Geometry and Computing

For further volumes:
www.springer.com/series/7580

Efi Fogel • Dan Halperin • Ron Wein

CGAL Arrangements and Their Applications

A Step-by-Step Guide

 Springer

Efi Fogel
Tel Aviv University
The Blavatnik School of Computer Science
Schreiber Building
69978 Tel Aviv
Israel
efif@post.tau.ac.il

Ron Wein
Tel Aviv University
The Blavatnik School of Computer Science
Schreiber Building
69978 Tel Aviv
Israel
wein_r@yahoo.com

Dan Halperin
Tel Aviv University
The Blavatnik School of Computer Science
Schreiber Building
69978 Tel Aviv
Israel
danha@post.tau.ac.il

ISSN 1866-6795   e-ISSN 1866-6809
ISBN 978-3-662-50712-4   ISBN 978-3-642-17283-0 (eBook)
DOI 10.1007/978-3-642-17283-0
Springer Heidelberg Dordrecht London New York

ACM Computing Classification: I.3.5, F.2.2

Mathematics Subjects Classification (2000): 65Dxx, 68U05

Printed on acid-free paper

Springer is part of Springer Science+Business Media (www.springer.com)

To Ilil, Adam, Shira, and Tom.
E.F.

To Gili and Naomi.
D.H.

To Hélène, Nogah, and Shir.
R.W.

Contents

Preface

What This Book Contains

This book is about how to use the CGAL *2D Arrangements* package to solve problems. It will teach you the functionality of this package and a few related CGAL packages. Every feature of the package is demonstrated by a small example program. Even the basic tools are sufficient to solve problems, as shown in Chapter 2, which includes the first *application*. We use the word application here to refer to a complete standalone program (written on top of CGAL arrangements) to solve a meaningful problem. The applications presented in the book include finding the minimum-area triangle defined by a set of points, planning the motion of a polygon translating amidst polygons in the plane, computing the offset polygon, constructing the farthest-point Voronoi diagram, coordinating the motion of two discs moving amidst obstacles in the plane, performing Boolean set operations on curved polygons, and more. The programs are designed such that you can use each of them as is, provided that the definition of your problem is the same as the definition of the problem the application program was set out to solve. If your problem is similar (but not exactly the same) it may require modification of the application program presented in the book. The book contains material that will help you in this modification process. As the book progresses, the applications become more involved and the range of problems solved by the applications widens. Moreover, the book was designed such that about halfway through, you will have accumulated sufficient knowledge to write your own application programs. You can download the code of the programs listed in the book along with appropriate input data-files from the book's website `http://acg.cs.tau.ac.il/cgal-arrangement-book`.

The book is about using CGAL arrangements; it is *not* about developing CGAL packages. The latter requires more in-depth knowledge of C++ and of generic programming than is required for reading this book, in addition to familiarity with the CGAL design rules and programming standards. If you are interested in developing basic CGAL code, you will find some of the necessary information on the CGAL website `http://www.cgal.org`. You will also find many pointers in the book to papers describing the design principles and implementation details of various related packages of CGAL. The CGAL code is open source. In particular, you can browse the entire code of the *2D Arrangements* package. In the book we barely discuss how the package was implemented.

One of the guiding principles in developing the CGAL *2D Arrangements* package is the separation of the combinatorial algorithms (e.g., plane sweep) on the one hand and the algebraic operations they use (e.g., intersecting two curves) on the other. This book is *not* about algebraic computation. We concentrate the algebra in so-called traits classes (see Chapter 5). The package comes with several ready-made traits-classes for the most common types of curves in use. We refer the interested reader to the book *Effective Computational Geometry of Curves and Surfaces* [29], which deals intensively with the algebra of the underlying algorithms for arrangements. More references to arrangement-related algebraic computation can be found in Section 5.7.

What You Should Know before Reading This Book

CGAL in general, and the *2D Arrangements* package in particular, rigorously adhere to the *generic programming* paradigm. The code of the library is written in C++, a programming language that is well equipped for writing software according to the generic programming paradigm through the extensive use of class templates and function templates. We assume that the reader is familiar with the C++ programming language and with the basics of the generic programming paradigm. Some information on the required material is provided in the Introduction chapter.

Most of the underlying algorithmic material was developed within the field of computational geometry. We made an effort throughout the book to supply sufficient details for the problems we introduce and the techniques that we use to solve those problems, so that you can get started without first reading a computational geometry book or taking a computational geometry course. This was however a nontrivial task, and we may have not accomplished it in full. Computational geometry is a broad and deep field. Quite often we sketch an algorithm or a technique, without providing all the fine details, extensions, or variations. We attempted to give adequate references in the Bibliographic Notes and Remarks section at the end of each chapter, so that you can learn more about the topics if you wish.

While the theory of geometric algorithms has been studied for several decades and is very well developed, the practice of implementing geometric algorithms is far less advanced. Even if you master the theory and know what your options are to attack a problem, the theoretical measures of asymptotic time and space consumption may be misleading when an algorithm is put to the test of actual work. The budding area of *algorithm engineering* is about figuring out the better strategies in practice. At least for arrangements, this type of information is still scarce. If you wish to develop your own *efficient* applications, you will need to know more about algorithm engineering; see more in the Introduction chapter.

How to Obtain and Install CGAL

The Computational Geometry Algorithm Library (CGAL) comprises a large collection of packages. It has a small set of thin library objects (e.g., `.lib` files on WINDOWS) executables are linked with, and other components. In this book we only discuss the *2D Arrangements*, *2D Intersection of Curves*, *2D Regularized Boolean Set-Operations*, *2D Minkowski Sums*, *Envelopes of Curves in 2D*, and *Envelopes of Surfaces in 3D* packages. The data structures and operations provided by the *2D Arrangements* package serve as a foundation layer for the other packages.[1] You must obtain the entire collection and install the header files and at least the main CGAL library object in order to use any one of its packages.[2] The remaining library objects are optional, but three of them, i.e., `CGAL_Core`, `CGAL_Qt`, and `CGAL_Qt4`, are relevant in the context of this book. The `CGAL_Core` library object provides a custom version of the CORE library and depends on GMP and MPFR. (All three supply multi-precision number types.) The `CGAL_Qt` and `CGAL_Qt4` library objects provide convenient components for demonstration programs that have visual effects. They depend on QT Version 3 and Version 4, respectively; additional information about all these dependencies is provided in the next subsection.

When you develop an application based on the *2D Arrangements* package or related packages, you introduce #`include` statements in your code that refer to header files, e.g., `CGAL/Arrangement_2.h`. The compiled objects linked with the CGAL library object (and perhaps with other library objects) form the application executable. There is no trace of the *2D Arrangements* code in the library object, as the entire code resides only in header files. As a matter of fact, the library object itself is very thin. Thus, the advantage of creating and using a shared library object[3] diminishes. Nevertheless, such an object is installed by default as part of the installation procedure.

The CGAL distribution consists of various packages. Each package comes with (i) header files

---

[1] The *2D Intersection of Curves*, *2D Regularized Boolean Set-Operations*, *2D Minkowski Sums*, and *Envelopes of Surfaces in 3D* packages depend on the *2D Arrangements* package and in particular on its main data structure `Arrangement_2`. The *Envelopes of Curves in 2D* package does not depend on the `Arrangement_2` data structure but it does depend on other components of the *2D Arrangements* package.

[2] In the future CGAL may be divided into interdependent components with a clear set of dependencies between them. This will allow you to obtain and install only a portion of the entire collection, or upgrade only those components that you may find necessary.

[3] It is called a dynamic-link library (DLL) on WINDOWS platforms, and a shared object (SO) on UNIX platforms.

that consist not only of the interface but also of the generic implementation of the package code, (ii) comprehensive and didactic documentation divided into two parts, namely, the programming guide and the reference manual, (iii) a set of non-interactive standalone example programs, and optionally (iv) an interactive demonstration program with visual effects.[4] The packages described in this book are no exception. The programming guide of the *2D Arrangements* package, for example, consists of roughly 100 pages. There are more than 50 (non-interactive) examples that exercise various operations on arrangements, and there is an interactive program included in the package that demonstrates its features. The *2D Regularized Boolean Set-Operations* and *Envelopes of Surfaces in 3D* packages also come with demonstration programs.

Libraries That CGAL Uses

BOOST: The CGAL library depends on other libraries, which must be installed a priori. First, it depends on BOOST. BOOST provides free peer-reviewed portable C++ source libraries that work well with, and are in the same spirit as, the C++ Standard Template Library (STL). Several BOOST libraries are implemented using only templates defined in header files; other libraries include source code that needs to be built and installed, such as the `Program_options` and `Thread` libraries. The latter is a prerequisite, and, as such, it must be installed before you commence the build process of CGAL. The former is required to build some of the CGAL demonstration programs. On WINDOWS, for example, you can download a WINDOWS installer for the current BOOST release (1.45.0 at the time this book was written), available at `http://www.boostpro.com/download`. Popular LINUX and UNIX distributions such as FEDORA, DEBIAN, and NETBSD include pre-built BOOST packages. Source code, documentation, and other related material can be obtained from `http://www.boost.org/`.

GMP and MPFR: GMP and MPFR are two libraries that provide types for multi-precision integers, rational numbers, and floating-point numbers. They are highly recommended but optional, as CGAL has built-in multi-precision number types used when none of the external libraries that provide multi-precision number types is installed on your system. (CORE and LEDA—see the next paragraph—also provide multi-precision rational-number types.) Do not use the CGAL built-in number types if you want to get optimal performance. LINUX distributions include the GMP and MPFR libraries by default. The MPFR library can be obtained from `http://www.mpfr.org/`. The page `http://gmplib.org/` provides useful information regarding the installation of GMP. The CGAL distribution includes precompiled versions of GMP and MPFR for Visual C++ users who find directly obtaining and installing these libraries less convenient. Notice that these libraries make use of processor-specific instructions when compiled with maximum optimization enabled. It is reasonable to assume that downloaded precompiled objects do not contain these instructions. If it turns out that this is a performance bottleneck for your application, you can always download the source package and compile it locally.

CORE and LEDA: If you want to handle arrangements of non-linear curves in a robust manner, you must use support for algebraic numbers to carry out some of the arithmetic operations involved. In some of those cases the CORE or LEDA libraries come to the rescue. In particular, constructing an arrangement of conic curves or of Bézier curves requires the CORE library. A custom version of the CORE library that has its own license is distributed with CGAL for your convenience. More information on the CORE library can be found at `http://www.cs.nyu.edu/exact/core/`. Information on LEDA can be found at `http://www.algorithmic-solutions.com/leda/`. There are free and commercial editions for this library.

---

[4] A noteworthy component every package includes (but not distributed as part of public releases) is a collection of functional and regression tests. The tests of all packages combined compose the CGAL test suite. All the tests in the test suite run daily, and their results are automatically assembled and analyzed.

QT: Most CGAL demos, including the ones provided by the *2D Arrangements*, *2D Regularized Boolean Set-Operations*, and *Envelopes of Surfaces in 3D* packages, are based on the QT cross-platform library, developed by Trolltech, and currently owned by Nokia. Some of the CGAL demos are still based on QT Version 3—an older version of QT, which was not licensed free of charge for WINDOWS. Unfortunately, some of the demos of the packages above are included in this subset. However, they were being ported at the time this book was written to Version 4, the latest version of QT. Editions with open-source licenses of QT Version 4 are available for WINDOWS as well as for LINUX. QT is included by default in LINUX distributions, and can be obtained from `http://qt.nokia.com/products`.

Supported Platforms[5]

Table 1: CGAL-supported platforms.

Compiler	Operating System
GNU g++	LINUX, Mac OS, Solaris, XP, XP64, Vista, WINDOWS 7
MS Visual C++ (.NET)	XP, XP64, Vista, WINDOWS 7
Intel Compiler	LINUX

CGAL Version 3.8, the latest version of CGAL at the time this book was written, is supported by the platforms listed in Table 1. If your platform is not on this list, it does not necessarily mean that CGAL does not work on your platform. Note that 64-bit variants of these platforms are also supported. For the specific versions of the supported operating systems and compilers consult the official Web page `http://www.cgal.org/platforms.html`.

Visual C++-Related Issues

Choosing the Runtime Library: Do not mix static and dynamic versions of the C runtime (CRT) libraries, and make sure that all static and dynamic link libraries are built with the same version of the CRT libraries. More than one copy of a (CRT) library in a process can cause problems because static data in one copy is not shared with other copies. The linker prevents you from linking with both static and dynamic versions within one executable file (`.exe`), but you can still end up with two (or more) copies of CRT libraries. For example, a dynamic-link library linked with the static (non-DLL) versions of the CRT libraries can cause problems when used with an executable file that was linked with the dynamic (DLL) version of the CRT libraries. (You should also avoid mixing the debug and non-debug versions of the libraries in one process, because the debug version might have different definitions of objects, as is the case, for example, with the STL containers and iterators in VC9.)

Automatic Library Selection: Starting from CGAL Version 3.3, BOOST-style automatic library selection (auto-linking) is used for all CGAL library objects (including **CGAL**, **CGAL_Qt**, **CGAL_Qt4**, and **CGAL_Core**) and the third-party libraries GMP and MPFR installed via the WINDOWS installer. This feature allows you to choose any build configuration for your application without worrying about a matching precompiled library (`.lib`) to link against. Auto-linking also prevents mixing different runtime libraries.

Runtime Type Information: You must enable runtime type information. This is forced with the /GR compiler option.

Compiling: When compiling some of the source code listed in this book (and also included on the book's website `http://acg.cs.tau.ac.il/cgal-arrangement-book`) using Visual C++, the following warning is issued by the compiler:

---

[5]The term "platform" refers to the combination of an operating system and a compiler.

warning C4503: ... decorated name length exceeded, name was truncated.

This implies that the decorated name was longer than the compiler limit (4,096 according to MSDN), and thus was truncated. You must reduce the number of arguments or name length of identifiers used to avoid the truncation. For example, instead of

typedef CGAL::Cartesian<Number_type> Kernel;

define

struct Kernel : **public** CGAL::Cartesian<Number_type> {};

In many cases, though, the truncation is harmless. You can suppress only these warning messages by adding the following pragma statement at the beginning of the file that contains the source code:

#**pragma warning**(disable : 4503)

Alternatively, you may use the /w or the /W compiler options to suppress all warnings or to set the appropriate warning-level, respectively.[6]

Installing CGAL

There are ongoing efforts to make the CGAL installation process as easy as possible. In spite of these efforts, and mainly due to the dependence of CGAL on other libraries, you may encounter difficulties while trying to install CGAL on your computer for the first time. We next give some tips on the installation and refer you below to sources containing more information. Answers to frequently asked questions appear at `http://www.cgal.org/FAQ.html` and an archive of a discussion forum can be found at `https://lists-sop.inria.fr/sympa/arc/cgal-discuss`.

All versions of CGAL, including the latest version of the library, that is, Version 3.8 at the time this book was written, can be obtained online. Precompiled versions of CGAL library-objects, source files required for development, demonstration programs, and documentation are available as **deb** packages for LINUX distributions based on DEBIAN, such as DEBIAN and UBUNTU.

There are no precompiled versions for WINDOWS. A self-extracting executable that installs the CGAL sources, and that allows you to select and download some precompiled third-party libraries is available at `http://gforge.inria.fr/projects/cgal/`. It obtains the source files required for the configuration and compilation of the various CGAL library-objects, and all other components required for the development of applications. Alternatively, you can find an archive (a UNIX "tar ball"), e.g., `CGAL-3.7.tar.gz`, that contains all the code (but not necessarily the manuals) at `http://www.cgal.org/download.html`. Download the archive, and follow the instructions described at `http://www.cgal.org/Manual/last/doc_html/installation_manual/Chapter_installation_manual.html`, which are summarized below, to configure, build, and install CGAL. This process, commonly referred to as the *build* process, is carried out by CMAKE—a cross-platform build system generator. If CMAKE is not installed already on your system, you can obtain it from `http://www.cmake.org/`.

The archive can be expanded using the following command:

tar zxvf *tar-ball*

As a result, the CGAL root directory will be created, e.g., `CGAL-3.8`.

Ideally, building CGAL and then building an example that depends on CGAL amounts to

```
cd CGAL-3.7                              # go to CGAL directory
cmake .                                  # configure CGAL
make                                     # build the CGAL library objects
cd examples/Arrangement_on_surface_2     # go to an example directory
cmake -DCGAL_DIR=$HOME/CGAL-3.7 .        # configure the examples
make io                                  # build the io example
```

---

[6]Suppressing all warnings is a risky practice, as warnings may convey crucial information about real problems.

When you build CGAL, you are free to choose which generator to use. The generator is determined by a specific compilier, a development environment, or other choices. On WINDOWS, for example, you may choose to generate `project` and `solution` files for a specific version of the common Visual C++ development environment. You may choose to generate `makefile` files by a specific version of Visual C++, or you may choose to generate `makefile` files for the CYGWIN port of g++. The CGAL code is written in ISO/IEC C++, and compiles with most C++ compilers; consult the website `http://www.cgal.org/platforms_frame.html` for an up-to-date list of supported compilers and operating systems.

License

The CGAL library is released under open-source licenses. Common parts of CGAL (i.e., the *kernel* and the *support library*; see Section 1.4.2) are distributed under the terms of the GNU Lesser General Public License (or GNU LGPL for short) and the remaining part (i.e., the *basic library*, which contains most of the software described in the book) is distributed under the terms of the Q Public License (QPL), a non-copyleft free software license created by Trolltech for its free edition of the QT toolkit. The QPL captures the general meaning of the GNU General Public Licence (GPL), but is incompatible with it, meaning that you cannot legally distribute products derived from both GPL'ed and QPL'ed code. Trolltech changed its policy, and released Version 4.0 of QT under the terms of GPL Version 2, abandoning QPL. As a consequence, the use of QPL became rare and is decreasing. Most probably, GPL will replace QPL for all relevant CGAL components. (This may have already taken place by the time you are reading this.) A copy of each license used by CGAL components is included in the distribution in the `LICENSE` file and its derivatives.

The CGAL library may be used freely in noncommercial applications. Commercial licenses for selected components of CGAL are provided by GEOMETRY FACTORY.[7] There are licenses for industrial and academic development and for industrial research.

Style Conventions

These style conventions are used in this book:

- **Bold** — command names.

- SMALL CAPS — special terms and names, e.g., CGAL.

- Sans Serif — generic programming concepts, e.g., CopyConstructible.

Code excerpts are set off from the text in a monospace font, for example:

```
#include <iostream>
int main() { std::cout << "Hello_World" << std::endl; return 0; }
```

Implicit association between identifiers in the code set off in a monospace font and the corresponding entity denoted in math font is assumed. For example, the identifier p, which explicitly refers to a point object (or a pointer to a point object) also implicitly refers to the point denoted by p.

Topics that are particularly complicated—and that you can skip, if, for example, you are new to CGAL—are surrounded by the *advanced* marks:

—————— *advanced* ——————

This can apply to a single paragraph or to an entire section.

—————— *advanced* ——————

---

[7]See `http://www.geometryfactory.com/`.

Simple tasks that are left for the reader to practice with are marked with the icon. For example, assuming that the listing of an imaginary program coded in `ex_compute_union.cpp` is given:

Try: Modify the program coded in `ex_compute_union.cpp` such that it computes the union of discs rather than the union of squares.

Example programs are marked with the ▲ icon.

Example: Typically, this consists of a description of the example followed by the program code.

Exercises

At the end of every chapter, with the exception of the Introduction chapter, you will find several exercises. These are meant to recite the material covered in that chapter, and altogether strengthen the practical direction this book has taken. The exercises are divided into two categories according to their difficulty levels. Solving an exercise marked as **Project** typically requires considerably more resources than solving a non-marked exercise. Several programming exercises (typically marked as **Project**) ask for the implementation of a useful feature that is currently not implemented, but could nicely fit into CGAL. If you work out such an exercise, and you believe that your solution meets the high standards of CGAL code, kindly send it to us. We will consider your contribution as a potential candidate for inclusion in a future release.

The Cover

The illustration on the cover of the book depicts an arrangement of Fibonacci spirals, which govern the layout of sunflower seeds. This is explained in detail in Exercise 5.7.

Errata

Errors and corrections will appear on `http://acg.cs.tau.ac.il/cgal-arrangement-book`, the book's website, as soon as errors are detected.

Acknowledgments

The CGAL *2D Arrangements* package is an integrated part of CGAL. The initial development of CGAL was funded by two European Union projects, CGAL and GALIA, over three years (1996–1999). Several sites kept on working on CGAL after the European Union funding stopped. The European Union projects ECG[8] (Effective Computational Geometry for curves and surfaces) and ACS[9] (Algorithms for Complex Shapes with certified topology and numerics) provided further partial support for new research and development in CGAL.

The work on the *2D Arrangements* package and its related packages took place mostly at Tel Aviv University, and has been supported in part by the Israel Science Foundation through several grants to Dan Halperin.

The code of the *2D Arrangements* and *2D Intersection of Curves* packages is the result of a long development process. Initially (and until Version 3.1), the code was spread among several components, namely, `Topological_map`, `Planar_map_2`, `Planar_map_with_intersections_2`,

---

[8]See `http://www-sop.inria.fr/prisme/ECG/`.
[9]See `http://acs.cs.rug.nl/`.

and `Arrangement_2`, that were developed by Ester Ezra, Eyal Flato, Efi Fogel, Dan Halperin, Iddo Hanniel, Idit Haran, Shai Hirsch, Eugene Lipovetsky, Oren Nechushtan, Sigal Raab, Ron Wein, Baruch Zukerman, and Tali Zvi.

In Version 3.2, as part of the ACS project, the packages have gone through a major redesign, resulting in improved and unified *2D Arrangements* and *2D Intersection of Curves* packages. The code of the two new packages was restructured and developed by Efi Fogel, Idit Haran, Ron Wein, and Baruch Zukerman. This version included for the first time a new geometry-traits class that handles circular and linear curves, and is based on the circular kernel. The circular kernel was developed by Monique Teillaud, Sylvain Pion, and Julien Hazebrouck.

A significant outcome of the redesign was the reduction of the geometry-traits concept, which was assisted by Monique Teillaud. Version 3.2 also exploited an optimized multi-set data structure implemented as a red-black tree, which was developed by Ron Wein. This feature was reviewed by Remco Veltkamp.

Version 3.2.1 featured arrangements of unbounded curves for the first time. The design and development of this feature required yet another restructuring of the entire package. All this was done by Eric Berberich, Efi Fogel, Dan Halperin, Ophir Setter, and Ron Wein. Michael Hemmer helped in tuning up parts of the geometry-traits concept related to unbounded curves. Sylvain Pion reviewed Version 3.2.1.

Version 3.7 of CGAL introduced a geometry-traits class that handles planar algebraic curves of arbitrary degree. It was developed by Michael Kerber and Eric Berberich and reviewed by Efi Fogel.

At the time this book was written an internal version of CGAL introduced a new geometry-traits class that handles rational arcs. It was developed by Oren Salzman and Michael Hemmer, reviewed by Eric Berberich, and was expected to replace in version 3.9 an old traits, which handles the same family of curves, developed by Ron Wein.

The code of the *2D Regularized Boolean Set-Operations* package was developed by Efi Fogel, Dan Halperin, Ophir Setter, Guy Zucker, Ron Wein, and Baruch Zukerman. Andreas Fabri reviewed this package. The code of the *2D Minkowski Sums* package was developed by Ron Wein. Michael Hoffmann reviewed this package and the two packages that we mention next. The code of the *Envelopes of Curves in 2D* package was developed by Ron Wein. The code of the *Envelopes of Surfaces in 3D* package was developed by Dan Halperin, Michal Meyerovitch, Ron Wein, and Baruch Zukerman.

The ongoing work on an extension of the *2D Arrangements* package to handle arrangements on surfaces (which is still at a prototype stage, and hence not described in the book) had a major influence on the quality of the planar-arrangement code. This work is carried out by the authors of this book together with Eric Berberich, Michael Hemmer, Michael Kerber, Kurt Mehlhorn, and Ophir Setter.

The quality of the book in general and the example programs and applications listed in it in particular were immensely improved following a diligent review and code revisiting and rewriting conducted by Ophir Setter.

The *2D Arrangements* package depends on other components of the library, which have been developed, and are still being developed, by many people. The package benefits from many services provided collectively to all parts of CGAL by groups and individuals, many of whom are not explicitly acknowledged here. Nonetheless, we would like to thank the CGAL Board (formerly the CGAL Editorial Board) and its current and past members, Pierre Alliez, Eric Berberich, Andreas Fabri, Bernd Gärtner, Michael Hemmer, Susan Hert, Michael Hoffmann, Menelaos Karavelas, Lutz Kettner, Sylvain Pion, Marc Pouget, Laurent Rineau, Monique Teillaud, Mariette Yvinec, and Remco Veltkamp, GEOMETRY FACTORY and its members, Andreas Fabri, Fernando Cacciola, Sébastien Loriot, and Laurent Rineau, and all members of the CGAL developers' community for providing those vital components and services and for sharing their wisdom through many rich discussions.

Over the years the *2D Arrangements* package benefitted from intensive discussions in private meetings, workshops, conferences, and over electronic means, with many folks, including Ioannis Z. Emiris, Peter Hechenberger, Eli Packer, and Stefan Schirra. We are indebted to them for their invaluable insights and critiques.

Last but not least we owe a special debt to the creators of CGAL and the people who nursed and nurtured CGAL during its early stages. This book would not have been written if it had not been for these visionaries.

Tel Aviv, 2011

Efi Fogel
Dan Halperin
Ron Wein

Chapter 1

Introduction

1.1 Arrangements

Geometric arrangements, or arrangements for short, are subdivisions of some space induced by geometric objects. The figure to the right shows an arrangement of two curves C_1 and C_2 in the plane. It has three faces—two bounded faces f_1 and f_2 (filled with diagonal-stripe patterns) and an unbounded face. The arrangement has seven vertices—four represent the endpoints of C_1

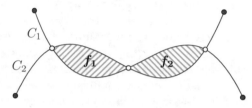

and C_2 (drawn as small discs), and three represent the intersection points of the two curves (drawn as small rings). The arrangement also has eight edges, each of which is a maximal portion of one curve not intersecting the other.

This sounds simple (and is easy to visualize), and indeed it is. In spite of their seeming simplicity, arrangements turn out to be an extremely powerful tool in understanding and solving a large variety of problems.

Arrangements are not restricted to curves in the plane. There are useful arrangements in three and higher dimensions (these are not so easy to visualize) and they can be induced by geometric objects of any type, such as spheres, simplices, polytopes, or Bézier surfaces. In this book we focus on *two-dimensional arrangements* of curves. Here are a few problems efficiently solvable using two-dimensional arrangements:

Boolean set operations: Given two sets of regions in the plane, which we refer to as the *blue* regions and the *red* regions, compute the *purple* regions, namely regions of the plane covered by both a red region and a blue region. The purple regions are the *intersection* of the two sets. Similarly, compute the *union*, the *symmetric difference*, and other operations on the planar sets.

Minimum-area triangle: Given a finite set of points in the plane, find the three of them defining the smallest-area triangle out of all possible triples of points in the set.[1]

Translational motion-planning: A robot is translating in a room cluttered with obstacles. Determine whether the robot can get out of the room through the door without colliding with the obstacles in the room.

Farthest-point: Given a set of points $P = \{p_1, \ldots, p_n\}$ in the plane, devise a data structure that can efficiently answer queries of the following form: Locate the farthest point in P from a given query point q.

---

[1]This problem can be trivially solved in $O(n^3)$ time by examining all triples. With arrangements we can solve it in only $O(n^2)$ time.

E. Fogel et al., *CGAL Arrangements and Their Applications*, Geometry and Computing 7, DOI 10.1007/978-3-642-17283-0_1, © Springer-Verlag Berlin Heidelberg 2012

Throughout this book we give complete solutions, including source code, to the problems above, and many more.

Arrangements are subdivisions; thus, it seems natural that planar arrangements help in solving problems on maps in Geographic Information Systems (GIS) or in computing Boolean operations on planar regions. However, their range of applications is much broader. What makes arrangements such a useful tool is that they enable the discretization of continuous problems without giving up the correctness or exactness of the solution. This is explained and demonstrated through the various applications presented in the book.

We postpone a formal definition of arrangements to the next chapter. In fact, we put arrangements aside for a short while to describe more general background material both in programming (Section 1.2) and in geometric computing (Section 1.3). We get back to arrangements when we outline the contents of the book in Section 1.5.

1.2 Generic Programming

Several definitions of the term *generic programming* have been proposed since it was first coined around the early 1960s, along with the introduction of the LISP programming language. Here we confine ourselves to the classic notion first described by Musser and Stepanov [166], who considered generic programming as a discipline that consists of the gradual lifting of concrete algorithms abstracting over details, while retaining the algorithm semantics and efficiency. Within this context, several approaches have been put into trial through the introduction of new features in existing computer languages, or even entirely new computer languages. The software described in this book is written in C++, a programming language that is well equipped for writing software according to the generic programming paradigm through the extensive use of class templates and function templates.

1.2.1 Concepts and Models

One crucial abstraction supported by all contemporary computer languages is the subroutine (also known as procedure or function, depending on the programming language). Another abstraction supported by C++ is that of abstract data typing, where a new data type is defined together with its basic operations. C++ also supports object-oriented programming, which emphasizes packaging data and functionality together into units within a running program, and is manifested in hierarchies of polymorphic data-types related by inheritance. It allows referring to a value and manipulating it without the need to specify its exact type. As a consequence, one can write a single function that operates on a number of types within an inheritance hierarchy. Generic programming identifies a more powerful abstraction (perhaps less tangible than other abstractions). It is a formal hierarchy of polymorphic abstract requirements on data types, referred to as *concepts*, and a set of classes that conform precisely to the specified requirements, referred to as *models*. Models that describe behaviors are referred to as *traits* classes [168]. Traits classes typically add a level of indirection in template instantiation to avoid accreting parameters to templates.

A generic algorithm has two parts—the actual instructions that describe the steps of the algorithm, and a set of requirements that specify the properties its argument types must satisfy. The following swap() function is an example of the first part of a generic algorithm:

template <**typename** T> **void** swap(T& a, T& b) { T tmp = a; a = b; b = tmp; }

When the code is compiled, each function call is instantiated, and the template parameter T is substituted with a data type that must have an assignment operator. A data type that fulfils this requirement is a model of a concept commonly called Assignable [12]. The substituted data-type must have a copy constructor; namely, it must also be a model of the concept CopyConstructible. The int data type, for example, is a model of these two concepts, so it can substitute the T template parameter when the function template is instantiated.[2]

---

[2]See http://www.sgi.com/tech/stl/ for a complete specification of the SGI STL, where the concepts Assignable

Concept Requirements

A concept is a set of requirements divided into four categories, namely, associated types, valid expressions, runtime characteristics, and complexity guarantees. When a type meets all requirements of a concept, the type is considered a *model* of the concept. When a concept extends the requirements of another concept, the former is said to be a *refinement* of the latter, and the latter is said to be a generalization of the former.

Associated Types are auxiliary types. For example, a type that represents a two-dimensional point, namely `Point_2`, is required by every arrangement-traits concept; see Section 5.

Valid Expressions are C++ expressions that must compile successfully. For example, `p = q`, where `p` and `q` are both objects of type `Point_2`. Valid expressions identify the set of operations that a model of the concept must support.

Runtime Characteristics are characteristics of the variables involved in the valid expressions that apply during the variables' lifespans. The runtime characteristics often take the form of preconditions and postconditions, which must always be satisfied; see Section 1.4.2 for a brief explanation about condition handling. For example, a condition that requires that an input point p lies on an input curve c on invocation of a predicate that accepts both p and c as arguments. Having preconditions typically minimizes the concept, as the operations provided by a model must operate only on restricted arguments. Formally, removing preconditions from, and introducing postconditions to, a requirement set results in a refined concept.

Complexity Guarantees are maximum limits on the computing resources consumed by the valid expressions.

1.2.2 Traits Classes

The use of traits classes to gain software flexibility is central to CGAL. The name "traits class" comes from a standard C++ design pattern [168], which provides a way of associating information with a compile-time entity (typically a type). For example, the standard class-template `std::iterator_traits<T>` looks roughly like this:

```
template <typename Iterator>
struct iterator_traits {
    typedef ... iterator_category;
    typedef ... value_type;
    typedef ... difference_type;
    typedef ... pointer;
    typedef ... reference;
};
```

Iterators play an important role in generic programming. A function that operates on a range of objects usually accepts two iterators that specify this range. The `value_type` nested in the iterator traits specifies the type of object the iterators are pointing at, while the `iterator_category` can be used to select more efficient algorithms depending on the iterator capabilities.

A key property of trait classes is that they are non-intrusive. Namely, they allow us to associate information with arbitrary types, without interfering with the internal representation of these types. Thus, it is possible to define a traits class also for built-in types and types defined in third-party libraries.

Within the context of CGAL, we give a slightly broader interpretation to the term "traits class," by requiring it not only to define nested types of geometric objects, but also requiring it to support predicates involving objects of these types, and allowing it to carry state. Some algorithms also require the provision of constructions of geometric objects by the traits class; see Section 5.4.5.

---

and CopyConstructible are defined, among others.

Let's begin with an easy geometric example. Consider a function that accepts a set of points given by the range [pts_begin, pts_end),[3] and computes the minimal axis-parallel rectangle that contains all points in the range. It does so by locating the points with extremal x- or y-coordinates, and then constructs the bounding iso-rectangle accordingly.

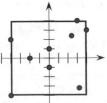

```
template <typename Input_iterator, typename Traits>
typename Traits::Iso_rectangle_2
bounding_rectangle (Input_iterator pts_begin, Input_iterator pts_end)
{
  Traits          traits;
  Input_iterator left, right, top, bottom, curr = pts_begin;
  left = right = top = bottom = curr;
  while (++curr; curr != pts_end; ++curr) {
    if (traits.compare_x(*curr, *left) == CGAL::SMALLER) left = curr;
    else if (traits.compare_x(*curr, *right) == CGAL::LARGER) right = curr;
    if (traits.compare_y(*curr, *bottom) == CGAL::SMALLER) bottom = curr;
    else if (traits.compare_y(*curr, *top) == CGAL::LARGER) top = curr;
  }

  return traits.construct_iso_rectangle(*left, *right, *bottom, *top);
}
```

The compile-time requirements that a substituted traits class must satisfy in this case are as follows: First, it must have a default constructor; namely, it must model the concept Default-Constructible. Secondly, it has to define the nested type Iso_rectangle_2 (and, implicitly, also a point type, say, Point_2).[4] Thirdly, it should supply two three-valued predicates[5] that compare two points by their x-coordinates and by their y-coordinates, respectively. Finally, it should also support the construction of an axis-parallel iso-rectangle from four points that specify its extremal x- and y-coordinates. Note, however, that the actual representation of points and rectangles (the coordinate system, the number type used to represent the coordinates, etc.) and the implementation of the traits-class operations are entirely decoupled from the function template bounding_rectangle(). Another requirement, which cannot be enforced during compile time, and thus falls under the runtime-characteristic category, is that the set of input points given by the range [pts_begin, pts_end) must be nonempty.

Consider an imaginary generic implementation of a data structure that handles geometric arrangements induced by planar curves; its prototype follows.

```
template <typename Traits> class Arrangement_2 { ... };
```

When the Arrangement_2 class template is instantiated, the Traits parameter must be substituted with a class that defines the type of the inducing curves, and some functions that operate on curves of this type. Such a traits-class models a concept that is expected to be much more involved than the concept of the traits of the bounding_rectangle() function template above. As a matter of fact, there is an entire refinement hierarchy of traits concepts, models of which can be used to instantiate the real arrangement data structure to obtain instances with different capabilities. A precise definition of this hierarchy is given in Chapter 5.

One important objective is to minimize the set of requirements the traits concept imposes. A tight traits concept may save tremendously in terms of the cost of analysis and programming of classes that model the concept. Another important reason for reaching the minimal set of requirements is to avoid computing the value of the same algebraic expression in different ways.

---

[3]This notation means that pts_begin points to the first point in the range, while pts_end points beyond the last point of the range (it is therefore called a *past-the-end iterator*, and need not point to any valid point object).

[4]CGAL prescribes the _2 suffix for two-dimensional data structures and algorithms, and the _3 suffix for three-dimensional structures.

[5]The predicate return value is CGAL::SMALLER, CGAL::EQUAL, or CGAL::LARGER.

The importance of this is amplified in the context of computational geometry, as a non-tight model that consists of slightly different implementations of the same algebraic entity can lead to superficial degenerate conditions, which in turn can drastically impair the performance.

1.2.3 Generic and Object-Oriented Programming

Generic programming is based on *parametric polymorphism*, while object-oriented programming is based on *subtype polymorphism*. In C++ these paradigms are implemented with templates and inheritance, respectively.[6] An algorithm implemented according to the standard object-oriented programming paradigm alone may resort to using dynamic cast to achieve flexibility, is forced to have tight coupling through the inheritance relationship, may require additional memory for each object to accommodate the virtual-function table-pointer, and adds for each call to a virtual-member function an overhead, as the call must indirectly pass through the virtual-function table.

An algorithm implemented according to the generic programming paradigm does not suffer from these disadvantages. The set of requirements on data types is not tied to a specific C++ language feature. Therefore, it might be more difficult to grasp. In return, a generic implementation gains stronger type checking at compile time and a higher level of flexibility, without loss of efficiency. In fact, it may expedite the computation. Many articles and a few books have been written on the subject. We refer the reader to [12] for a complete introduction.

Following the object-oriented programming paradigm it is easy to extend data types by defining new classes or new functions. Supporting both directions of extensibility creates a problem, known as the expression problem [169]. We end this section with an example that demonstrates the strength of generic programming for solving a variant of this problem.

Let `Line_2` and `Point_2` be two types of some library that are used to represent a line and a point in the plane, respectively. The `Point_2` type defines a function operator that accepts two lines, and the `Line_2` type defines a function operator that accepts two points. The semantics of these two operations is irrelevant. Nevertheless, to give you an authentic feel, assume that the function operator of the `Point_2` type computes the coordinates of the intersection point of the given lines and stores them, resetting its privately stored data-members. Similarly, assume that the function operator of the `Line_2` type computes the coefficients of the line that passes through the given points and stores them, resetting its privately stored data-members.

```
// Forward declarations.
struct Point_2;
struct Line_2;

struct Point_2 {
  virtual void operator()(Line_2& line1, Line_2& line2) { ... }
};
struct Line_2 {
  virtual void operator()(Point_2& point1, Point_2& point2) { ... }
};
```

Now, consider a prospective user of the library who needs to attach color data to the types above. Assume that the color type defines a blending operation that constructs a new color given two colors.

```
struct Color {
  int v;
  void blend(Color& color1, Color& color2) { v = (color1.v + color2.v) / 2; }
};
```

Using virtual functions to express polymorphism, our user will attempt to derive two new types, `Color_point_2` and `Color_line_2`, from the `Point_2` and `Line_2` base types, respectively. The

---

[6]Other languages provide different mechanisms.

parameter type in the member operations of a derived construct must be the same type as (or a base type of) the type of the parameter in the base class. This implies, for example, that for the operations in the derived constructs to override operations in the base constructs the parameters to `Color_line_2::operator()` must be of type `Point_2` (and not `Color_point_2`).

```
struct Color_point_2 : public Point_2 {
  Color color;
  virtual void operator()(Line_2& line1, Line_2& line2);
};
struct Color_line_2 : public Line_2 {
  Color color;
  virtual void operator()(Point_2& point1, Point_2& point2);
};
```

Unfortunately, the arguments passed to the operations above must be downcast, in order to extract the color field each type is extended with:

```
void Color_point_2::operator()(Line_2& line1, Line_2& line2)
{
  Point_2::operator()(line1, line2);
  color.blend(dynamic_cast<Color_line_2&>(line1).color,
              dynamic_cast<Color_line_2&>(line2).color);
}
void Color_line_2::operator()(Point_2& point1, Point_2& point2)
{
  Line_2::operator()(point1, point2);
  color.blend(dynamic_cast<const Color_point_2&>(point1).color,
              dynamic_cast<const Color_point_2&>(point2).color);
}
```

This means that programs using `Color_point_2` or `Color_line_2` are no longer type-safe. An object of an unmatching type could be passed to any one of the extended operators above in error, causing an exception at runtime.

Using templates to express polymorphism, the extended constructs remain type-safe. The point and line types can be defined as follows:

```
struct Point_2 {
  template <typename Line> void operator()(Line& line1, Line& line2)
  { ... }
};
struct Color_point_2 : public Point_2 {
  Color color;
  template <typename Line> void operator()(Line& line1, Line& line2)
  {
    Point_2::operator()(line1, line2);
    color.blend(line1.color, line2.color);
  }
};

struct Line_2 {
  template <typename Point> void operator()(Point& point1, Point& point2)
  { ... }
};
struct Color_line_2 : public Line_2 {
  Color color;
  template <typename Point> void operator()(Point& point1, Point& point2)
  {
```

```
    Line_2::operator()(point1, point2);
    color.blend(point1.color, point2.color);
  }
};
```

1.2.4 Libraries

Alexander Stepanov began exploring the potential of compile-time polymorphism for revolutionizing software development in 1979. With the help of several other researchers his work evolved into a prime generic programming library—the Standard Template Library (STL). This library had became part of the C++ standard library in 1994, approximately one year before early development of CGAL started; see Section 1.4.1 for details about the evolution of CGAL.

Through the years a few other generic programming libraries emerged. One notable library in the context of computational geometry is LEDA (Library of Efficient Data Types and Algorithms), a library of combinatorial and geometric data-types and algorithms [151].[7] Early development of LEDA started in 1988, ten years before the first public release of CGAL became available [64].[8] In some sense LEDA is a predecessor of CGAL, although the two libraries are headed in different directions. While LEDA is mostly a large collection of fundamental graph-related and general purpose data structures and algorithms, CGAL is a large collection of data structures and algorithms focusing on geometry.

A noticeable influence on generic programming is wielded by the BOOST online community, which encourages the development of free C++ software gathered in the BOOST library collection.[9] It is a large set of portable and high-quality C++ libraries that work well with, and are in the same spirit as, the C++ Standard Template Library. The BOOST Graph Library (BGL) [198], which consists of generic graph-algorithms, serves a particularly important role in our context. It can be used, for example, to implement the underlying topological data structure of an arrangement instance, that is, a model of the concept ArrangementDcel. The role of this concept is explained in Section 2.2. Using some generic programming techniques, an arrangement instance can be adapted as a BGL graph, and passed as input to generic algorithms already implemented in the BGL, such as Dijkstra's shortest path algorithm; Chapter 7 explains in detail how this is done.

1.3 Geometric Computing

Implementing geometric algorithms and data structures is notoriously difficult, much harder than may seem when just considering the algorithm as described in a paper or a book. In the traditional computational geometry literature two assumptions are usually made to simplify the design and analysis of geometric algorithms. First, inputs are in "general position." That is, degenerate input (e.g., three curves intersecting at a common point) is precluded. Secondly, operations on real numbers yield accurate results (the "real RAM" model [182], which also assumes that each basic operation takes constant time). Unfortunately, these assumptions do not hold in practice, as degenerate input is commonplace in practical applications and numerical errors are inevitable while using standard computer arithmetic. Finding roots of polynomials is a much more demanding operation than elementary arithmetic operations, and exact operations on real numbers carried out to eliminate numerical errors do not take constant time. Thus, an algorithm implemented without keeping this in mind may yield incorrect results, get into an infinite loop, or just crash while running on a degenerate, or nearly degenerate, input; see [132, 187] for examples. The estimate of the time consumption of such an algorithm may be completely off target. These pitfalls have become well known, and have been the subject of intensive research [187, 219].

Indeed, the last decade has seen significant progress in the development of software for computational geometry. The mission of such a task, which Kettner and Näher [133] call *geometric*

---

[7]See LEDA's homepage at http://www.algorithmic-solutions.com/leda/.
[8]See CGAL's homepage at http://www.cgal.org/.
[9]See the homepage of the BOOST C++ libraries at http://www.boost.org/.

programming, is to develop software that is correct, efficient, flexible (namely, adaptable and extendible[10]), and easy to use.

1.3.1   Separation of Topology and Geometry

The use of the generic programming paradigm enables a convenient separation of the topology and the geometry of data structures.[11] This is a key aspect in the design of geometric software, and is put into practice, for example, in the design of CGAL polyhedra, CGAL triangulations, and our CGAL arrangements. This separation allows the convenient abstraction of algorithms and data structures in combinatorial and topological terms, regardless of the specific geometry of the objects at hand. This abstraction is realized through class and function templates that represent specific data structures and algorithmic frameworks, respectively. Consider again our imaginary `Arrangement_2` class template from the previous section; its improved prototype is listed below. It is instantiated with yet two other types. The first, referred to as a *traits* class, defines the set of geometric-object types and the operations on objects of these types. The second defines the topological-object types and the operations required to maintain the incidence relations among objects of these types. A natural choice is a doubly-connected edge list (DCEL) data structure, on which we elaborate in Section 2.1.

template <**typename** Traits , **typename** Dcel> **class** Arrangement_2 { ... };

An immediate advantage of the separation of the topology and the geometry of data structures is that users with limited expertise in computational geometry can employ the data structure with their own special type of objects. They must, however, supply the relevant traits class, which mainly involves algebraic computations. A traits class also encapsulates the number types used to represent coordinates of geometric objects and to carry out algebraic operations on them. It encapsulates the type of coordinate system used (e.g., Cartesian and homogeneous), and the geometric or algebraic computation methods themselves. Naturally, a prospective user of the package that develops a traits class would like to face as few requirements as possible in terms of traits development.

1.3.2   Exact Geometric Computation

A geometric algorithm is written in terms of a well-chosen set of basic questions called predicates, such as "is a point above, on, or below a line?" and constructions of new basic geometric objects, such as the construction of a line segment from two endpoints or of the intersection point of two segments. Each computational step is either a construction step or a conditional step based on the evaluation of a predicate. The latter is typically carried out through the computation of the sign of an expression. Different computational paths lead to results with different combinatorial characteristics. Although numerical errors can sometimes be tolerated and interpreted as small perturbations of the input, they may break some invariants, such as the planarity of a set of points, and lead to invalid combinatorial structures or inconsistent state during a program execution. Thus, for algorithms that do not construct new geometric objects (except perhaps at the last step), it suffices to ensure that all predicates are evaluated correctly to eliminate inconsistencies and guarantee combinatorially correct results. This is easier said than done, but nowadays possible, as explained below. For algorithms that do construct new geometric objects the task of certifying correctness becomes even harder, as the input to predicates may be new geometric objects constructed during the execution of the algorithms.

The need for robust software implementations of computational geometry algorithms has driven many researchers over the last decade to develop variants of the classic algorithms that are less

---

[10] *Adaptability* refers to the ability to incorporate existing user code, and *extendibility* refers to the ability to enhance the software with more code in the same style.

[11] In this context, we sometimes say *combinatorics* instead of topology, and say *algebra* or *numerics* instead of geometry. We always mean the same thing—the separation of the abstract, graph-like structure (the topology) from the actual embedding in the plane (the geometry).

susceptible to degenerate inputs. The approaches taken to overcome the difficulties in robustly implementing geometric algorithms can be roughly divided into two categories: (i) Exact computing and (ii) fixed-precision approximation. In the latter approach the algorithms are modified so that they can consistently cope with the limited precision of computer arithmetic. In the former, which is the approach taken by CGAL in general and the *2D Arrangements* package in particular, ideal computer arithmetic is emulated for the specific type of objects being manipulated, and the code is prepared for successfully handling degenerate input.

Advances in computer algebra enabled the development of efficient software libraries that offer exact arithmetic manipulations of unbounded integers, rational numbers (GMP, GNU's multi-precision integer and rational-number library[12]), unbounded floating-point numbers (MPFR, GNU's multi-precision floating-point library[13]) and algebraic numbers (the CORE library [124][14] and the numerical facilities of LEDA [151, Chapter 4]; these multi-precision number types serve as fundamental building blocks in the robust implementation of many geometric applications in general (see [219] for a review) and of those that employ arrangements in particular, including the *2D Arrangements* package.

Exact Geometric Computation (EGC), as summarized by Yap [219], simply amounts to ensuring that we never err in predicate evaluations. EGC represents a significant relaxation of the naive concept of numerical exactness. We only need to compute to sufficient precision to make the correct predicate evaluation. This has led to the development of several techniques such as precision-driven computation, lazy evaluation, adaptive computation, and floating-point filters, some of which are implemented in CGAL, such as numerical filtering. Here, computation is carried out using a number type that supports only inexact arithmetic (e.g., double-precision floating-point arithmetic), while keeping a bound on the computation error. The error bound is used by a filter that issues an event, referred to as a *filter failure* in the hacker's jargon, when the computation reaches a stage of uncertainty. When a filter failure occurs, the computation is redone using exact arithmetic. However, in cases where such a state is never reached, expensive computation is avoided, while the result is still certified.

Once the set of basic operations that a high-level algorithm is based on is determined, we have to make sure that they return correct results (possibly allowing intermediate roundoff errors). Given this, the algorithms on top of the basic operations just work—not in most cases, but always! The important thing for you to know is that you don't have to know about it. It all works automatically behind the scenes. Well, almost. If you look at an example program in CGAL, it typically starts with a long sequence of typedef statements. This is to some extent a consequence of the exact computation paradigm.

The chain of typedefs is necessary in order to choose the appropriate parameters for the algorithm. The handling of numerical robustness does not happen in the algorithm itself, but in its basic predicates and constructions. CGAL offers many different ways of getting these basics right, and which one is the best depends on the concrete application. That's why you have to choose. But in most cases, if you simply work with the settings that you find in the example program closest to your intended application, you will be fine. Switching between number types and exact computation techniques, and choosing the appropriate components that best suit the application needs, typically requires only a minor code change reflected in the instantiation of just a few data types. This convenient handling of numerical issues to expedite exact geometric computation is another advantage gained by the use of generic programming.

1.3.3 Well-Behaved Curves

What constitutes valid curves that can be handled by the *2D Arrangements* package is discussed in detail in Chapter 5, where the models of the traits classes are described. However, when we cite combinatorial complexity bounds or bounds on the resources (running time, storage) required by algorithms, we often postulate stricter assumptions on the input curves. The prevalent term

---

[12]See http://gmplib.org/.

[13]See http://www.mpfr.org/.

[14]See http://cs.nyu.edu/exact/core_pages/.

in use is that the curves are *well behaved*, which may have different interpretations in different settings. If we are concerned with combinatorial complexity bounds for curves in the plane, then the standard assumptions are that (i) each curve is non-self-intersecting (so-called Jordan arc) and (ii) every pair of curves intersects in at most some constant number of points. For algorithmic purposes we need to require more since we assume that any operation on a small constant number of curves takes unit time. In this sense arcs of algebraic curves of degree bounded by a constant (namely the zero set of bivariate polynomials of constant maximum total degree) are well behaved. Naturally, what are typically considered well-behaved surfaces in $\mathbb{R}^3$ is even more complicated to state. See Section 1.6 for more information.

Remarks. (i) From the complexity-bound perspective, most of the arrangements that we deal with in the book can be regarded as defined by well-behaved curves. Even though the package allows for self-intersecting curves, for most types each curve can be decomposed into a constant number of well-behaved curves, thus having no effect on the asymptotic bounds that we cite. (ii) One type of curves that we deal with is special in this sense: *polylines*, namely concatenations of an unlimited number of line segments. A polyline is not considered well behaved, as it cannot be decomposed into a constant number of constant-descriptive complexity subcurves. Informative bounds for arrangements of polylines are expressed by other parameters in addition to the number of polylines, for example, the total number of segments in all the polylines together.

1.4 The Computational Geometry Algorithm Library

1.4.1 CGAL Chronicles

Several research groups in Europe started to develop small geometry libraries on their own in the early 1990s. A consortium of several sites in Europe and Israel was founded in 1995 to cultivate the labor of these groups and gather their produce in a common library called CGAL — the Computational Geometry Algorithms Library. The goal was to promote research in computational geometry and translate the results into useful, reliable, and efficient programs for industrial and academic applications [64,133,173,202], the very same goal that governs CGAL development efforts to date.

An INRIA startup, GEOMETRY FACTORY,[15] was founded in January 2003. The company sells CGAL commercial licenses, support for CGAL, and customized development based on CGAL.

In November 2003, when Version 3.0 was released, CGAL became an open-source project,[16] allowing new contributions from various resources. Most parts of CGAL are now distributed under the GNU Lesser General Public License (GNU LGPL).

CGAL has evolved through the years and now represents the state of the art in computational geometry software. The implementations of the CGAL software modules described in this book are complete and robust, as they handle all degenerate cases. They rigorously adhere to the generic programming paradigm to overcome problems encountered when effective computational geometry software is implemented. A glimpse at the structure of CGAL is given in the following section.

1.4.2 CGAL Content

CGAL is written in C++ according to the generic programming paradigm described above. It has a common programming style, which is very similar to that of the STL. Its application programming interface (API) is homogeneous, and allows for convenient and consistent interfacing with other software packages and applications. The library consists of about 900,000 lines of code divided among approximately 4,000 files. CGAL also comes with numerous examples and demos. The didactic manual comprises about 3,500 pages. There are approximately 65 chapters arranged in

---

[15]See http://www.geometryfactory.com/.
[16]See http://www.cgal.org/.

14 parts. The `Arrangement_2` package, for example, consists of about 140,000 lines of code divided among approximately 300 files, described in a manual of about 300 pages.

One salient piece of CGAL consists of the geometric kernels [64]. A geometric kernel consists of types of constant-size non-modifiable geometric primitive objects (e.g., points, lines, triangles, and circles) and operations on objects of these types.

Another distinguishable piece, referred to as the "Support Library," consists of non-geometric facilities, such as circulators,[17] random generators, and I/O support for interfacing CGAL with various visualization tools (i.e., input and output streams). An important contribution of this piece is the number type support, on which we elaborate in the next subsection. This part also contains extensive debugging utilities that handle warnings and errors that may result from unfulfilled conditions.

Recall that a set of a traits-concept requirements may include runtime characteristics in the form of preconditions and postconditions. The code of CGAL in general, and the code of the *2D Arrangements* package in particular, follows specific rules regarding unfulfilled conditions, as follows. By default an unfulfilled condition is ignored when the code is compiled in *release* mode. When the code is compiled in *debug* mode, an unfulfilled condition leads to an assertion, which is the default, leads to an exception, or is completely ignored, based on your choice. You can control expensive (in terms of computing resources) conditions separately; consult the CGAL manual for further details.

The rest of the library offers a collection of geometric data structures and algorithms such as convex hulls, polygons and polyhedra, and operations on them (Boolean operations, polygon offsetting), 2D and 3D triangulations, Voronoi diagrams, surface meshing, subdivision, and reconstruction, search structures, geometric optimization, interpolation, and kinetic data structures. The 2D arrangements and its related data structures naturally fit in. These data structures and algorithms are parameterized by traits classes that define the interface between them and the primitives they use. In many cases, a kernel can be used as a traits class, or at least the subtypes of a kernel can be used as components of traits classes for these data structures and algorithms.

1.4.3   The Algebraic Foundations of CGAL

CGAL is aiming towards exact computation with non-linear objects, such as objects defined by algebraic curves and surfaces, as well as linear objects. As a consequence, types representing polynomials, algebraic extensions, and finite fields play an important role in related applications. To this end, CGAL defines a refinement hierarchy of algebraic concepts. Models of these concepts differ by their algebraic structures and by the property of whether they are embeddable on the real axis. In general, the requirements of a concept in this hierarchy specify what arithmetic operations, including their precisions, are supported by a model of the concept (e.g., exact division or square root).

Two compound concepts are relevant to you, namely, FieldNumberType and RingNumberType. The former combines the requirements of the concepts Field and RealEmbeddable. This roughly means that a model of this concept defines the operations $+$, $-$, $*$, and $/$ with semantics (approximately) corresponding to those of a field in the algebraic-structure sense. Note that, strictly speaking, the built-in storage type `int` does not fulfill the requirements on a field type. `int` objects correspond to elements of a ring rather than a field, as the operation $/$ is not the inverse of the $*$ operation when applied to numbers of the `int` type. The RingNumberType concept combines the requirements of the concepts IntegralDomainWithoutDivision and RealEmbeddable. This roughly means that a model of this concept defines the operations $+$, $-$, and $*$ with semantics (approximately) corresponding to those of a ring in the algebraic-structure sense. For more information on this hierarchy consult the manual of the Algebraic Foundations package. This package also introduces the notion of interoperable types, which allows you to perform operations with mixed types. For reasons that will become evident later, a kernel class must provide two nested types, namely,

---

[17] A circulator models the concept Circulator. It is quite similar to the Iterator concept, but it is used to iterate over circular ranges.

`Kernel::FT` and `Kernel::RT`, that model the concepts FieldNumberType and RingNumberType, respectively.

1.4.4 The Geometry Kernel of CGAL

The list of types of constant-size non-modifiable geometric primitive objects and operations on objects of these types establishes a concept referred to as the Kernel concept. Algorithms and data structures in the basic library of CGAL are parameterized by a traits class that subsumes the types of objects on which the algorithm or data structure operates as well as the operations on objects of these types. For many algorithms and data structures in the basic library you can use a model of the Kernel concept as a traits class. For some algorithms you do not even have to specify the kernel; it is detected automatically using generic programming techniques based on the types of the geometric objects passed to the algorithm. In some other cases, the algorithms or data structures need more than is provided by the Kernel concept. In these cases, a kernel cannot be used as a traits class, but can certainly aid in developing one [115]. This is, for example, the case with CGAL arrangements.

CGAL provides several models of the Kernel concept divided into two main families that differ in the way points are represented. The two families consist of models that use *Cartesian* and *homogeneous* point representations, respectively.[18] However, the interface of the kernel objects is designed such that it works well with both Cartesian and homogeneous representations. For example, a planar point can be constructed from three arguments as well (the three homogeneous coordinates of the point). The common interfaces allow you to develop code that is independent of the chosen representation.

All models are parameterized by number types representing the coordinates of the points. Instantiating a kernel with a particular number type is a trade-off between efficiency and accuracy. Naturally, availability also plays a role. The choice depends on the algorithm implementation and the expected input data to be handled. Evidently, number types have semantic constraints. That is, they should be meaningful in the sense that they approximate some subfield of the real numbers. CGAL provides several models of its number type concepts; some of them implement techniques to expedite exact computation mentioned in previous sections.

The `Cartesian<NumberType>` class template models the Kernel concept. It uses a Cartesian representation of coordinates. The template parameter `NumberType` determines the type of the coordinates of the kernel objects. The injected number type used with the Cartesian representation class-template models the concept FieldNumberType as described above. The built-in type `int`, also mentioned above, does not model the FieldNumberType concept. However, for some computations with Cartesian representation, no division operation is needed; that is, a model of the RingNumberType concept is sufficient in this case. Both types, `Cartesian<NumberType>::FT` and `Cartesian<NumberType>::RT` are defined as `NumberType`.

The `Cartesian<NumberType>` class template uses reference counting of the geometric objects. CGAL also provides the class template `Simple_cartesian<NumberType>` a kernel that uses Cartesian representation but no reference counting. Debugging is easier with `Simple_cartesian<NumberType>`, since the coordinates are stored within the class, and hence direct access to the coordinates is possible. Depending on the algorithm, it can also be slightly more or less efficient than `Cartesian<NumberType>`. Both `Simple_cartesian<NumberType>::FT` and `Simple_cartesian<NumberType>::RT` are mapped to `NumberType`.

Using homogeneous coordinates obviates the need for division operations in numerical computations, since the additional coordinate serves as a common denominator. Avoiding divisions can be useful for exact geometric computation. The `Homogeneous<RingNumberType>` class template uses a homogeneous representation for the coordinates of the kernel objects. As for the Cartesian representation, you must instantiate the type used to store the coordinates. Since the homogeneous representation does not use division, the number type associated with an instance of the homogeneous representation class-template must model the weaker concept RingNumberType

---

[18] A point in the plane in *homogeneous* representation is given by (x, y, w). The corresponding Cartesian representation is $(x/w, y/w)$.

only. However, some operations provided by this kernel involve division, for example computing Cartesian coordinates. To keep the requirements on the number type parameter of `Homogeneous` low, the number type `Quotient<RingNumberType>` is used for operations that require division. This number type can be viewed as an adaptor, which turns a model of the RingNumberType concept into a model of the FieldNumberType concept. It maintains numbers as quotients, i.e., a numerator and a denominator. The nested type `Homogeneous<RingNumberType>::FT` is defined as `Quotient<RingNumberType>`; the nested type `Homogeneous<RingNumberType>::RT` is defined as `RingNumberType`.

As with the `Simple_cartesian<NumberType>` class template, CGAL also provides the class template `Simple_homogeneous<RingNumberType>`—a kernel that uses homogeneous representation but no reference counting. Depending on the algorithm, it can also be slightly more or less efficient than `Homogeneous<RingNumberType>`. Again, the nested type `Simple_homogeneous< RingNumberType>::FT` is defined as `Quotient<RingNumberType>`, while the nested type `Simple_ homogeneous<RingNumberType>::RT` is defined as `RingNumberType`.

CGAL also provides a glue layer that adapts number type classes implemented by external libraries as models of its number type concepts. For example, `CGAL::Gmpq` is a wrapper class for the GNU multi-precision rational-number type, and models the concept FieldNumberType.

In summary, the instantiation of a kernel typically succeeds the instantiation of a particular number type, and precedes the instantiation of a traits class and the arrangement data structure type as follows:

```
typedef CGAL::Gmpq                           Number_type;
typedef CGAL::Cartesian<Number_type>         Kernel;
typedef CGAL::Arr_linear_traits_2<Kernel>    Traits_2;
typedef CGAL::Arrangement_2<Traits_2>        Arrangement_2;
```

CGAL kernels apply filtering techniques based on interval arithmetic [32, 153] to achieve exact and efficient predicate evaluation. Some kernels also apply lazy evaluations [65], where the exact computations are delayed at runtime until they are actually needed, if at all, to achieve exact and efficient geometric-object constructions as well as to further expedite predicate evaluation.

For your convenience, CGAL provides some predefined types of useful kernels. All predefined kernels are Cartesian. They all support constructions of points from double-precision floating-point (`double`) Cartesian coordinates. They all provide exact geometric predicates, but they handle geometric constructions differently:

- `Exact_predicates_exact_constructions_kernel` provides exact geometric constructions, in addition to exact geometric predicates.

- `Exact_predicates_inexact_constructions_kernel` provides exact geometric predicates, but geometric constructions may be inexact due to roundoff errors. Naturally, it is faster than the `Exact_predicates_exact_constructions_kernel` kernel; thus, it is preferable for geometric algorithms that just issue predicates involving their input entities and do not perform construction of new geometric objects (e.g., computing the convex hull of a set of points). In the context of arrangements, we usually have to construct points of intersection between input curves, so this kernel should be used with discretion.

When using a predefined kernel, replace the first two statements in the code excerpt above with the following single one:

```
typedef CGAL::Exact_predicates_exact_constructions_kernel Kernel;
```

Types appear as standalone classes encapsulated by the kernel. For example, a two-dimensional point is defined as follows:

```
Kernel::Point_2 p(1,2), q(-2,1);
```

The Kernel concept lists over 100 predicates and over 150 constructors. It is possible to extend any primitive-object type of every kernel model or even to exchange it for a different type, without the need to redefine any one of the operations with the help of generic programming. Operations appear either as free functions or as function objects (also known as functors) encapsulated by the kernel. This interface presents a good face to alternative styles of programming. Formally, these are types nested in the kernel; each such type refines a Functor concept variant[19] and models some concept. For example, to compare the x-coordinates of two points p and q you may use a free function:

```
if (CGAL::compare_x_2(p, q)) { ... }
```

or you may use a functor:

```
if (kernel.compare_x_2_object()(p, q)) { ... }
```

Using the second alternative enables encapsulating both types and operations by the kernel. This allows for the types and the operations to be adapted and exchanged at the same time—for example, it is possible to extend the types defined by the kernel with some auxiliary data (e.g., adding color attributes to points).

A geometric kernel of CGAL can be used as a traits class for many purposes. Let us revise the example listed in Section 1.2.2, and implement the function **bounding_rectangle()** in terms of kernel functors:

```
template <typename Input_iterator, typename Traits>
typename Traits::Iso_rectangle_2
bounding_rectangle (Input_iterator pts_begin, Input_iterator pts_end)
{
  Traits                        traits;
  typename Traits::Compare_x_2  comp_x = traits.compare_x_2_object();
  typename Traits::Compare_y_2  comp_y = traits.compare_y_2_object();
  Input_iterator                left, right, top, bottom, curr = pts_begin;
  left = right = top = bottom = curr;
  while (++curr; curr != pts_end; ++curr) {
    if (comp_x(*curr, *left) == CGAL::SMALLER) left = curr;
    else if (comp_x(*curr, *right) == CGAL::LARGER) right = curr;
    if (comp_y(*curr, *bottom) == CGAL::SMALLER) bottom = curr;
    else if (comp_y(*curr, *top) == CGAL::LARGER) top = curr;
  }

  return
    traits.construct_iso_rectangle_2_object()(*left, *right, *bottom, *top);
}
```

1.4.5 The State of CGAL

The above accurately describes the model of CGAL at the time this book was written. Constant and persistent improvement to the source code and the didactic manuals, review of packages by the CGAL Board,[20] and exhaustive testing through the years led to excellent quality internationally recognized as unrivaled in its field. CGAL has a foothold in many domains related to computational geometry and can be found in many academic and research institutes as well as commercial products. Release 3.5 was downloaded more than 15,000 times, and the public discussion list counts more than 1,000 subscribed users.

---

[19] In general, a model of a Functor concept must define the operator(); see, for example, http://www.sgi.com/tech/stl/functors.html for more details.

[20] The CGAL Board (formerly the CGAL Editorial Board) is a directing group responsible for guiding the development of the library, developers, and the user community.

1.4.6 The Namespace of CGAL

All names introduced by CGAL, especially those documented in the manuals, are scoped in a namespace called `CGAL`, which is in global scope. You can either qualify names from CGAL by adding `CGAL::`, e.g., `CGAL::Point_2<CGAL::Cartesian<int> >`, make a single name from CGAL visible in a scope via a `using` statement, e.g., `using CGAL::Cartesian;`, and then use this name unqualified in this scope, or even make all names from namespace `CGAL` visible in a scope with `using namespace CGAL;`. The latter, however, may cause naming collisions, and is recommended for use with caution. Argument-dependent name lookup (ADL, also known as Koenig lookup), which is applied by advanced C++ compilers to look up unqualified function-names based on the types of the arguments passed to functions, is widely recognized as an essential feature. However, it makes namespaces less strict, and so can require the use of fully qualified names when this would not otherwise be needed.

The code listed in this book includes references to different namespaces, i.e., `std`, `boost`, `CORE`, `leda`, and `CGAL`. However, it does not rely on ADL beyond the normal dependency-level. Nevertheless, to remove all doubt, most references to names defined in these namespaces are fully qualified. In the text itself we gave ourselves the freedom to remove the qualification in many cases (at least for names of functions and constructs defined in the `CGAL` namespace).

1.5 Outline of the Book

In the following chapters we introduce the *2D Arrangements* package of CGAL [214] step by step. We use ample programming *examples* to explain and demonstrate how to write code and use the capabilities of the package. These examples are typically small and serves to present one feature at a time. Along with these demonstrative examples, the book contains a collection of larger scale examples, which we call *applications*. We present a problem (of the type listed in Section 1.1), a solution to the problem using arrangements, and a complete program that implements the solution. We are hopeful that the application programs can be used as is, or that after reading (parts of) this book, it will be easy for you to modify the application programs and adapt them to your specific needs. All programs and examples can be found on the book's website `http://acg.cs.tau.ac.il/cgal-arrangement-book`.

Chapter 2 introduces the data structure that represents the incidence relations among the features of a two-dimensional arrangement, namely the doubly-connected edge list (DCEL). It exposes the fundamental `Arrangement_2` class template, which deals with the combinatorial part of the arrangement algorithms. You will learn how to construct an arrangement, traverse it, and locally modify it.

Chapter 3 introduces utility classes and free (global) functions that operate on arrangements, and encapsulate more involved algorithmic frameworks. Reading this chapter you will learn how to issue queries on an arrangement and how to apply global changes to it, such as the insertion of a set of curves into a nonempty arrangement.

To simplify the exposition of the *2D Arrangements* package, Chapters 2 and 3 focus on arrangements of line segments, which are perhaps the simplest form of *bounded* planar curves. However, the arrangement package can also handle *unbounded* curves. In Chapter 4 we show how to construct and maintain an arrangement of linear curves, namely, lines, rays, and line segments. At the end of this chapter we solve the **minimum-area triangle problem**, which requires the construction and traversal of an arrangement of lines.

We recommend that you read Chapters 2, 3, and 4 in full. For many purposes these may be the only chapters you really need to read before working with CGAL arrangements. Most of the ensuing chapters elaborate and extend the basic capabilities presented in these chapters.

Recall that it is possible to construct arrangements of diverse families of planar curves, and not just linear curves. The geometry of the curves is encapsulated in a so-called *traits class*. Chapter 5 reviews the various traits classes supplied with the *2D Arrangements* package. You will find ready-made traits classes for line segments (even several types of these), for conic arcs, for Bézier curves, and more. Reading this chapter you will also learn how to attach additional information of your needs and preferences to the curves, through *traits-class decorators*.

Adding information to curves is one way to enhance the arrangements with your own application data. At other times it is natural to enhance arrangement faces, edges,[21] or vertices with data. In Chapter 6 we show how to extend the DCEL features with the desired data. In addition, we present the *notification* mechanism that enables external (possibly user-defined) classes to be notified on modifications to the arrangement. This mechanism is important for keeping the application data up-to-date as the arrangement undergoes a change. Also in this chapter, we present the map-overlay operation, which lays two arrangements, one on top of the other, to produce a third arrangement. This operation typically applies to extended arrangements.

In many applications arrangements are used to discretize continuous problems as stated above. The discretization is through an *invariant*, which is a property of each point in space. In the arrangement, all the points in every cell of any dimension have the same invariant. A typical algorithm for solving a problem with arrangements goes over all (or some of) the cells of an arrangement and computes the invariant in every cell. Often, the invariant changes only slightly between adjacent cells, so traversing the cells in some contiguous order (say breadth first or depth first) and only updating the invariant as it changes from one cell to its neighbor could be much more efficient than traversing the cells in an arbitrary order, independently computing the invariant for each cell. This calls for using graph algorithms on top of the arrangements. The BOOST Graph Library supplies a wealth of algorithms, and is written in the same spirit as STL and CGAL, following the generic programming paradigm. Chapter 7 describes two adaptors of arrangements to BGL graphs provided by our package. The *primal adaptor* regards the vertices and edges of the arrangement as nodes and arcs of a planar graph. The *dual adaptor* assigns a graph node to each arrangement face. Two nodes in the graph are connected by an arc if the boundaries of their respective faces share a common edge. Having a convenient way to traverse arrangement features paves the way for easier problem solving. Indeed, we give an application in this chapter to exemplify the usefulness of the BOOST adaptors, which asks for the largest common point sets under ϵ-congruence: Given two finite sets of points A and B in the plane, we wish to find equally sized subsets $A' \subseteq A$ and $B' \subseteq B$ of maximal cardinality, such that points in A' match points in B' under translation, up to distance less than some given ϵ.

Boolean operations on polygons are often required in computer graphics (CG), computer-aided design and manufacturing (CAD/CAM), electronic design automation (EDA), as well as in many other areas. Chapter 8 presents a layer developed on top of the basic arrangement software, which supports intersection, union, and other Boolean operations on polygons. Moreover, it supports other types of operations, the polygons need not be linear, and the operations can be applied to more than two operands. We call the latter type *multiway* operations. One of the applications presented in the chapter applies multiway operations on *generalized* polygons, namely, simple polygons whose boundaries consist of straight line segments and circular arcs.

One of the most frequently used operations in domains that employ geometric computing is constructing the *Minkowski sums* of two objects. The Minkowski sum of two sets A and B in the plane is the set $\{a + b \mid a \in A, b \in B\}$, denoted by $A \oplus B$. Chapter 9 presents functions that compute the sum of two polygons and the sum of a polygon and a disc of radius r; the latter operation is often referred to as *offsetting* the polygon by r. In a manner that is explained in the chapter, Minkowski sums arise in various layout, placement and material-cutting problems, as well as in the case of translational collision-free motion-planning amidst obstacles. We provide two motion-planning applications: (i) We give a complete solution to the **translational motion-planning problem** for a polygon robot moving amidst polygonal obstacles in the plane. (ii) We give a hybrid solution to the **two-disc coordination problem** for robots translating amidst polygonal obstacles in the plane, combining arrangements with sampling-based techniques.

In Chapter 10 we take (half) a step toward three-dimensional arrangements: We compute lower and upper *envelopes* of surfaces in three-dimensional space. This is only half a step, since such

---

[21] Notice that arrangement *edges* are different from the underlying input curves that induce the arrangement. The edges correspond to maximal portions of the input curves that are x-monotone and do not intersect other curves.

envelopes are most suitably described by special two-dimensional arrangements, which we call the minimization diagrams for lower envelopes or the maximization diagrams for upper envelopes. Envelopes have many applications in their own right. For example, one can compute fairly general *Voronoi diagrams* in the plane using envelopes of surfaces. The chapter starts with the much simpler case of envelopes of curves in the plane.

We conclude the book by pointing out future directions for developing software for arrangements. Chapter 11 discusses two-dimensional arrangements on curved surfaces, higher-dimensional arrangements, fixed-precision geometric computing (in contrast to the exact geometric paradigm followed in this book), and more applications.

1.6   Further Reading

The book brings together two broad topics: arrangements on the one hand and geometric programming on the other hand. Arrangements have been intensively studied for decades. Some of the earlier publications include [88, Chapter 18], [89], and [90] by Grünbaum, and the monograph by Zaslavsky [221]. The computational study of arrangements of hyperplanes till the mid-1980s is summarized in the book by Edelsbrunner [53]. The combinatorics and algorithmics of arrangements of curves and surfaces till the mid-1990s are summarized in the book by Sharir and Agarwal [196]. Many books have one or more chapters on arrangements, including [30, 149, 164, 171]. We mention in particular the general computational geometry book by de Berg et al. [45], since in addition to the chapter on arrangements of lines, it has several chapters that could come in handy for you while reading this book, covering topics such as the plane-sweep paradigm, trapezoidal decomposition, point location and other search structures, Minkowski sums, and more.

There are (at least) two expansive surveys on arrangements, which beyond their common topics are distinguished by different foci: One [4] carefully discusses what constitutes "well-behaved" curves and surfaces and has a more in-depth coverage of the theory of arrangements, while the other [98] discusses the robustness issues as addressed in the context of arrangements and surveys software developed for arrangements. The first chapter of the book *Effective Computational Geometry for Curves and Surfaces* [29] is dedicated to arrangements, with an emphasis on the numerical and algebraic computation underlying the arrangement construction. It is complementary to this book, since here we hardly discuss how the algebraic methods, which are encapsulated in the traits classes, are implemented.

The rigorous study of geometric programming got a big boost in the mid-1990s, and several dozens of papers on the topic have been published since in the applied track of the European Symposium on Algorithms (ESA) and its predecessor the Workshop on Algorithm Engineering (WAE), in the workshop on Algorithm Engineering and Experiments (ALENEX), and in several related journals, such as the ACM Journal on Experimental Algorithmics. The major conferences on computational geometry and in particular the Symposium on Computational Geometry (SoCG) occasionally have papers on the topic, and the journal *Computational Geometry, Theory & Applications*, in addition to publishing papers on geometric programming, dedicated a special issue to CGAL. Good starting points on the topic are the chapter on CGAL and LEDA in the *Handbook of Discrete and Computational Geometry* [133], and the book on LEDA [151].

More generally, Austern, in his book [12], introduces you to the generic programming paradigm. He explains central ideas underlying generic programming and shows how these ideas lead to the fundamental concepts of STL. Musser et al. [167] provide a comprehensive and practical introduction to STL. Stroustrup, the inventor of the C++ programming language, describes key programming and design techniques supported by modern standard C++ [200]. Vandevoorde et al. [204] cover all the ins and outs of programming with templates. The BOOST Graph Library is presented in the book by Siek et al. [198], which also provides excellent insights into generic programming in general. We often refer to software *design-patterns*. While the implementations of the design patterns described in this book are generic, and thus slightly differ from the object-oriented implementations as described in Gamma et al. [83], their intents are identical. Finally, our concise notes would not be complete without our mentioning Alexandrescu's book [7], which offers programming techniques that combine design patterns, generic programming, and C++.

Chapter 2

Basic Arrangements

We start with a formal definition of two-dimensional arrangements, and proceed with an introduction to the data structure used to represent the incidence relations among features of two-dimensional arrangements, namely, the *doubly-connected edge list*, or DCEL for short. Then we describe the main class of the *2D Arrangements* package and the functions it supports. This chapter contains the basic material you need to know in order to use CGAL arrangements.

2.1 Representation of Arrangements: The DCEL

Given a set $\mathcal{C}$ of planar curves, the *arrangement* $\mathcal{A}(\mathcal{C})$ is the subdivision of the plane into zero-dimensional, one-dimensional, and two-dimensional *cells*,[1] called *vertices*, *edges*, and *faces*, respectively, induced by the curves in $\mathcal{C}$.

The curves in $\mathcal{C}$ can intersect each other (a single curve may also be self-intersecting or may comprise several disconnected branches) and are not necessarily x-monotone.[2] We construct in two steps a collection $\mathcal{C}''$ of x-monotone subcurves that are pairwise disjoint in their interiors as follows. First, we decompose each curve in $\mathcal{C}$ into maximal x-monotone subcurves and possibly isolated points, obtaining the collection $\mathcal{C}'$. Note that an x-monotone curve cannot be self-intersecting. Then, we decompose each curve in $\mathcal{C}'$ into maximal connected subcurves not intersecting any other curve (or point) in $\mathcal{C}'$ in its interior. The collection $\mathcal{C}''$ contains isolated points if the collection $\mathcal{C}'$ contains such points. The arrangement induced by the collection $\mathcal{C}''$ can be conveniently embedded as a planar graph, the vertices of which are associated with curve endpoints or with isolated points, and the edges of which are associated with subcurves. It is easy to see that the faces of $\mathcal{A}(\mathcal{C})$ are the same as the faces of $\mathcal{A}(\mathcal{C}'')$. There are possibly more vertices in $\mathcal{A}(\mathcal{C}'')$ than in $\mathcal{A}(\mathcal{C})$—the vertices where curves were cut into x-monotone (non-intersecting) pieces; accordingly there may also be more edges in $\mathcal{A}(\mathcal{C}'')$. This graph can be represented using a *doubly-connected edge list* (DCEL) data structure, which consists of containers of vertices, edges, and faces and maintains the incidence relations among these cells. It is one of a family of combinatorial data structures called *halfedge data structures* (HDS), which are edge-centered data structures capable of maintaining incidence relations among cells of, for example, planar subdivisions, polyhedra, or other orientable, two-dimensional surfaces embedded in space of an arbitrary dimension. Geometric interpretation is added by classes built on top of the halfedge data structure.

The DCEL data structure represents each edge using a pair of directed *halfedges*, one going from the xy-lexicographically smaller (left) endpoint of the curve towards the xy-lexicographically

---

[1] We use the term *cell* to describe the various dimensional entities in the induced subdivision. Sometimes, the term *face* is used for this purpose in the literature. However, in this book, we use the term *face* to describe a *two-dimensional* cell.

[2] A continuous planar curve C is *x-monotone* if every vertical line intersects it at most once. For example, a non-vertical line segment is always x-monotone, and so is the graph of any continuous function $y = f(x)$. A circle of radius r centered at (x_0, y_0) is not x-monotone, as the vertical line $x = x_0$ intersects it at $(x_0, y_0 - r)$ and at $(x_0, y_0 + r)$. For convenience, we always deal with *weakly x-monotone* curves, which include vertical linear curves.

E. Fogel et al., *CGAL Arrangements and Their Applications*, Geometry and Computing 7,
DOI 10.1007/978-3-642-17283-0_2, © Springer-Verlag Berlin Heidelberg 2012

Fig. 2.1: An arrangement of interior-disjoint line-segments with some of the DCEL records that represent it. The unbounded face f_0 has a single connected component that forms a hole inside it, and this hole consists of several faces. The halfedge e is directed from its source vertex v_1 to its target vertex v_2. This halfedge, together with its twin e', corresponds to a line segment that connects the points associated with v_1 and v_2 and separates the face f_1 from f_2. The predecessor e_{prev} and successor e_{next} of e are part of the chain that forms the boundary of the face f_2. The face f_1 has a more complicated structure, as it contains two holes in its interior: One hole contains two faces f_3 and f_4, while the other hole consists of just two edges. f_1 also contains two isolated vertices u_1 and u_2 in its interior.

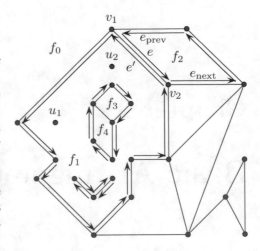

larger (right) endpoint, and the other, known as its *twin* halfedge, going in the opposite direction.

As each halfedge is directed, it has a *source* vertex and a *target* vertex. Halfedges are used to separate faces and to connect vertices, with the exception of *isolated vertices* (representing isolated points), which are disconnected. If a vertex v is the target of a halfedge e, we say that v and e are *incident* to each other. The halfedges incident to a vertex v form a circular list sorted in clockwise order around this vertex; see the figure to the right. (An isolated vertex has no incident halfedges.)

An *edge* of an arrangement is a maximal portion of a curve between two vertices of the arrangement. Each edge is represented in the DCEL by a pair of twin halfedges. Each halfedge e stores a pointer to its *incident face*, which is the face lying to its left. Moreover, every halfedge is followed by another halfedge sharing the same incident face, such that the target vertex of the halfedge is the same as the source vertex of the next halfedge. The halfedges around faces form

circular chains, such that all halfedges of a chain are incident to the same face and wind along its boundary; see the figure above. We call such a chain a *connected component of the boundary*, or CCB for short.

The unique CCB of halfedges winding in a counterclockwise orientation along a face boundary is referred to as the *outer CCB* of the face. For the time being, let us consider only arrangements of bounded curves, such that exactly one unbounded face exists in every arrangement. The unbounded face does not have an outer boundary. Any other connected component of the boundary of the face is called a *hole*, or *inner CCB*, and can be represented as a circular chain of halfedges winding in a clockwise orientation around it. Note that a hole does not necessarily correspond to a face at all, as it may have no area, or alternatively it may contain several faces inside it. Every face can have several holes in its interior, or may contain no holes at all. In addition, every face may contain isolated vertices in its interior. See Figure 2.1 for an illustration of the various DCEL features.

So much on the abstract description of the DCEL. This is the underlying data structure of the CGAL arrangement class. Occasionally, this is the only data structure we need, especially when we are only concerned with traversing the arrangement.

2.2   The Main Arrangement Class

The main component in the *2D Arrangements* package is the `Arrangement_2<Traits,Dcel>` class template. An instance of this template is used to represent planar arrangements. The class template provides the interface needed to construct such arrangements, traverse them, and maintain them.

The design of the *2D Arrangements* package is guided by two aspects of modularity as follows: (i) the separation of the representation of the arrangements and the various geometric algorithms that operate on them, and (ii) the separation of the topological and geometric aspects of the planar subdivision. The latter separation is exhibited by the two template parameters of the `Arrangement_2` class template; their description follows.

- The `Traits` template parameter should be substituted with a model of one of the geometry traits concepts, for example, the ArrangementBasicTraits_2 concept. A model of this traits concept defines the types of x-monotone curves and two-dimensional points, `X_monotone_curve_2` and `Point_2`, respectively, and supports basic geometric predicates on them.

  In this chapter we always use `Arr_non_caching_segment_traits_2` as our traits-class model in order to construct arrangements of line segments. In Chapter 3 we also use `Arr_segment_traits_2` as our traits-class model. In Chapter 4 we use `Arr_linear_traits_2` to construct arrangements of linear curves (i.e., lines, rays, and line segments). The *2D Arrangements* package contains several other traits classes that can handle other types of curves, such as polylines (continuous piecewise-linear curves), conic arcs, and arcs of rational functions. We exemplify the usage of these traits classes in Chapter 5. A few additional models have been developed by other groups of researchers.

- The `Dcel` template parameter should be substituted with a class that models the ArrangementDcel concept, which is used to represent the topological layout of the arrangement. This parameter is substituted with `Arr_default_dcel<Traits>` by default, and we use this default value in this and in the following three chapters. However, in many applications it is necessary to extend the DCEL features. This is done by substituting the `Dcel` parameter with a different type; see Section 6.2 for further explanation and examples.

The function template `print_arrangement_size()` listed below prints out quantitative measures of a given arrangement. While in what follows it is used only by examples, it demonstrates well the use of the member functions `number_of_vertices()`, `number_of_edges()`, and `number_of_faces()`, which return the number of vertices, edges, and faces of an arrangement, respectively. The function template is defined in the header file `Arr_print.h`.

```
template <typename Arrangement>
void print_arrangement_size(const Arrangement& arr)
{
  std::cout << "The arrangement size:" << std::endl
            << "   |V| = " << arr.number_of_vertices()
            << ",  |E| = " << arr.number_of_edges()
            << ",  |F| = " << arr.number_of_faces() << std::endl;
}
```

You can also obtain the number of halfedges of an arrangement using the member function `number_of_halfedges()`. Recall that the number of halfedges is always twice the number of edges.

Example: The simple program listed below constructs an arrangement of three connected line-segments forming a triangle. It uses the CGAL *Cartesian kernel* (see Section 1.4.4) with an integral-number type to instantiate the `Arr_segment_traits_2` class template. The resulting arrangement consists of two faces, a bounded triangular face and the unbounded face. Constructing and maintaining arrangements using limited-precision numbers, such as `int`, works properly only under severe restrictions, which in many cases render the program not very useful. In this example, however, the points are far apart, and constructions of new geometric objects do not occur. Thus, it is safe to use `int` after all. The program constructs an arrangement induced by three line segments that are pairwise disjoint in their interior, prints out the number of faces, and ends. It uses the `insert()` free-function, which inserts the segments into

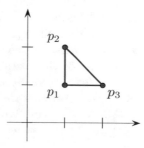

the arrangement; see Section 3.4. It uses the member function **number_of_faces()** to obtain the number of faces (two in this case). We give more elaborate examples in the rest of this chapter. The programs in those examples rely on computing with numbers of arbitrary precision, which guarantees robust execution and correct results.

```
// File: ex_triangle.cpp

#include <CGAL/Cartesian.h>
#include <CGAL/Arr_non_caching_segment_traits_2.h>
#include <CGAL/Arrangement_2.h>

typedef int                                              Number_type;
typedef CGAL::Cartesian<Number_type>                     Kernel;
typedef CGAL::Arr_non_caching_segment_traits_2<Kernel>   Traits;
typedef Traits::Point_2                                  Point;
typedef Traits::X_monotone_curve_2                       Segment;
typedef CGAL::Arrangement_2<Traits>                      Arrangement;

int main()
{
  Point      p1(1, 1), p2(1, 2), p3(2, 1);
  Segment    cv[] = {Segment(p1, p2), Segment(p2, p3), Segment(p3, p1)};
  Arrangement  arr;
  insert(arr, &cv[0], &cv[sizeof(cv)/sizeof(Segment)]);
  std::cout << "Number_of_faces:_" << arr.number_of_faces() << std::endl;
  return 0;
}
```

 Try: Modify the program above so that it inserts the number of vertices and halfedges as well as the number of faces into the standard output-stream.

2.2.1 Traversing the Arrangement

The simplest and most fundamental arrangement operations are the various traversal methods, which allow you to systematically go over the relevant features of the arrangement at hand.

Since the arrangement is represented as a DCEL, which stores containers of vertices, halfedges, and faces, the **Arrangement_2** class template supplies iterators for these containers. For example, if **arr** is an **Arrangement_2** object, the calls **arr.vertices_begin()** and **arr.vertices_end()** return iterators of the nested **Arrangement_2::Vertex_iterator** type that define the valid range of arrangement vertices. The value type of this iterator is **Arrangement_2::Vertex**. Moreover, the vertex-iterator type is convertible to **Arrangement_2::Vertex_handle**, which serves as a pointer to a vertex. As we show next, all functions related to arrangement features accept handle types as input parameters and return handle types as their output. A handle models the STL concept TrivialIterator.[3] Throughout this book, we use the identifiers **v**, **he**, and **f** to refer to a vertex handle, a halfedge handle, and a face handle, respectively.

In addition to the iterators for arrangement vertices, halfedges, and faces, the **Arrangement_2** class template also provides an iterator for edges, namely **Arrangement_2::Edge_iterator**. The value type of this iterator is **Arrangement_2::Halfedge**, which is used to represent one of the twin halfedges associated with the edge. The calls **arr.edges_begin()** and **arr.edges_end()** return iterators that define the valid range of arrangement edges.

All iterator, circulator,[4] and handle types also have non-mutable (**const**) counterparts. These non-mutable iterators are useful for traversing an arrangement without changing it. For example,

---

[3] A handle is a lightweight object that behaves like a pointer; hence, it is more efficient to pass handles around.

[4] A *circulator* is used when traversing a circular list, such as the list of halfedges incident to a vertex.

the arrangement has a mutable member-function called `arr.vertices_begin()` that returns an `Arrangement_2::Vertex_iterator` object and another non-mutable member-function that returns an `Arrangement_2::Vertex_const_iterator` object. In fact, all methods listed in this section that return an iterator, a circulator, or a handle have non-mutable counterparts. It should be noted that, for example, an `Arrangement_2::Vertex_handle` is convertible into an `Arrangement_2::Vertex_const_handle`, but not the other way around.

Conversions of non-mutable handles to the corresponding mutable handles are nevertheless possible. They can be performed using the overloaded member-function `non_const_handle()`. There are three variants that accept a non-mutable handle to a vertex, a halfedge, or a face, respectively. Only mutable objects of type `Arrangement_2` can call the `non_const_handle()` method; see, e.g., Section 3.1.1.

Traversal Methods for an Arrangement Vertex

A vertex v of an arrangement induced by bounded curves is always associated with a geometric entity, namely, with a `Point_2` object, which can be obtained by `v->point()`. Recall that `v` identifies a vertex handle; hence, we treat it as a pointer.

The call `v->is_isolated()` determines whether the vertex v is isolated or not. Recall that the halfedges incident to a non-isolated vertex, namely the halfedges that share a common target vertex, form a circular list around this vertex. The call `v->incident_halfedges()` returns a circulator of the nested type `Arrangement_2::Halfedge_around_vertex_circulator` that enables the traversal of this circular list around a given vertex v in a clockwise order. The value type of this circulator is `Arrangement_2::Halfedge`. By convention, the target of the halfedge is v. The call `v->degree()` evaluates to the number of the halfedges incident to v.

Example: The function below prints all the halfedges incident to a given arrangement vertex (assuming that the `Point_2` type can be inserted into the standard output-stream using the << operator). The arrangement type is the same as in the simple example (coded in `ex_triangle.cpp`) above.

```
template <typename Arrangement>
void print_incident_halfedges(typename Arrangement::Vertex_const_handle v)
{
  if (v->is_isolated()) {
    std::cout << "The vertex (" << v->point() << ") is isolated" << std::endl;
    return;
  }
  std::cout << "The neighbors of the vertex (" << v->point() << ") are:";
  typename Arrangement::Halfedge_around_vertex_const_circulator  first, curr;
  first = curr = v->incident_halfedges();
  do std::cout << " (" << curr->source()->point() << ")";
  while (++curr != first);
  std::cout << std::endl;
}
```

If v is an isolated vertex, the call `v->face()` obtains the face that contains v.

Traversal Methods for an Arrangement Halfedge

A halfedge e of an arrangement induced by bounded curves is associated with an `X_monotone_curve_2` object, which can be obtained by `he->curve()`, where `he` identifies a handle to e.

The calls `he->source()` and `he->target()` return handles to the halfedge source-vertex and target-vertex, respectively. You can obtain a handle to the twin halfedge using `he->twin()`. Note that from the definition of halfedges in the DCEL structure, the following invariants always hold:
- `he->curve()` is equivalent to `he->twin()->curve()`,
- `he->source()` is equivalent to `he->twin()->target()`, and

- `he->target()` is equivalent to `he->twin()->source()`.

Every halfedge has an incident face that lies to its left, which can be obtained by `he->face()`. Recall that a halfedge is always one link in a connected chain (CCB) of halfedges that share the same incident face. The `he->prev()` and `he->next()` calls return handles to the previous and next halfedges in the CCB, respectively.

As the CCB is a circular list of halfedges, it is only natural to traverse it using a circulator. Indeed, `he->ccb()` returns an `Arrangement_2::Ccb_halfedge_circulator` object for traversing all halfedges along the connected component of `he`. The value type of this circulator is `Arrangement_2::Halfedge`.

Example: The function template `print_ccb()` listed below prints all x-monotone curves along a given CCB (assuming that the `Point_2` and the `X_monotone_curve_2` types can be inserted into the standard output-stream using the `<<` operator).

```
template <typename Arrangement>
void print_ccb (typename Arrangement::Ccb_halfedge_const_circulator circ)
{
  std::cout << "(" << circ->source()->point() << ")";
  typename Arrangement::Ccb_halfedge_const_circulator  curr = circ;
  do {
    typename Arrangement::Halfedge_const_handle he = curr;
    std::cout << "   [" << he->curve() << "]   "
              << "(" << he->target()->point() << ")";
  } while (++curr != circ);
  std::cout << std::endl;
}
```

Traversal Methods for an Arrangement Face

An `Arrangement_2` object `arr` that identifies an arrangement of bounded curves always has a single unbounded face. The call `arr.unbounded_face()` returns a handle to this face. Note that an empty arrangement contains nothing *but* the unbounded face.

Given a handle to a face f, you can issue the call `f->is_unbounded()` to determine whether the face f is unbounded. Bounded faces have an outer CCB, and the `outer_ccb()` method returns a circulator of type `Arrangement_2::Ccb_halfedge_circulator` for traversing the halfedges along this CCB. Note that the halfedges along this CCB wind in a *counterclockwise* order around the outer boundary of the face.

A face can also contain disconnected components in its interior, namely, holes and isolated vertices. You can access these components as follows:

- You can obtain a pair of `Arrangement_2::Hole_iterator` iterators that define the range of holes inside a face f by calling `f->holes_begin()` and `f->holes_end()`.

  The value type of this iterator is `Arrangement_2::Ccb_halfedge_circulator`, defining the CCB that winds in a *clockwise* order around a hole.

- The calls `f->isolated_vertices_begin()` and `f->isolated_vertices_end()` return `Arrangement_2::Isolated_vertex_iterator` iterators that define the range of isolated vertices inside the face f. The value type of this iterator is `Arrangement_2::Vertex`.

Example: The function template `print_face()` listed below prints the outer and inner boundaries of a given face. It uses the function template `print_ccb()` listed above.

```
template <typename Arrangement>
void print_face (typename Arrangement::Face_const_handle f)
{
  // Print the outer boundary.
```

```cpp
  if (f–>is_unbounded()) std::cout << "Unbounded_face._" << std::endl;
  else {
    std::cout << "Outer_boundary:_";
    print_ccb<Arrangement>(f–>outer_ccb());
  }

  // Print the boundary of each of the holes.
  int                                              index = 1;
  typename Arrangement::Hole_const_iterator   hole;
  for (hole = f–>holes_begin(); hole != f–>holes_end(); ++hole, ++index) {
    std::cout << "____Hole_#" << index << ":_";
    print_ccb<Arrangement>(*hole);
  }

  // Print the isolated vertices.
  typename Arrangement::Isolated_vertex_const_iterator   iv;
  for (iv = f–>isolated_vertices_begin(), index = 1;
       iv != f–>isolated_vertices_end(); ++iv, ++index)
    std::cout << "____Isolated_vertex_#" << index << ":_"
              << "(" << iv–>point() << ")" << std::endl;
}
```

Example: The function template `print_arrangement()` listed below prints the features of a given arrangement. The file `arr_print.h`, includes the definitions of this function, as well as the definitions of all other functions listed in this section. This concludes the preview of the various traversal methods.

```cpp
template <typename Arrangement>
void print_arrangement(const Arrangement& arr)
{
  CGAL_precondition(arr.is_valid());

  // Print the arrangement vertices.
  typename Arrangement::Vertex_const_iterator   vit;
  std::cout << arr.number_of_vertices() << "_vertices:" << std::endl;
  for (vit = arr.vertices_begin(); vit != arr.vertices_end(); ++vit) {
    std::cout << "(" << vit–>point() << ")";
    if (vit–>is_isolated()) std::cout << "_-_Isolated." << std::endl;
    else std::cout << "_-_degree_" << vit–>degree() << std::endl;
  }

  // Print the arrangement edges.
  typename Arrangement::Edge_const_iterator      eit;
  std::cout << arr.number_of_edges() << "_edges:" << std::endl;
  for (eit = arr.edges_begin(); eit != arr.edges_end(); ++eit)
    std::cout << "[" << eit–>curve() << "]" << std::endl;

  // Print the arrangement faces.
  typename Arrangement::Face_const_iterator     fit;
  std::cout << arr.number_of_faces() << "_faces:" << std::endl;
  for (fit = arr.faces_begin(); fit != arr.faces_end(); ++fit)
    print_face<Arrangement>(fit);
}
```

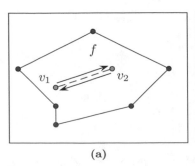
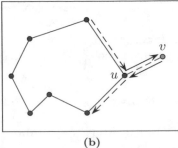
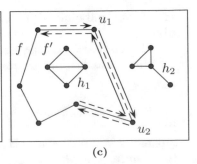

(a)   (b)   (c)

Fig. 2.2: Illustrations of the various specialized insertion procedures. The inserted x-monotone curve is drawn as a dashed line, surrounded by two solid arrows that represent the pair of twin halfedges added to the DCEL. Existing vertices are drawn as dark discs, while new vertices are drawn as light discs. Existing halfedges that are affected by the insertion operations are drawn as dashed arrows. (a) Inserting a curve that induces a new hole inside the face f. (b) Inserting a curve from an existing vertex u that corresponds to one of its endpoints. (c) Inserting an x-monotone curve, the endpoints of which correspond to existing vertices u_1 and u_2. In this case the new pair of halfedges close a new face f'. The hole h_1, which belonged to f before the insertion, becomes a hole in this new face.

2.2.2   Modifying the Arrangement

In this section we review the various member functions of the `Arrangement_2` class template that allow you to modify the topological structure of the arrangement through the introduction of new edges or vertices or the modification or removal of existing edges or vertices.

The arrangement member-functions that insert new x-monotone curves into the arrangement, thus enabling the construction of a planar subdivision, are rather specialized, as they assume that the interior of the inserted curve is disjoint from all existing arrangement vertices and edges, and in addition require a priori knowledge of the location of the inserted curve. Indeed, for most purposes it is more convenient to construct an arrangement using the free (global) insertion functions, which relax these restrictions. However, as these free functions are implemented in terms of the specialized insertion functions, we start by describing the fundamental functionality of the arrangement class, and describe the operation of the free functions in Chapter 3.

Inserting Non-Intersecting x-Monotone Curves

The most trivial functions that allow you to modify the arrangement are the specialized functions for the insertion of an x-monotone curve the interior of which is disjoint from the interior of all other curves in the existing arrangement and does not contain any point of the arrangement. In addition, these functions require that the location of the curve in the arrangement be known.

The rather harsh restrictions on the inserted curves enable an efficient implementation. While inserting an x-monotone curve, the interior of which is disjoint from all curves in the existing arrangement, is quite straightforward, as we show next, (efficiently) inserting a curve that intersects with the curves already in the arrangement is much more complicated and requires the application of nontrivial geometric algorithms. The decoupling of the topological arrangement representation from the various algorithms that operate on it dictates that the general insertion operations be implemented as free functions that operate on the arrangement and the inserted curve(s); see Section 3.4 for more details and examples.

When an x-monotone curve is inserted into an existing arrangement, such that the interior of this curve is disjoint from the interior of all curves in the arrangement, only the following three scenarios are possible, depending on the status of the endpoints of the inserted curve:

1. If both curve endpoints do not correspond to any existing arrangement vertex we have to create two new vertices, corresponding to the curve endpoints, and connect them using a pair of twin halfedges. This halfedge pair forms a new hole inside the face that contains the

curve in its interior; see Figure 2.2a for an illustration.

2. If exactly one endpoint corresponds to an existing arrangement vertex (we distinguish between a vertex that corresponds to the left endpoint of the inserted curve and one that corresponds to its right endpoint), we have to create a new vertex that corresponds to the other endpoint of the curve and to connect the two vertices by a pair of twin halfedges that form an "antenna" emanating from the boundary of an existing connected component; see Figure 2.2b. (Note that if the existing vertex used to be isolated, this operation is actually equivalent to forming a new hole inside the face that contains this vertex.)

3. If both endpoints correspond to existing arrangement vertices, we connect these vertices using a pair of twin halfedges. (If one or both vertices are isolated, this case reduces to case (2) or case (1), respectively.) The two following subcases may occur:

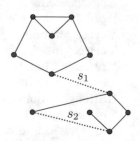

   - Two disconnected components are merged into a single connected component (as is the case with the segment s_1 in the figure to the right).

   - A new face is created, which is split from an existing arrangement face. In this case we also have to examine the holes and isolated vertices in the existing face and move the relevant ones to belong to the new face (as is the case with the segment s_2 in the figure to the right); see also Figure 2.2c.

The `Arrangement_2` class template offers insertion functions that perform the special insertion procedures listed above, namely `insert_in_face_interior()`, `insert_from_left_vertex()`, `insert_from_right_vertex()`, and `insert_at_vertices()`. The first function accepts an x-monotone curve c and a handle to an arrangement face f that contains this curve in its interior. The other functions accept an x-monotone curve c and handles to the existing vertices that correspond to the curve endpoint(s). Each of the four functions returns a handle to one of the twin halfedges that have been created; more precisely:

- `insert_in_face_interior(c, f)` returns a handle to the halfedge directed from the vertex corresponding to the left endpoint of c towards the vertex corresponding to its right endpoint.

- `insert_from_left_vertex(c, v)` and `insert_from_right_vertex(c, v)` each returns a handle to the halfedge, the source of which is the vertex v, and the target of which is the new vertex that has just been created.

- `insert_at_vertices(c, v1, v2)` returns a handle to the halfedge directed from v_1 to v_2.

Example: The program below demonstrates the usage of the four specialized insertion functions. It creates an arrangement of five line segments $s_1, \ldots, s_5$, as depicted in the figure to the right.[5] The first line segment s_1 is inserted in the interior of the unbounded face, while the four succeeding line segments $s_2, \ldots, s_5$ are inserted using the vertices created by the insertion of preceding segments. The arrows in the figure mark the direction of the halfedges $e_1, \ldots, e_5$ returned from the insertion functions, to make it easier to follow the flow of the program. The resulting arrangement consists of three faces, where the two bounded faces form together a hole in the unbounded face.

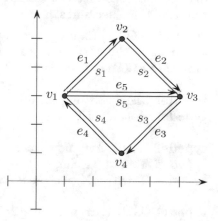

Two header files are included in the code, in order to make this and the following examples more compact. The

---

[5]Notice that in all figures in this book the coordinate axes are drawn only for illustrative purposes and are *not* part of the arrangement.

file `arr_inexact_construction_segments.h` is listed immediately after the program. The file `arr_print.h` is introduced in Section 2.2.1.

// File: ex_edge_insertion.cpp

```cpp
#include "arr_inexact_construction_segments.h"
#include "arr_print.h"

int main()
{
  Point           p1(1, 3), p2(3, 5), p3(5, 3), p4(3, 1);
  Segment         s1(p1, p2), s2(p2, p3), s3(p3, p4), s4(p4, p1), s5(p1, p3);

  Arrangement     arr;
  Halfedge_handle e1 = arr.insert_in_face_interior(s1, arr.unbounded_face());
  Vertex_handle   v1 = e1->source();
  Vertex_handle   v2 = e1->target();
  Halfedge_handle e2 = arr.insert_from_left_vertex(s2, v2);
  Vertex_handle   v3 = e2->target();
  Halfedge_handle e3 = arr.insert_from_right_vertex(s3, v3);
  Vertex_handle   v4 = e3->target();
  Halfedge_handle e4 = arr.insert_at_vertices(s4, v4, v1);
  Halfedge_handle e5 = arr.insert_at_vertices(s5, v1, v3);

  print_arrangement(arr);
  return 0;
}
```

As mentioned above, in all examples listed in this chapter and some of the examples listed in the following chapter the `Traits` parameter of the `Arrangement_2<Traits, Dcel>` class template is substituted with an instance of the `Arr_segment_traits_2<Kernel>` class template. In these examples the `Arr_segment_traits_2` class template is instantiated with the predefined CGAL kernel that evaluates predicates in an exact manner, but constructs geometric objects in an inexact manner, as none of these examples construct new geometric objects. In the remaining examples listed in the next chapter, as well as in most other examples listed in the book, the traits class-template is instantiated with a kernel that evaluates predicates and constructs geometric objects, both in an exact manner; see Section 1.4.4 for more details about the various kernels. The statements below define the types for arrangements of line segments common to all examples that do not construct new geometric objects. They are kept in the header file `arr_inexact_construction_segments.h`.

```cpp
#include <CGAL/Exact_predicates_inexact_constructions_kernel.h>
#include <CGAL/Arr_non_caching_segment_traits_2.h>
#include <CGAL/Arrangement_2.h>

typedef CGAL::Exact_predicates_inexact_constructions_kernel Kernel;
typedef Kernel::FT                                          Number_type;

typedef CGAL::Arr_non_caching_segment_traits_2<Kernel>      Traits;
typedef Traits::Point_2                                     Point;
typedef Traits::X_monotone_curve_2                          Segment;

typedef CGAL::Arrangement_2<Traits>                         Arrangement;
typedef Arrangement::Vertex_handle                          Vertex_handle;
typedef Arrangement::Halfedge_handle                        Halfedge_handle;
typedef Arrangement::Face_handle                            Face_handle;
```

Manipulating Isolated Vertices

Isolated points are in general simpler geometric entities than curves, and indeed the member functions that manipulate them are easier to understand.

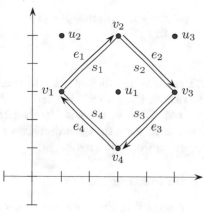

The call `arr.insert_in_face_interior(p, f)` inserts an isolated point p, located in the interior of a given face f, into the arrangement and returns a handle to the arrangement vertex associated with p it has created. Naturally, this function has the precondition that p is an isolated point; namely, it does not coincide with any existing arrangement vertex and does not lie on any edge. As mentioned in Section 2.2.1, it is possible to obtain the face containing an isolated vertex v by calling `v->face()`. The member function `remove_isolated_vertex(v)` accepts a handle to an isolated vertex v as input and removes it from the arrangement.

Example: The program below demonstrates the usage of the arrangement member-functions for manipulating isolated vertices. It first inserts three isolated vertices, u_1, u_2, and u_3, located inside the unbounded face of the arrangement. Then it inserts four line segments, $s_1, \ldots, s_4$, that form a square hole inside the unbounded face; see the figure above for an illustration. Finally, it traverses the vertices and removes those isolated vertices that are still contained in the unbounded face (u_2 and u_3 in this case).

```cpp
// File: ex_isolated_vertices.cpp

#include "arr_inexact_construction_segments.h"
#include "arr_print.h"

int main()
{
  // Insert isolated points.
  Arrangement      arr;
  Face_handle      uf = arr.unbounded_face();
  Vertex_handle    u1 = arr.insert_in_face_interior(Point(3, 3), uf);
  Vertex_handle    u2 = arr.insert_in_face_interior(Point(1, 5), uf);
  Vertex_handle    u3 = arr.insert_in_face_interior(Point(5, 5), uf);

  // Insert four segments that form a square-shaped face.
  Point            p1(1, 3), p2(3, 5), p3(5, 3), p4(3, 1);
  Segment          s1(p1, p2), s2(p2, p3), s3(p3, p4), s4(p4, p1);

  Halfedge_handle e1 = arr.insert_in_face_interior(s1, uf);
  Vertex_handle   v1 = e1->source();
  Vertex_handle   v2 = e1->target();
  Halfedge_handle e2 = arr.insert_from_left_vertex(s2, v2);
  Vertex_handle   v3 = e2->target();
  Halfedge_handle e3 = arr.insert_from_right_vertex(s3, v3);
  Vertex_handle   v4 = e3->target();
  Halfedge_handle e4 = arr.insert_at_vertices(s4, v4, v1);

  // Remove the isolated vertices located in the unbounded face.
  Arrangement::Vertex_iterator curr, next = arr.vertices_begin();
  for (curr = next++; curr != arr.vertices_end(); curr = next++) {
    // Keep an iterator to the next vertex, as curr might be deleted.
```

```
    if (curr->is_isolated() && curr->face() == uf)
      arr.remove_isolated_vertex(curr);
  }

  print_arrangement(arr);
  return 0;
}
```

 Try: A more efficient way to remove the isolated vertices that lie inside the unbounded face is to obtain the unbounded face, and then traverse its isolated vertices, removing them as they are visited. Replace the code in the program above that traverses all vertices and removes the isolated ones with code that obtains the unbounded face and then traverses its isolated vertices, removing them as they are visited.

Manipulating Halfedges

While reading the previous subsection you learned how to insert new points that induce isolated vertices into the arrangement. You may wonder now how you can insert a new point that lies on an x-monotone curve that is associated with an existing arrangement edge.

The introduction of a vertex, the geometric mapping of which is a point p that lies on an x-monotone curve, requires the splitting of the curve in its interior at p. The two resulting subcurves induce two new edges, respectively. In general, the `Arrangement_2` class template relies on the geometry traits to perform such a split. As a matter of fact, it relies on the geometry traits to perform all geometric operations. To insert a point p that lies on an x-monotone curve associated with an existing edge e into the arrangement $\mathcal{A}$, you must first construct the two curves c_1 and c_2, which are the two subcurves that result from splitting the x-monotone curve associated with the edge e at p. Then, you have to issue the call `arr.split_edge(he, c1, c2)`, where `arr` identifies the arrangement $\mathcal{A}$ and `he` is a handle to one of the two halfedges that represent the edge e. The function splits the two halfedges that represent e into two pairs of halfedges, respectively. Two new halfedges are incident to the new vertex v associated with p. The function returns a handle to the new halfedge, the source of which is the source vertex of the halfedge handled by `he`, and the target of which is the new vertex v. For example, if the halfedge drawn as a dashed line at the top in the figure above is passed as input, the halfedge drawn as a dashed line at the bottom is returned as output.

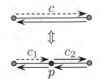

The reverse operation is also possible. Consider a vertex v of degree 2 that has two incident edges e_1 and e_2 associated with two curves c_1 and c_2, respectively, such that the union of c_1 and c_2 results in a single continuous x-monotone curve c of the type supported by the traits class in use. To merge the edges e_1 and e_2 into a single edge associated with the curve c, essentially removing the vertex v from the arrangement identified by `arr`, you need to issue the call `arr.merge_edge(he1, he2, c)`, where `he1` and `he2` are handles to halfedges representing e_1 and e_2, respectively.

Finally, the call `arr.remove_edge(he)` removes the edge e from the arrangement, where `he` is a handle to one of the two halfedges that represents e. Note that this operation is the reverse of an insertion operation, so it may cause a connected component to split into two, or two faces to merge into one, or a hole to disappear. By default, if the removal of e causes one of its end vertices to become isolated, this vertex is removed as well. However, you can control this behavior and choose to keep the isolated vertices by supplying additional Boolean flags to `remove_edge()` indicating whether the source or the target vertices are to be removed should they become isolated.

 Example: The example program below shows how the edge-manipulation functions can be used. The program works in three steps, as demonstrated in Figure 2.3. Note that the program uses the fact that `split_edge()` returns one of the new halfedges (after the split) that has the same direction as the original halfedge (the first parameter of the function) and is directed towards the split point. Thus, it is easy to identify the vertices u_1 and u_2 associated with the split points.

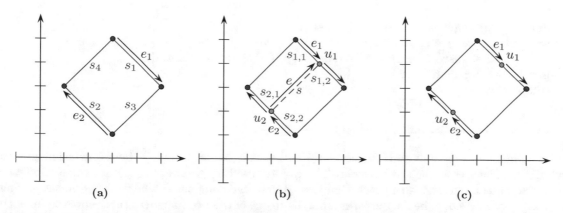

Fig. 2.3: The three steps of the example program `ex_edge_manipulation.cpp`. In Step (a) it constructs an arrangement of four line segments. In Step (b) the edges e_1 and e_2 are split, and the split points are connected with a new segment s that is inserted into the arrangement. This operation is undone in Step (c), where e is removed from the arrangement, rendering its end vertices u_1 and u_2 redundant. We therefore remove these vertices by merging their incident edges and go back to the arrangement depicted in (a).

```cpp
// File: ex_edge_manipulation.cpp

#include "arr_inexact_construction_segments.h"
#include "arr_print.h"

int main()
{
  // Step (a) - construct a rectangular face.
  Point         q1(1, 3), q2(3, 5), q3(5, 3), q4(3, 1);
  Segment       s4(q1, q2), s1(q2, q3), s3(q3, q4), s2(q4, q1);

  Arrangement    arr;
  Halfedge_handle e1 = arr.insert_in_face_interior(s1, arr.unbounded_face());
  Halfedge_handle e2 = arr.insert_in_face_interior(s2, arr.unbounded_face());

  e2 = e2->twin();       // as we wish e2 to be directed from right to left
  arr.insert_at_vertices(s3, e1->target(), e2->source());
  arr.insert_at_vertices(s4, e2->target(), e1->source());
  std::cout << "After step (a):" << std::endl;
  print_arrangement(arr);

  // Step (b) - split e1 and e2 and connect the split points with a segment.
  Point   p1(4,4), p2(2,2);
  Segment s1_1(q2, p1), s1_2(p1, q3), s2_1(q4, p2), s2_2(p2, q1), s(p1, p2);

  e1 = arr.split_edge(e1, s1_1, s1_2);
  e2 = arr.split_edge(e2, s2_1, s2_2);
  Halfedge_handle e = arr.insert_at_vertices(s, e1->target(), e2->target());
  std::cout << std::endl << "After step (b):" << std::endl;
  print_arrangement(arr);

  // Step (c) - remove the edge e and merge e1 and e2 with their successors.
  arr.remove_edge(e);
```

```
arr.merge_edge(e1, e1->next(), s1);
arr.merge_edge(e2, e2->next(), s2);
std::cout << std::endl << "After_step_(c):" << std::endl;
print_arrangement(arr);
return 0;
}
```

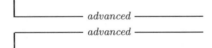

The member functions `modify_vertex()` and `modify_edge()` modify the geometric mappings of existing features of the arrangement. The call `arr.modify_vertex(v, p)` accepts a handle to a vertex v and a reference to a point p, and sets p to be the point associated with the vertex v. The call `arr.modify_edge(he, c)` accepts a handle to one of the two halfedges that represent an edge e and a reference to a curve c, and sets c to be the x-monotone curve associated with e. (Note that both halfedges are modified; that is, both expressions `he->curve()` and `he->twin()->curve()` evaluate to c after the modification.) These functions have preconditions that p is geometrically equivalent to `v->point()` and c is equivalent to `e->curve()`, respectively.[6] If these preconditions are not met, the corresponding operation may invalidate the structure of the arrangement. At first glance it may seem as if these two functions are of little use. However, you should keep in mind that there may be extraneous data (probably non-geometric) associated with the point objects or with the curve objects, as defined by the traits class. With these two functions you can modify this data; see more details in Section 5.5.

In addition, you can use these functions to replace a geometric object (a point or a curve) with an equivalent object that has a more compact representation. For example, if we use some simple rational-number type to represent the point coordinates, we can replace the point $\left(\frac{20}{40}, \frac{99}{33}\right)$ associated with some vertex v with an equivalent point with normalized coordinates, namely $\left(\frac{1}{2}, 3\right)$.

Advanced Insertion Functions

Assume that the specialized insertion function `insert_from_left_vertex(c,v)` is given a curve c, the left endpoint of which is already associated with a non-isolated vertex v. Namely, v has already several incident halfedges. It is necessary in this case to locate the exact place for the new halfedge mapped to the newly inserted curve c in the circular list of halfedges incident to v. More precisely, in order to complete the insertion, it is necessary to locate the halfedge e_{pred} directed towards v such that c is located between the curves associated with e_{pred} and the next halfedge in the clockwise order in the circular list of halfedges around v; see the figure

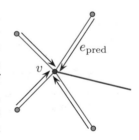

to the right. This may take $O(d)$ time, where d is the degree of the vertex v.[7] However, if the halfedge e_{pred} is known in advance, the insertion can be carried out in constant time, and without performing any geometric comparisons.

The `Arrangement_2` class template provides advanced versions of the specialized insertion functions for a curve c, namely `insert_from_left_vertex(c,he_pred)` and `insert_from_right_vertex(c,he_pred)`. These functions accept a handle to the halfedge e_{pred} as specified above, instead of a handle to the vertex v. They are more efficient, as they take constant time and do not perform any geometric operations. Thus, you should use them when the halfedge e_{pred} is

---

[6] Roughly speaking, two curves are equivalent iff they have the same graph. In Section 5.1.1 we give a formal definition of curves and curve equivalence.

[7] We can store the handles to the halfedges incident to v in an efficient search structure to obtain $O(\log d)$ access time. However, as d is usually very small, this may lead to a waste of storage space without a meaningful improvement in running time in practice.

known. In cases where the vertex v is isolated or the predecessor halfedge for the newly inserted curve is not known, the simpler versions of these insertion functions should be used. Similarly, the member function `insert_at_vertices()` is overloaded with two additional versions as follows. One accepts two handles to the two predecessor halfedges around the two vertices v_1 and v_2 that correspond to the curve endpoints. The other one accepts a handle to one vertex and a handle to the predecessor halfedge around the other vertex.

Example: The program below shows how to construct an arrangement of eight pairwise interior-disjoint line-segments $s_1, \ldots, s_8$, as depicted in the figure to the right, using the specialized insertion functions that accept predecessor halfedges. The corresponding halfedges $e_1, \ldots, e_8$ are drawn as arrows. Note that the point p_0 is initially inserted as an isolated point and later on is connected to the other four vertices to form the four bounded faces of the final arrangement.

// File: ex_special_edge_insertion.cpp

```
#include "arr_inexact_construction_segments.h"
#include "arr_print.h"

int main()
{
  Point            p0(3, 3), p1(1, 3), p2(3, 5), p3(5, 3), p4(3, 1);
  Segment          s1(p1, p2), s2(p2, p3), s3(p3, p4), s4(p4, p1);
  Segment          s5(p1, p0), s6(p0, p3), s7(p4, p0), s8(p0, p2);

  Arrangement      arr;
  Vertex_handle    v0 = arr.insert_in_face_interior(p0, arr.unbounded_face());
  Halfedge_handle  e1 = arr.insert_in_face_interior(s1, arr.unbounded_face());
  Halfedge_handle  e2 = arr.insert_from_left_vertex(s2, e1);
  Halfedge_handle  e3 = arr.insert_from_right_vertex(s3, e2);
  Halfedge_handle  e4 = arr.insert_at_vertices(s4, e3, e1->twin());
  Halfedge_handle  e5 = arr.insert_at_vertices(s5, e1->twin(), v0);
  Halfedge_handle  e6 = arr.insert_at_vertices(s6, e5, e3->twin());
  Halfedge_handle  e7 = arr.insert_at_vertices(s7, e4->twin(), e6->twin());
  Halfedge_handle  e8 = arr.insert_at_vertices(s8, e5, e2->twin());

  print_arrangement(arr);
  return 0;
}
```

It is possible to perform even more refined operations on an `Arrangement_2` object given specific topological information. As most of these operations are very fragile and do not test preconditions on their input in order to gain efficiency, they are not included in the public interface of the `Arrangement_2` class template. Instead, the `Arr_accessor<Arrangement>` class template enables access to these internal arrangement operations; see more details in the Reference Manual.

⌐──────── *advanced* ──────────

2.2.3 Input/Output Functions

In some cases, you may like to save an arrangement object constructed by some application, so that later on it can be restored. In other cases you may like to create nice drawings that represent arrangements constructed by some application. These drawings can be printed or displayed on a computer screen and dynamically change as the arrangement itself changes.

Input/Output Stream

Consider an arrangement that represents a very complicated geographical map, and assume that there are applications that need to answer various queries on this map. Naturally, you can store the set of curves that induce the arrangement, but this implies that you would need to construct the arrangement from scratch each time you wish to reuse it. A more efficient solution is to write the arrangement to a file in a format that other applications can read.

The *2D Arrangements* package provides an *inserter* operator (<<), which inserts an arrangement object into an output stream, and an *extractor* operator (>>), which extracts an arrangement object from an input stream. The arrangement is written using a simple predefined plain-text format that encodes the arrangement topology, as well as all geometric entities associated with vertices and edges.

The ability to use the input and output operators requires that the `Point_2` type and the `X_monotone_curve_2` type defined by the traits class both support the << and >> operators. Only traits classes that handle linear objects are guaranteed to provide these operators for the geometric types they define. Thus, you can safely write and read arrangements of line segments, polylines, or unbounded linear objects.[8]

Example: The example below constructs the arrangement depicted on Page 27 and writes it to an output file. It also demonstrates how to reread the arrangement from a file.

```cpp
// File: ex_io.cpp

#include <fstream>

#include <CGAL/basic.h>
#include <CGAL/IO/Arr_iostream.h>

#include "arr_inexact_construction_segments.h"
#include "arr_print.h"

int main()
{
  // Construct the arrangement.
  Point           p1(1, 3), p2(3, 5), p3(5, 3), p4(3, 1);
  Segment         s1(p1, p2), s2(p2, p3), s3(p3, p4), s4(p4, p1), s5(p1, p3);

  Arrangement     arr1;
  Halfedge_handle e1 = arr1.insert_in_face_interior(s1, arr1.unbounded_face());
  Vertex_handle   v1 = e1->source();
  Vertex_handle   v2 = e1->target();
  Halfedge_handle e2 = arr1.insert_from_left_vertex(s2, v2);
  Vertex_handle   v3 = e2->target();
  Halfedge_handle e3 = arr1.insert_from_right_vertex(s3, v3);
  Vertex_handle   v4 = e3->target();
  Halfedge_handle e4 = arr1.insert_at_vertices(s4, v4, v1);
  Halfedge_handle e5 = arr1.insert_at_vertices(s5, v1, v3);

  // Write the arrangement to a file.
  std::cout << "Writing" << std::endl;
  print_arrangement_size(arr1);
```

---

[8]Traits classes that handle non-linear objects use algebraic-number types. The inserter (<<) and extractor (>>) operators for these non-linear objects can be provided only if these operators are available for algebraic numbers.

```
std::ofstream  out_file("arr_ex_io.dat");
out_file << arr1;
out_file.close();

// Read the arrangement from the file.
Arrangement        arr2;
std::ifstream    in_file("arr_ex_io.dat");
in_file >> arr2;
in_file.close();
std::cout << "Reading" << std::endl;
print_arrangement_size(arr2);

return 0;
}
```

The inserter and extractor operators utilize the free functions **write()** and **read()**. These functions use a *formatter* object, which defines the I/O format. Both **read()** and **write()** functions use the **Arr_text_formatter** formatter class, which ignores auxiliary data that might be attached to the arrangement features. If you wish to write or read arrangements extended with auxiliary data, use other plain-text formats, such as Postscript, XML, and IPE,[9] or even use binary formats, you must call the **read()** and **write()** functions and pass an appropriate formatter. Section 6.2 describes how you can write or read an arrangement with auxiliary data stored with its features. The *2D Arrangements* package also comes with formatters that write and read arrangements that maintain cross-mappings between input curves and the arrangement edges they induce, referred to as arrangements-with-history objects; see Section 6.4 for details.

Output QT-Widget Stream

The *2D Arrangements* package includes an interactive program that demonstrates its features. As mentioned in the preface of this book, this demonstration program is still based on QT 3, an older version of QT, and like many other CGAL programs based on QT 3, it uses a QT stream called **Qt_widget**. You can display the drawings of arrangements in a graphical window using **Qt_widget** streams just like the demonstration program does. All you need to do is follow the guidelines for handling **Qt_widget** objects, and apply the *inserter*, which inserts an arrangement into a **Qt_widget** stream to complete the drawing. The ability to use this output operator requires that the **Point_2** and **X_monotone_curve_2** types defined by the traits class both support the inserter (<<) operator that inserts the respective geometric object into a **Qt_widget** stream. The **Arr_rational_arc_traits_2** class template (see Section 5.4.3) and the **Arr_linear_traits_2** class template (see Section 5.2) currently do not provide this operator for the geometric types they define. Thus, only arrangements of line segments, polylines, or conic arcs can be drawn this way without additional code. The << operator for polylines and conic arcs is defined in **CGAL/IO/Qt_widget_Polyline_2.h** and **CGAL/IO/Qt_widget_Conic_arc_2.h**, respectively. These files must be explicitly included to insert polylines or conic arcs into **Qt_widget** streams.

All CGAL programs based on QT Version 3, including the arrangement demonstration-program, were being ported to Version 4, the latest version of QT, at the time this book was written. In the new setup the visualization of two-dimensional CGAL objects is done with the QT Graphics View Framework. This framework enables managing and interacting with a large number of custom-made 2D graphical items, and provides a view widget for visualizing the items, with support for zooming and rotation.[10]

---

[9]http://tclab.kaist.ac.kr/ipe/.

[10]See http://doc.qt.nokia.com/latest/graphicsview.html for more information on this framework. The package "CGAL and the QT Graphics View Framework" [66] provides the means to integrate CGAL and the Graphics View Framework.

2.3   Application: Obtaining Silhouettes of Polytopes

We conclude this chapter with a small application that obtains the silhouettes of bounded convex polyhedra in $\mathbb{R}^3$, commonly referred to as convex *polytopes*. Given a convex polytope P, the program obtains the outline of the shadow of P cast on the xy-plane, where the scene is illuminated by a light source at infinity directed along the negative z-axis. The silhouette is represented as an arrangement that contains two faces—an unbounded face and a single hole inside the unbounded face. The silhouette is the outer boundary of the latter.

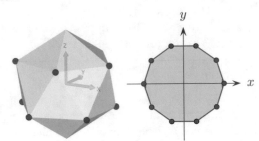

The figure to the right shows an icosahedron and its silhouette. The corresponding input file located in the example folder is called `icosahedron.dat`.

The program first constructs the input convex polytope and stores the result in a temporary object, the type of which is an instance of the CGAL `Polyhedron_3` class template. Then, it traverses the convex polytope facets. For each facet facing upwards (namely, whose outer normal has a positive z component), it traverses the edges on the facet boundary. The traversal of the facets and the traversal of the halfedges around each facet are done in a way similar to the traversals of arrangement cells as described in Section 2.2.1; see the Reference Manual for the exact interface of the `Polyhedron_3` class template. (Notice that the positive normal z-coordinate requirement rules out faces that are parallel to the z-axis.) Let e be the current polytope edge being processed, and let e' be the vertical projection of e (onto the xy-plane). The program inserts e' into the arrangement using one of the specialized insertion member-functions listed in Section 2.2.2. Once the traversal of the temporary polytope is completed, it is discarded. Finally, the program traverses the arrangement edges, and removes all edges that are not incident to the unbounded face.

We must ensure that each edge is inserted into the arrangement exactly once to avoid overlaps. To this end, we maintain a set E of handles to polytope edges. It contains the edges of the polytope, the projections of which have already been inserted into the arrangement. For each edge e being processed, if e is not in E, we insert the projection of e into the arrangement and we also insert e into E. We use the `std::set` data structure to maintain the set E. It requires the provision of a model of the StrictWeakOrdering STL concept that compares handles. We use the functor `Less_than_handle` listed below, and defined in the header file `Less_than_handle.h`. The functor compares the addresses of the handled objects. Thus, (different) handles to the same object are evaluated as equal. (The induced ordering is consistent, but arbitrary and irrelevant.)

```
struct Less_than_handle {
  template <typename Type>
  bool operator()(Type s1, Type s2) const { return (&(*s1) < &(*s2)); }
};
```

——— *advanced* ———

An alternative technique to avoid duplicate insertion is to extend each record of the polyhedron data structure that represents a halfedge with a Boolean value that indicates whether it has been inserted into the arrangement. This technique is more efficient, as it does not use the set auxiliary data structure. The technique used to extend the polyhedron data structure is similar to the technique used to extend the arrangement data structure, as both are based on the same halfedge data structure. The extending technique is discussed only in Chapter 6.

——— *advanced* ———

We must determine the appropriate insertion routines to insert the segments into the arrangement. To this end, we use yet another auxiliary data structure, namely a map, which maps polyhedron vertices to corresponding arrangement vertices. Before inserting a segment into the arrangement we search the map for the arrangement vertices that correspond to the segment endpoints, and dispatch the appropriate insertion routine.

┌──────── *advanced* ────────────

An alternative technique is to extend each record of the polyhedron data structure that represents a vertex with the handle to the corresponding arrangement vertex. This field is initialized when the polyhedron vertex is processed for the first time, and is used during subsequent encounters with the vertex. Again, this technique is more efficient, as it does not use the map auxiliary data structure, but also more advanced. Moreover, the processing can be expedited further using the more efficient insertion methods that accept halfedges as operands.

└──────── *advanced* ────────────

The functor template **Arr_inserter<Polyhedron, Arrangement>** listed below, and defined in the header file **Arr_inserter.h**, is used to insert a segment, which is the projection of a polytope edge, into the arrangement. When it is instantiated its template parameters **Polyhedron** and **Arrangement** must be substituted with instances of the polyhedron and arrangement types, respectively. Its function operator accepts four parameters as follows: the target arrangement object, the polytope edge, and the two points that are the projection of the endpoints of the polytope edge onto the xy-plane.

```
#include <CGAL/basic.h>

#include "Less_than_handle.h"

template <typename Polyhedron, typename Arrangement> class Arr_inserter {
private:
  typedef typename Polyhedron::Halfedge_const_handle
    Polyhedron_halfedge_const_handle;
  typedef std::map<typename Polyhedron::Vertex_const_handle,
                   typename Arrangement::Vertex_handle, Less_than_handle>
                                                         Vertex_map;
  typedef typename Vertex_map::iterator                  Vertex_map_iterator;
  typedef typename Arrangement::Point_2                  Point_2;
  typedef typename Arrangement::X_monotone_curve_2       X_monotone_curve_2;

  Vertex_map m_vertex_map;
  std::set<Polyhedron_halfedge_const_handle, Less_than_handle> m_edges;
  const typename Arrangement::Traits_2::Compare_xy_2 m_cmp_xy;
  const typename Arrangement::Traits_2::Equal_2 m_equal;

public:
  Arr_inserter(const typename Arrangement::Traits_2& traits) :
    m_cmp_xy(traits.compare_xy_2_object()),
    m_equal(traits.equal_2_object())
  {}

  void operator()(Arrangement& arr, Polyhedron_halfedge_const_handle he,
                  Point_2& prev_arr_point, Point_2& arr_point)
  {
    // Avoid the insertion if he or its twin have already been inserted.
    if ((m_edges.find(he) != m_edges.end()) ||
        (m_edges.find(he->opposite()) != m_edges.end()))
      return;

    // Locate the arrangement vertices, which correspond to the projected
    // polyhedron vertices, and insert the segment corresponding to the
    // projected-polyhedron edge using the proper specialized insertion
```

```
    // function.
  m_edges.insert(he);
  X_monotone_curve_2 curve(prev_arr_point, arr_point);
  Vertex_map_iterator it1 = m_vertex_map.find(he->opposite()->vertex());
  Vertex_map_iterator it2 = m_vertex_map.find(he->vertex());
  if (it1 != m_vertex_map.end()) {
    if (it2 != m_vertex_map.end())
      arr.insert_at_vertices(curve, (*it1).second, (*it2).second);
    else {
      typename Arrangement::Halfedge_handle arr_he =
        (m_cmp_xy(prev_arr_point, arr_point) == CGAL::SMALLER) ?
        arr.insert_from_left_vertex(curve, (*it1).second) :
        arr.insert_from_right_vertex(curve, (*it1).second);
      m_vertex_map[he->vertex()] = arr_he->target();   // map the new vertex
    }
  }
  else if (it2 != m_vertex_map.end()) {
    typename Arrangement::Halfedge_handle arr_he =
      (m_cmp_xy(prev_arr_point, arr_point) == CGAL::LARGER) ?
      arr.insert_from_left_vertex(curve, (*it2).second) :
      arr.insert_from_right_vertex(curve, (*it2).second);
    // map the new vertex.
    m_vertex_map[he->opposite()->vertex()] = arr_he->target();
  }
  else {
    typename Arrangement::Halfedge_handle arr_he =
      arr.insert_in_face_interior(curve, arr.unbounded_face());
    // map the new vertices.
    if (m_equal(prev_arr_point, arr_he->source()->point())) {
      m_vertex_map[he->opposite()->vertex()] = arr_he->source();
      m_vertex_map[he->vertex()] = arr_he->target();
    } else {
      m_vertex_map[he->opposite()->vertex()] = arr_he->target();
      m_vertex_map[he->vertex()] = arr_he->source();
    }
  }
  }
};
```

As we are interested in focusing on the arrangement construction and manipulation, and not on the polytope construction, we simplify the code that constructs the polytope, perhaps at the account of its performance. The program reads only the boundary points of the input polytope from a given input file and computes their convex hull instead of parsing an input file that contains a complete representation of the input polytope as a polyhedral mesh, for example.[11] In Chapter 8 we introduce a similar program that obtains the silhouettes of bounded polyhedra, which are not necessarily convex. In this case we are compelled to parse complete representations of the input polyhedra. Regardless of the construction technique, the normals to all facets are computed in a separate loop using the functor template `Normal_equation` listed below, and defined in the header file `Normal_equation.h`. An instance of the functor template is applied to each facet. It computes the cross product of two vectors that correspond to two adjacent edges on the boundary of the input facet, thus obtaining a normal to the facet underlying plane.

```
struct Normal_equation {
```

---

[11] A polyhedral mesh representation consists of an array of boundary vertices and the set of boundary facets, where each facet is described by an array of indices into the vertex array.

```
template <typename Facet> typename Facet::Plane_3 operator()(Facet & f) {
    typename Facet::Halfedge_handle h = f.halfedge();
    return CGAL::cross_product(h->next()->vertex()->point() -
                                   h->vertex()->point(),
                               h->next()->next()->vertex()->point() -
                               h->next()->vertex()->point());
  }
};
```

By default, each record that represents a facet of the polytope is extended with the underlying plane of the facet. The plane is defined with four coefficients. You can override the default and store only the normal to the plane, which is defined by three coordinates instead of the plane four coefficients. In this application we are interested only in the normal to the plane. The listing of the main function follows.

```
// File: ex_polytope_projection.cpp

#include <set>
#include <map>

#include "arr_inexact_construction_segments.h"
#include "read_objects.h"
#include "arr_print.h"
#include "Normal_equation.h"
#include "Arr_inserter.h"

#include <CGAL/convex_hull_3.h>
#include <CGAL/Polyhedron_traits_with_normals_3.h>

typedef CGAL::Polyhedron_traits_with_normals_3<Kernel>  Polyhedron_traits;
typedef CGAL::Polyhedron_3<Polyhedron_traits>           Polyhedron;
typedef Kernel::Point_3                                 Point_3;

int main(int argc, char* argv[])
{
  // Read a sequence of 3D points.
  const char* filename = (argc > 1) ? argv[1] : "polytope.dat";
  std::list<Point_3> points;
  read_objects<Point_3>(filename, std::back_inserter(points));

  // Construct the polyhedron.
  Polyhedron polyhedron;
  CGAL::convex_hull_3(points.begin(), points.end(), polyhedron);
  // Compute the normals to all polyhedron facets.
  std::transform(polyhedron.facets_begin(), polyhedron.facets_end(),
                 polyhedron.planes_begin(), Normal_equation());

  // Construct the projection: go over all polyhedron facets.
  Traits traits;
  Arrangement arr(&traits);
  Kernel kernel;
  Kernel::Compare_z_3 cmp_z = kernel.compare_z_3_object();
  Kernel::Construct_translated_point_3 translate =
    kernel.construct_translated_point_3_object();
  Point_3 origin = kernel.construct_point_3_object()(CGAL::ORIGIN);
```

```
Arr_inserter<Polyhedron, Arrangement> arr_inserter(traits);
Polyhedron::Facet_const_iterator it;
for (it = polyhedron.facets_begin(); it != polyhedron.facets_end(); ++it) {
  // Discard facets whose normals have a non-positive z-components.
  if (cmp_z(translate(origin, it->plane()), origin) != CGAL::LARGER)
    continue;

  // Traverse the halfedges along the boundary of the current facet.
  Polyhedron::Halfedge_around_facet_const_circulator hit=it->facet_begin();
  const Point_3& prev_point = hit->vertex()->point();
  Point prev_arr_point = Point(prev_point.x(), prev_point.y());
  for (++hit; hit != it->facet_begin(); ++hit) {
    const Point_3& point = hit->vertex()->point();
    Point arr_point = Point(point.x(), point.y());
    arr_inserter(arr, hit, prev_arr_point, arr_point);
    prev_arr_point = arr_point;
  }
  const Point_3& point = hit->vertex()->point();
  Point arr_point = Point(point.x(), point.y());
  arr_inserter(arr, hit, prev_arr_point, arr_point);
  prev_arr_point = arr_point;
}
polyhedron.clear();

// Remove internal edges.
Face_handle unb_face = arr.unbounded_face();
Arrangement::Edge_iterator eit;
for (eit = arr.edges_begin(); eit != arr.edges_end(); ++eit) {
  Halfedge_handle he = eit;
  if ((he->face() != unb_face) && (he->twin()->face() != unb_face))
    arr.remove_edge(eit);
}

print_arrangement(arr);
return 0;
}
```

The program described above can be optimized, but even in its current state it is relatively efficient due to the good performance of the specialized insertion-functions. In addition, it is simple and, like all other programs presented in this book, it is robust and it produces exact results. It demonstrates well the use of the specialized insertion-functions.

As you might have noticed, the code above contains a call to an instance of a generic function-template called **read_objects()**. It reads the description of geometric objects from a file and constructs them. It accepts the name of an input file that contains the plain-text description of the geometric objects and an output iterator for storing the newly constructed objects. When the function is instantiated, the first template parameter, namely **Type**, must be substituted with the type of objects to read. It is assumed that an extractor operator (>>) that extracts objects of the given type from the input stream is available. The listing of the function template, which is defined in the file **read_objects.h**, is omitted here.

2.4   Bibliographic Notes and Remarks

The doubly-connected edge list (DCEL) as we use it in the *2D Arrangements* package enhances the structure described by de Berg et al. [45, Chapter 2]. They, in turn describe a variant of a structure originally suggested by Muller and Preparata [163]. Many structures to describe two-dimensional subdivisions were proposed over the years. The DCEL we use in particular is an evolution of the halfedge data structure (HDS) designed and implemented by Kettner [129]. This implementation of HDS is provided through the *Halfedge Data Structures* package of CGAL [131], and is directly used by the *3D Polyhedral Surfaces* package of CGAL [130]. We only use limited facilities provided by the *Halfedge Data Structures* package. An overview and comparison of different data structures together with a thorough description of the design of the HDS implemented in CGAL can be found in [129]. These structures are generalized by a topological model, called *combinatorial maps*, which enables the representation of subdivided objects in any fixed dimension; see, e.g., [146].

There is an alternative, very different, way to represent two-dimensional arrangements, via *trapezoidal decomposition* (a.k.a., vertical decomposition). We study this alternative in Section 3.6.

The preliminary design of CGAL's *2D Arrangements* package is reported in [69,107,108]. Over the years it has been significantly modified and enhanced. The major innovations in the current design are described in [77,213].

2.5 Exercises

2.1 Write a function template that accepts a face (which is not necessarily convex) of an arrangement induced by line segments, and returns a point located inside the face. Use the following prototype:

template <**typename** Arrangement, **typename** Kernel>
typename Kernel::Point_2
point_in_face(**typename** Arrangement::Face_const_handle f,
                const Kernel& ker);

with the following preconditions: (i) The traits class used supports line segments, (ii) the types `Arrangement::Traits_2::Point_2` and `Kernel::Point_2` are convertible to one another, and (iii) the `Kernel` has a nested functor called `Construct_midpoint_2`, the function operator of which accepts two points p and q and returns the midpoint of the segment pq.

2.2 Optimize the program coded in the files `polytope_projection.cpp` and `Arr_inserter.h` as much as possible. Use the specialized insertion-methods that accept handles to halfedges as operands instead of the insertion methods that accept handles to vertices.

2.3 In this exercise you are asked to gradually develop an interactive system that creates and modifies an arrangement. Optionally, equip the system with the ability to render the arrangement in a dedicated window, adding visual capabilities.

    (a) Write a function that accepts two positive integers m and n, and constructs a grid-like arrangement of $m \times n$ squared faces in the first quadrant of the Cartesian plane. The figure to the right depicts an arrangement of 3×2 squared faces. Use the specialized insertion methods that accept handles to vertices as operands to insert all the segments except the first one.

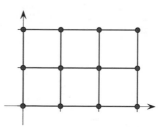

    (b) Apply an optimization to the function above, by using the specialized insertion methods that accept handles to halfedges as operands to insert all the segments except the first one.

    (c) Complete the development of the interactive system that creates and modifies an arrangement. When the system is ready to process a command, it prompts the user. The user as a response may issue one of the following commands:

        c m n — create an arrangement of $m \times n$ squared faces in the first quadrant of the Cartesian plane.

i m n — increase the number of columns by m and the number of rows by n.

d m n — decrease the number of columns by m and the number of rows by n.

m m n — multiply the number of columns and rows by m and n, respectively.

/ m n — divide the number of columns and rows by m and n, respectively.

t x y — translate the arrangement by the vector (x, y). Use the `modify_vertex()` and `modify_edge()` methods to perform the operation.

s x y — scale the arrangement by the scaling factors x and y.

h x y — add the vector (ix, jy) to every grid point $p_{i,j}$.

2.4 Given a set S of points in the plane such that no three of them lie on a common line, the triangulation of S is created by adding a maximal set of pairwise interior-disjoint segments connecting pairs of points in S. The result is an arrangement of segments whose vertices are the points in S, and which subdivides the convex hull of S into triangles. Write a program that reads an arrangement of line segments from a file and decides whether it is a triangulation of the vertices of the arrangement.

Remark: After reading the arrangement file, check that no three vertices of the arrangement lie on a line. You can do this naively in time that is cubic in the number of vertices. Carrying out this test efficiently is not trivial, and you will learn how to do it when you read Chapter 4.

2.5 Develop a function that constructs an arrangement of five interlocking rings that form the symbol of the Olympic Games, as illustrated in the figure to the right. Each ring must be represented as two polylines approximating the inner and outer circles of the ring. The number of segments that compose a polyline is $\lceil sr \rceil$, where r is the radius and s is some rational scale; two pairs, $\langle s_1, r_1 \rangle$ and $\langle s_2, r_2 \rangle$, are passed as input parameters to the function for the two polylines.

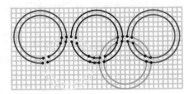

Develop a program that renders the arrangement that represents the symbol of the Olympic Games in a dedicated window with graphics. Assuming that the arrangement is displayed on a raster screen with a specific pixel resolution, the program should accept the window width and height, and the approximate length of a rendered segment that composes a polyline that represents a circle, given in terms of pixels. (Essentially, this length determines the quality of the rendered circles.)

Let r be the radius of a circle and let ℓ be the length of a segment of the polyline that represents the circle in terms of pixels. Set the rational scale that governs the number of segments that compose the polyline to $s = \frac{2\pi r}{\ell}$.

Chapter 3

Queries and Free Functions

Having reviewed the basic functionality of the member functions of the `Arrangement_2` class template and its nested types, we continue our tour with enhanced functionality implemented by utility classes and *free functions*.[1] We start with basic queries on arrangements, namely, *point location* and *vertical ray-shooting*, and proceed with the two central ways of inserting curves into an arrangement: (i) incrementally, one curve at a time using a zone-construction algorithm, or (ii) aggregately, using a plane-sweep algorithm.

3.1 Issuing Queries on an Arrangement

One of the most useful query types defined on arrangements is the *point-location* query: Given a point, find the arrangement cell that contains it. Typically, the result of a point-location query is one of the arrangement faces, but in degenerate situations the query point can lie on an edge, or it may coincide with a vertex.

Point-location queries are common in many applications, and also play an important role in the incremental construction of arrangements (and more specifically in the free insertion-functions described in Section 3.4). Therefore, it is crucial to have the ability to answer such queries effectively.

3.1.1 Point-Location Queries

Recall that the arrangement representation is decoupled from the geometric algorithms that operate on it. Thus, the `Arrangement_2` class template does not support point-location queries directly. Instead, the package provides a set of class templates that are capable of answering such queries; all are models of the concept ArrangementPointLocation_2. Each model employs a different algorithm or *strategy* for answering queries. A model of this concept must define the `locate()` member function, which accepts an input query-point and returns a polymorphic object representing the arrangement cell that contains this point. The returned object, which is of type `CGAL::Object`, can be assigned to a `Face_const_handle`, a `Halfedge_const_handle`, or a `Vertex_const_handle`, depending on whether the query point is located inside a face, on an edge, or on a vertex.

Note that the handles returned by the `locate()` functions are non-mutable (`const`). If necessary, such handles may be cast to mutable handles using the `non_const_handle()` methods provided by the arrangement class.

An object `pl` of any point-location class must be attached to an `Arrangement_2` object `arr` before it is used to answer point-location queries on `arr`. This attachment can be performed when `pl` is constructed or at a later time using the `pl.init(arr)` call.

---

[1] *Free functions* are functions that are free from any particular object. That is, they are not member functions of a specific class.

E. Fogel et al., *CGAL Arrangements and Their Applications*, Geometry and Computing 7, DOI 10.1007/978-3-642-17283-0_3, © Springer-Verlag Berlin Heidelberg 2012

Example: The function template `locate_point()` listed below accepts a point-location object, the type of which is a model of the ArrangementPointLocation_2 concept, and a query point. The function template issues a point-location query for the given point, and prints out the result. It is defined in the header file `point_location_utils.h`.

```
template <typename Point_location>
void locate_point(const Point_location& pl,
                  const typename Point_location::Arrangement_2::Point_2& q)
{
  CGAL::Object    obj = pl.locate(q);      // perform the point-location query

  // Print the result.
  print_point_location<typename Point_location::Arrangement_2> (q, obj);
}
```

The function template `locate_point()` calls an instance of the function template `print_point_location()`, which inserts the result of the query into the standard output-stream. It is listed below, and defined in the header file **point_location_utils.h**. Observe how the function `assign()` is used to cast the resulting **CGAL::Object** into a handle to an arrangement feature. The point-location object **pl** is assumed to be already attached to an arrangement.

```
template <typename Arrangement>
void print_point_location(const typename Arrangement::Point_2& q,
                          const CGAL::Object& obj)
{
  typename Arrangement::Vertex_const_handle    v;
  typename Arrangement::Halfedge_const_handle  e;
  typename Arrangement::Face_const_handle      f;

  std::cout << "The point (" << q << ") is located ";
  if (CGAL::assign (f, obj)) {                    // q is located inside a face
    if (f->is_unbounded())
      std::cout << "inside the unbounded face." << std::endl;
    else std::cout << "inside a bounded face." << std::endl;
  }
  else if (CGAL::assign(e, obj)) {                // q is located on an edge
    std::cout << "on an edge: " << e->curve() << std::endl;
  }
  else if (CGAL::assign(v, obj)) {                // q is located on a vertex
    if (v->is_isolated())
      std::cout << "on an isolated vertex: " << v->point() << std::endl;
    else std::cout << "on a vertex: " << v->point() << std::endl;
  }
  else CGAL_error_msg( "Invalid object!");        // this should never happen
}
```

Choosing a Point-Location Strategy

Each of the various point-location class templates employs a different algorithm or *strategy*[2] for answering queries.

---

[2]The term *strategy* is borrowed from the design-pattern taxonomy [83, Chapter 5]. A *strategy* provides the means to define a family of algorithms, each implemented by a separate class. All classes that implement the various algorithms are made interchangeable, letting the algorithm in use vary according to the user choice.

- `Arr_naive_point_location<Arrangement>` employes the *naive* strategy. It locates the query point naively, exhaustively scanning all arrangement cells.

- `Arr_walk_along_a_line_point_location<Arrangement>` employs the *walk-along-a-line* (or *walk* for short) strategy. It simulates a traversal, in reverse order, along an imaginary vertical ray emanating from the query point. It starts from the unbounded face of the arrangement and moves downward towards the query point until it locates the arrangement cell containing it.

- `Arr_landmarks_point_location<Arrangement,Generator>` uses a set of "landmark" points, the location of which in the arrangement is known. It employs the *landmark* strategy. Given a query point, it uses a nearest-neighbor search-structure (a KD-tree is used by default) to find the nearest landmark, and then traverses the straight-line segment connecting this landmark to the query point.

  There are various ways to select the landmark set in the arrangement. The selection is governed by the `Generator` template parameter. The default generator class, namely `Arr_landmarks_vertices_generator`, selects all the vertices of the attached arrangement as landmarks. Additional landmark generators that select the set in other ways, such as by sampling random points or choosing points on a grid, are also available; see the Reference Manual for more details.

  The landmark strategy requires that the type of the attached arrangement be an instance of the `Arrangement_2<Traits,Dcel>` class template, where the `Traits` parameter is substituted with a geometry-traits class that models the ArrangementLandmarkTraits_2 concept, which refines the basic ArrangementBasicTraits_2 concept; see Section 5.1 for details. Most traits classes included in the *2D Arrangements* package are models of this refined concept.

- `Arr_trapezoid_ric_point_location<Arrangement>` implements a logarithmic (expected) query-time algorithm. The arrangement faces are decomposed into simpler cells, each of constant complexity, known as *pseudo trapezoids*, and a search structure (a directed acyclic graph) is constructed on top of these cells, facilitating the search of the pseudo trapezoid (hence the arrangement cell) containing a query point in expected logarithmic time. The trapezoidal map and the search structure are built by a randomized incremental construction algorithm (RIC).

The first two strategies do not require any extra data. The class templates that implement them store a pointer to an arrangement object and operate directly on it. Attaching such point-location objects to an existing arrangement has virtually no running-time cost at all, but the query time is linear in the size of the arrangement (the performance of the walk strategy is much better in practice, but its worst-case performance is linear). Using these strategies is therefore recommended only when a relatively small number of point-location queries are issued by the application, or when the arrangement is constantly changing. (That is, changes in the arrangement structure are more frequent than point-location queries.)

On the other hand, the landmark and the trapezoidal map RIC strategies require auxiliary data structures on top of the arrangement structure, which they need to construct once they are attached to an arrangement object and need to keep up-to-date as this arrangement changes. The data structures needed by both strategies can be constructed in $O(N \log N)$ time, where N is the overall number of edges in the arrangement, but the constant hidden in the $O()$ notation for the trapezoidal map RIC strategy is much larger. Thus, construction needed by the landmark algorithm is in practice significantly faster than the construction needed by the trapezoidal map RIC strategy. In addition, although both data structures are asymptotically linear in size, using a KD-tree as the nearest-neighbor search-structure that the landmark algorithm stores significantly reduces memory consumption. The trapezoidal map RIC algorithm has expected logarithmic query-time, while the query time for the landmark strategy may be as large as linear. In practice however, the query times of both strategies are competitive.

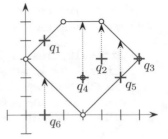

Fig. 3.1: The arrangement of line segments, as constructed in `ex_point_location.cpp`, `ex_vertical_ray_shooting.cpp`, and `ex_batched_point_location.cpp`. The arrangement vertices are drawn as small rings, while the query points $q_1, \ldots, q_6$ are drawn as crosses.

Updating the auxiliary data structures of the trapezoidal map RIC algorithm is done very efficiently. On the other hand, updating the nearest-neighbor search-structure of the landmark algorithm may consume more time when the arrangement changes frequently, especially when a KD-tree is used, as it must be rebuilt each time the arrangement changes. It is therefore recommended that the **Arr_landmarks_point_location** class template be used when the application frequently issues point-location queries on an arrangement that only seldom changes. If the arrangement is more dynamic and is frequently going through changes, the **Arr_trapezoid_ric_point_location** class template should be the selected point-location strategy.

Example: The program listed below constructs a simple arrangement of five line segments that form a pentagonal face, with a single isolated vertex in its interior, as depicted in Figure 3.1. Notice that we use the same arrangement structure in the next three example programs. The arrangement construction is performed by the function **construct_segments_arr()** defined in the header file **point_location_utils.h**. (Its listing is omitted here.) The program employs the naive and landmark strategies to issue several point-location queries on this arrangement.

```
// File: ex_point_location.cpp

#include <CGAL/basic.h>
#include <CGAL/Arr_naive_point_location.h>
#include <CGAL/Arr_landmarks_point_location.h>

#include "arr_inexact_construction_segments.h"
#include "point_location_utils.h"

typedef CGAL::Arr_naive_point_location<Arrangement>        Naive_pl;
typedef CGAL::Arr_landmarks_point_location<Arrangement>    Landmarks_pl;

int main()
{
  // Construct the arrangement.
  Arrangement   arr;
  Naive_pl      naive_pl(arr);
  construct_segments_arr(arr);

  // Perform some point-location queries using the naive strategy.
  locate_point(naive_pl, Point(1, 4));         // q1
  locate_point(naive_pl, Point(4, 3));         // q2
  locate_point(naive_pl, Point(6, 3));         // q3

  // Attach the landmarks object to the arrangement and perform queries.
  Landmarks_pl landmarks_pl;
  landmarks_pl.attach(arr);
  locate_point(landmarks_pl, Point(3, 2));     // q4
  locate_point(landmarks_pl, Point(5, 2));     // q5
```

```
locate_point(landmarks_pl, Point(1, 0));       // q6

    return 0;
}
```

Note that the program uses the `locate_point()` function template to locate a point and nicely print the result of each query; see Page 44.

3.1.2 Vertical Ray Shooting

Another query frequently issued on arrangements is the vertical ray-shooting query: Given a query point, which arrangement feature is encountered by a vertical ray shot upward (or downward) from this point? In the general case the ray hits an edge, but it is possible that it hits a vertex, or that the arrangement does not have any vertex or edge lying directly above (or below) the query point.

All point-location classes listed in the previous section are also models of the Arrangement-VerticalRayShoot_2 concept. That is, they all have member functions called `ray_shoot_up(q)` and `ray_shoot_down(q)` that accept a query point q. These functions output a polymorphic object of type `CGAL::Object`, which wraps a `Halfedge_const_handle`, a `Vertex_const_handle`, or a `Face_const_handle` object. The latter type is used for the unbounded face of the arrangement, in case there is no edge or vertex lying directly above (or below) q.

Example: The function template `shoot_vertical_ray()` listed below accepts a vertical ray-shooting object, the type of which models the ArrangementVerticalRayShoot_2 concept. It prints the result of the upward vertical ray-shooting operation from a given query point to the standard output-stream. The ray-shooting object `vrs` is assumed to be already attached to an arrangement. The function template is defined in the header file `point_location_utils.h`.

```
template <typename Ray_shooting>
void shoot_vertical_ray(const Ray_shooting& vrs,
                        const typename Ray_shooting::Arrangement_2::Point_2& q)
{
  // Perform the vertical ray-shooting query.
  CGAL::Object    obj = vrs.ray_shoot_up(q);

  // Print the result.
  typename Ray_shooting::Arrangement_2::Vertex_const_handle    v;
  typename Ray_shooting::Arrangement_2::Halfedge_const_handle  e;
  typename Ray_shooting::Arrangement_2::Face_const_handle      f;

  std::cout << "Shooting up from (" << q << ") : ";
  if (CGAL::assign(e, obj))                       // we hit an edge
    std::cout << "hit an edge: " << e->curve() << std::endl;
  else if (CGAL::assign(v, obj))                  // we hit a vertex
    std::cout << "hit " << ((v->is_isolated()) ? "an isolated" : "a")
              << " vertex: " << v->point() << std::endl;
  else if (CGAL::assign(f, obj)) {                // we did not hit anything
    CGAL_assertion(f->is_unbounded());
    std::cout << "hit nothing." << std::endl;
  }
  else CGAL_error();                              // this should never happen
}
```

Example: The program below uses the function template listed above to perform vertical ray-shooting queries on an arrangement. The arrangement and the query points are exactly the same as in `ex_point_location.cpp`; see Figure 3.1.

```cpp
// File: ex_vertical_ray_shooting.cpp

#include <CGAL/basic.h>
#include <CGAL/Arr_walk_along_line_point_location.h>
#include <CGAL/Arr_trapezoid_ric_point_location.h>

#include "arr_inexact_construction_segments.h"
#include "point_location_utils.h"

typedef CGAL::Arr_walk_along_line_point_location<Arrangement>  Walk_pl;
typedef CGAL::Arr_trapezoid_ric_point_location<Arrangement>    Trap_pl;

int main()
{
  // Construct the arrangement.
  Arrangement  arr;
  Walk_pl      walk_pl(arr);
  construct_segments_arr(arr);

  // Perform some vertical ray-shooting queries using the walk strategy.
  shoot_vertical_ray(walk_pl, Point(1, 4));          // q1
  shoot_vertical_ray(walk_pl, Point(4, 3));          // q2
  shoot_vertical_ray(walk_pl, Point(6, 3));          // q3

  // Attach the trapezoidal-RIC object to the arrangement and perform queries.
  Trap_pl trap_pl;  trap_pl.attach(arr);
  shoot_vertical_ray(trap_pl, Point(3, 2));          // q4
  shoot_vertical_ray(trap_pl, Point(5, 2));          // q5
  shoot_vertical_ray(trap_pl, Point(1, 0));          // q6

  return 0;
}
```

3.2   Two Algorithmic Frameworks: Plane Sweep and Zone Construction

In this section we describe two fundamental concepts in computational geometry, concepts that play a central role in the algorithmic study of arrangements. You can use the *2D Arrangements* package without knowing about the plane-sweep algorithm or what a zone of a curve is. However, being aware of these will help you in reading various parts of this book. Moreover, it may help you in choosing the right option for your needs when several options are available. Be warned that the description here is *brief*. See the Bibliographic Notes and Remarks at the end of the chapter for references to much more detailed discussions of the topics.

The Plane-Sweep Algorithm

The famous plane-sweep algorithm introduced by Bentley and Ottmann was originally formulated for sets of line segments in the plane. Let S be a set of input line segments in the plane. Consider the vertical line ℓ in the figure to the right. The intersections of the segments in S with ℓ define a one-dimensional arrangement $\mathcal{A}(S \cap \ell)$ on the line ℓ. It comprises three vertices and four one-dimensional cells, that is, two half-infinite intervals (rays) and two line segments. We can describe the

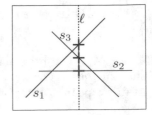

combinatorial structure of the arrangement $\mathcal{A}(\mathcal{S} \cap \ell)$ by listing the curves that intersect ℓ from bottom to top: $< s_2, s_3, s_1 >$. If we specify the equation $x = x_0$ of the line ℓ, then we also get the coordinates of the vertices of the one-dimensional arrangement.

If ℓ does not cross a vertex of the two-dimensional arrangement $\mathcal{A}(\mathcal{S})$, and it is slightly (infinitesimally) moved to the right or to the left, the combinatorial structure of the one-dimensional arrangement on ℓ remains intact. This structure changes only when ℓ hits a vertex of the arrangement, either an endpoint of a line segment in $\mathcal{S}$ or the intersection point of two (or more) segments in $\mathcal{S}$. Accordingly, we sweep the vertical line ℓ over the collection of curves in the plane from $x = -\infty$ to $x = +\infty$ to discover vertices, which are intersection points of curves, and other properties of the arrangement. We stop the sweep at a finite number of points, which we call *events*. The stopping points are endpoints and intersection points of the segments, since between any consecutive pair of events, the combinatorial structure of the one-dimensional arrangement on ℓ does not change.

We maintain two data structures to carry out the sweep efficiently. One structure describes the (dynamically changing) one-dimensional arrangement along the line ℓ; it is called the *status structure*. The second structure is a queue that keeps track of the events, sorted in increasing xy-lexicographic order; it is called the *event queue*. Events that are known in advance are inserted into the queue at the beginning of the algorithm. This is the case with events associated with the endpoints of the curves. Other events, such as intersection points between curves, are discovered as the algorithm proceeds. They are inserted into the queue when discovered. At each event, we dequeue the earliest, yet unhandled event from the queue. As the status structure and the event queue change constantly, they both should be implemented as data structure that can efficiently be updated; balanced binary trees (say red-black trees) will do the job. Sweeping n curves having k intersections takes $O((n + k) \log n)$ time. The status structure maintains at most n segments simultaneously, and each operation on it takes $O(\log n)$ time. The event queue may contain at most $O(n + k)$ events, where $k = O(n^2)$, and each operation on it takes $O(\log n)$ time.

In the literature you will find, in addition to a more detailed description of the algorithm, variants that handle degenerate cases and nuances about the time and space complexities of the algorithm. Notice that although the description above applies to line segments, the same procedure works seamlessly for any family of x-monotone curves.

The *2D Arrangements* package offers a generic implementation of the plane-sweep algorithm in the form of a class template called `Sweep_line_2`. It handles any set of arbitrary x-monotone curves, and serves as the foundation of a family of concrete operations described in the rest of this chapter, as well as several other operations described in later chapters, such as computing all points of intersections of a set of curves,[3] constructing an arrangement induced by a set of curves, aggregately inserting a set of curves into an existing arrangement, and computing the overlay of two arrangements. A concrete algorithm is realized through a plane-sweep visitor, a template parameter of `Sweep_line_2`, which resembles the *visitor* design-pattern [83, Chapter 5] and models the concept SweepLineVisitor_2. In this case, a visitor defines an operation based on the plane-sweep algorithm to be performed on a set of curves and points.[4]

Another parameter of the `Sweep_line_2` class template is the geometry-traits class, which must be substituted with a model of the ArrangementXMonotoneTraits_2 concept; see Section 5.1 for the precise definition of this concept. It defines the minimal set of geometric primitives, among other things, required to perform the plane-sweep algorithm briefly described above.

The Zone-Construction Algorithm

Given an arrangement of curves $\mathcal{A} = \mathcal{A}(\mathcal{C})$ in the plane, the *zone* of an additional curve $\gamma \notin \mathcal{C}$ in $\mathcal{A}$ is the union of the features of $\mathcal{A}$, whose closure is intersected by γ. In Figure 3.2, the shaded faces, the red edges, and the red vertex constitute the zone of the curve γ in an arrangement

---

[3]Strictly speaking, this operation is provided by the *2D Intersection of Curves* package, which is based on the *2D Arrangements* package.

[4]The Boost Graph Library, for example, uses visitors [198, Chapter 12.3] to support user-defined extensions to its fundamental graph-algorithms.

induced by line segments. The complexity of the zone is defined as the sum of the complexities of its constituents. (Notice that some vertices are counted multiple times.) The zone of a curve γ is computed by locating the left endpoint of C in the arrangement and then "walking" along the curve towards the right endpoint, keeping track of the vertices, edges, and faces crossed on the way.

The computational geometry literature abounds in interesting properties of zones, which have important algorithmic implications. For example, it is known that the complexity of the zone of a line in a planar arrangement of n lines is $O(n)$. This property is used in Section 4.2.1 in the analysis of the time complexity of the algorithm applied to solve the **minimum-area triangle problem**.

What we have found out during our work on the CGAL *2D Arrangements* package is that we repeatedly go over the zone of a curve in an arrangement in order to construct portions of, or compute various properties of, the arrangement, so much so that we have designed a zone algorithmic framework to unify these operations, and reuse as much code as possible. The

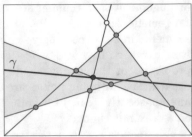

Fig. 3.2: The zone of a line γ in an arrangement of five lines.

single-curve insertion functions are examples of a zone algorithm, as are various other operations that appear in subsequent sections, for example, in the construction of lower envelopes of surfaces in 3-space; see Section 10.3.

The *2D Arrangements* package includes the `Arrangement_zone_2` class template, which computes the zone of an arrangement. Similarly to the `Sweep_line_2` template, the `Arrangement_zone_2` template is parameterized with a zone visitor, a model of the concept ZoneVisitor_2, and it serves as the foundation of a family of concrete operations, such as inserting a single curve into an arrangement and determining whether a query curve intersects with the curves of an arrangement.

It is sometimes necessary to compute the zone of a curve in an arrangement without actually inserting the curve. In other situations, the entire zone is not required, as in the case of a process that only checks whether a query curve crosses an existing arrangement vertex; if the answer is positive, the process can terminate as soon as such a vertex is located. While the plane-sweep algorithm operates on a set of input x-monotone curves and its visitors may only use the notifications they receive to construct their output structures, the zone-computation algorithm operates on an arrangement object and its visitors may modify the same arrangement object as the computation progresses. This makes the interaction of the main class with its visitors slightly more intricate.

After the location of one endpoint of a curve γ is found in the arrangement, it is possible to traverse the zone of γ in the arrangement in time proportional to the complexity of the traversed zone. For the case where the curves inducing the arrangement and γ are all lines, this is trivially achieved. However, for other types of curves (especially segments of curves) this is a less trivial task, and requires extra vertical segments to simplify the traversed zone. The code of the `Arrangement_zone_2` class template is a trivial implementation that uses only the DCEL. If the curves of the arrangement as well as γ are all well behaved (see Section 1.3.3 for a precise definition), the complexity of the zone is (near-)linear in the number of curves in $\mathcal{C}$.

3.3   Batched Point-Location

Here is our first example where the plane-sweep algorithm is put to use. Suppose that at a given moment our application has to issue a relatively large number m of point-location queries on a specific arrangement object. Naturally, it is possible to define a point-location object and use it to issue separate queries on the arrangement. However, as explained in Section 3.1.1, choosing a simple point-location strategy (either the naive or the walk strategy) means inefficient query-processing, while the more sophisticated strategies need to construct auxiliary structures that incur considerable overhead in running time.

Alternatively, the *2D Arrangements* package includes a free `locate()` function that accepts an arrangement and a range of query points as its input and sweeps through the arrangement to locate all query points in one pass. The function outputs the query results as pairs, where each pair consists of a query point and a polymorphic object of type `CGAL::Object`, which represents the cell containing the point; see Section 3.1.1. The output pairs are sorted in increasing *xy*-lexicographical order of the query point.

The batched point-location operation is carried out by sweeping the arrangement. Thus, it takes $O((m + N) \log (m + N))$ time, where N is the number of edges in the arrangement. Issuing separate queries exploiting a point-location strategy with logarithmic query time per query, such as the trapezoidal map RIC strategy (see Section 3.1.1), is asymptotically more efficient. However, experiments show that when the number m of point-location queries is of the same order of magnitude as N, the batched point-location operation is more efficient in practice. One of the reasons for the inferior performance of the alternative (asymptotically faster) procedures is the necessity to construct and maintain complex additional data structures.

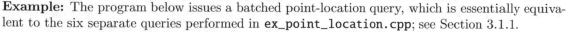

Example: The program below issues a batched point-location query, which is essentially equivalent to the six separate queries performed in **ex_point_location.cpp**; see Section 3.1.1.

```
// File: ex_batched_point_location.cpp

#include <list>

#include <CGAL/basic.h>
#include <CGAL/Arr_batched_point_location.h>

#include "arr_inexact_construction_segments.h"
#include "point_location_utils.h"

int main()
{
  // Construct the arrangement.
  Arrangement    arr;
  construct_segments_arr(arr);

  // Perform a batched point-location query.
  std::list<Point>        points;
  points.push_back(Point(1, 4));   points.push_back(Point(4, 3));
  points.push_back(Point(6, 3));   points.push_back(Point(3, 2));
  points.push_back(Point(5, 2));   points.push_back(Point(1, 0));
  std::list<std::pair<Point, CGAL::Object> >   results;
  locate(arr, points.begin(), points.end(), std::back_inserter(results));

  // Print the results.
  std::list<std::pair<Point, CGAL::Object> >::const_iterator   res_iter;
  for (res_iter = results.begin(); res_iter != results.end(); ++res_iter)
    print_point_location<Arrangement>(res_iter->first, res_iter->second);
  return 0;
}
```

3.4 Free Insertion Functions

The `Arrangement_2` class template is used to represent two-dimensional subdivisions induced by planar curves. Its interface is minimal in the sense that the member functions hardly perform any geometric algorithms and are mainly used for maintaining the topological structure of the subdivision. In this section we explain how to utilize the free functions that enhance that set of operations on arrangements. The implementation of these operations typically requires nontrivial geometric algorithms, and occasionally incurs additional requirements on the traits class.

3.4.1   Incremental Insertion Functions

Inserting Non-Intersecting Curves

Section 2.2 explains how to construct arrangements of x-monotone curves that are pairwise disjoint in their interior when the locations of the segment endpoints in the arrangement are known. Here we relax this constraint, and allow the locations of the inserted x-monotone curve endpoints to be arbitrary, as they may be unknown at the time of insertion. We retain, for the moment, the requirement that the interior of the inserted curve is disjoint from all existing arrangement edges and vertices.

The call `insert_non_intersecting_curve(arr, c, pl)` inserts the x-monotone curve c into the arrangement `arr`, with the precondition that the interior of c is disjoint from all existing edges and vertices of `arr`. The third argument `pl` is a point-location object attached to the arrangement; it is used to locate both endpoints of c in the arrangement. Each endpoint is expected to either coincide with an existing vertex or lie inside a face. It is possible to invoke one of the specialized insertion functions (see Section 2.2), based on the query results, and insert c at its proper location.[5] The insertion operation thus hardly requires any geometric operations on top on the ones needed to answer the point-location queries. Moreover, it is sufficient that the traits class that substitutes the `Traits` template parameter of the `Arrangement_2<Traits,Dcel>` class template when the latter is instantiated models the concept ArrangementBasicTraits_2 (or the concept ArrangementLandmarkTraits_2 if the landmark point-location strategy is used), which does not have to support the computation of intersection points between curves. This implies that using a kernel that provides exact geometric predicates, but potentially inexact geometric constructions due to roundoff errors, is still sufficient.

The free-function template `insert_non_intersecting_curve(arr, c)` is overloaded. There is a variant that instead of accepting a user-defined point-location object, it constructs a local object of the walk point-location class and uses it to insert the curve.

Inserting (Possibly) Intersecting x-Monotone Curves

The `insert_non_intersecting_curve()` function template is very efficient, but its preconditions on the input curves are still rather restricting. Let's assume that the traits class that substitutes the `Traits` template parameter of the `Arrangement_2<Traits,Dcel>` class template models the refined ArrangementXMonotoneTraits_2 concept and supports curve intersection computation; see Section 5.1 for the exact details. Given an x-monotone curve, it is sufficient to locate its left endpoint in the arrangement and to trace its *zone* (see Section 3.2 above) until the right endpoint is reached. Each time the new curve c crosses an existing vertex or edge, the curve is split into subcurves (in the latter case, we have to split the curve associated with the existing halfedge as well) and new edges are associated with the resulting subcurves. Recall that an edge is represented by a pair of twin halfedges, so we split it into two halfedge pairs.

The call `insert(arr, c, pl)` performs this insertion operation. The `insert()` function template accepts an x-monotone curve c, which may intersect some of the curves already in the arrangement `arr`, and inserts it into the arrangement by computing its zone. Users may supply a point-location object `pl` or use the default walk point-location strategy; namely, the variant `insert(arr, c)` is also available. The running time of this insertion function is proportional to the complexity of the zone of the curve c.

---

[5]The `insert_non_intersecting_curve()` function template, as well as all other functions reviewed in this section, is a function template parameterized by an arrangement type and a point-location type. (The latter must be substituted with a model of the ArrangementPointLocation_2 concept.)

─────── *advanced* ───────

In some cases users may have prior knowledge of the location of the left endpoint of the x-monotone curve c they wish to insert, so they can perform the insertion without issuing any point-location queries. This can be done by calling `insert(arr, c, obj)`, where `obj` is a polymorphic object of type CGAL::Object that represents the location of the left endpoint of c in the arrangement. In other words, it wraps a `Vertex_const_handle`, a `Halfedge_const_handle`, or a `Face_const_handle` object; see also Section 3.1.1.

─────── *advanced* ───────

Inserting General Curves

So far, all the examples have constructed arrangements of line segments, where the `Arrangement_2` template was instantiated with an instance of the `Arr_segment_traits_2` class template. In this case, the fact that `insert()` accepts an x-monotone curve does not seem to be a restriction, as all line segments are x-monotone. (Note that we always deal with *weakly* x-monotone curves, and we consider vertical line-segments to be *weakly* x-monotone.)

Consider an arrangement of circles. A circle is obviously not x-monotone, so `insert()` cannot be used in this case.[6] It is necessary to subdivide each circle into two x-monotone circular arcs, namely, its upper half and its lower half, and to insert the two individual x-monotone arcs.

The free function `insert()` is overloaded. It is possible to call this function and pass a curve that is not necessarily x-monotone, but this is subject to an important condition. Consider the call `insert(arr, c, pl)`, where c is not necessarily x-monotone. In this case the type of `arr` must be an instance of the `Arrangement_2` class template, where the `Traits` template parameter is substituted with a traits class that models the concept ArrangementTraits_2, which refines the ArrangementXMonotoneTraits_2 concept. It has to define an additional `Curve_2` type, which may differ from the `X_monotone_curve_2` type. It also has to support the subdivision of curves of this new type into x-monotone curves and possibly singular points; see the exact details in Chapter 5. The `insert(arr, c, pl)` function performs the insertion of a curve c that does not need to be x-monotone into the arrangement by subdividing it into x-monotone subcurves and inserting all individual x-monotone subcurves. Users may supply a point-location object `pl`, or use the default walk point-location strategy by calling `insert(arr, c)`.

Inserting Points

The `Arrangement_2` class template has a member function that inserts a point as an isolated vertex in a given face. The free function `insert_point(arr, p, pl)` inserts a vertex that corresponds to the point p into `arr` at an arbitrary location. It uses the point-location object `pl` to locate the point in the arrangement (by default, the walk point-location strategy is used), and acts according to the result as follows:
- If p is located inside a face, it is inserted as an isolated vertex inside this face.
- If p lies on an edge, the edge is split to create a vertex associated with p.
- Otherwise, p coincides with an existing vertex and no further actions are needed.

In all cases, the function returns a handle to the vertex associated with p.

The type of `arr` must be an instance of the `Arrangement_2` class template instantiated with a traits class that models the ArrangementXMonotoneTraits_2 concept, as the insertion operation may involve the splitting of curves.

Example: The program below constructs an arrangement of five intersecting line-segments $s_1, \ldots, s_5$. It is known that s_1 and s_2 do not intersect, so `insert_non_intersecting_curve()` is used to insert them into the empty arrangement. The rest of the segments are inserted using `insert()`. Using a kernel that constructs geometric objects in an inexact manner (due to

───────────────

[6]A key operation performed by `insert()` is to locate the left endpoint of the curve in the arrangement. A circle, however, does not have any endpoints!

roundoff errors) may yield a program that computes incorrect results and crashes from time to time. This is avoided by using a kernel that provides exact geometric-object constructions as well as exact geometric-predicate evaluations. The header file `arr_exact_construction_segments.h`, just like the header file `arr_inexact_construction_segments.h`, contains the definitions for arrangements of line segments. Unlike the latter, it uses a kernel suitable for arrangements induced by curves that intersect each other, namely a kernel that is exact all way. Note that we alternately use the naive point-location strategy, given explicitly to the insertion functions, and the default walk point-location strategy.

The resulting arrangement is depicted to the right, where the vertices that correspond to segment endpoints are drawn as dark discs, and the vertices that correspond to intersection points are drawn as circles. It consists of 13 vertices, 16 edges, and 5 faces. We also perform a point-location query on the resulting arrangement. The query point q is drawn as a plus sign. The face that contains it is drawn with a shaded texture. The program calls an instance of the function template `print_arrangement_size()`, which prints quantitative measures of the arrangement; see Page 21 for its listing.

```
// File: ex_incremental_insertion.cpp

#include <CGAL/basic.h>
#include <CGAL/Arr_naive_point_location.h>

#include "arr_exact_construction_segments.h"
#include "arr_print.h"

typedef CGAL::Arr_naive_point_location<Arrangement>   Naive_pl;

int main()
{
  // Construct the arrangement of five line segments.
  Arrangement   arr;
  Naive_pl      pl(arr);
  insert_non_intersecting_curve(arr, Segment(Point(1, 0), Point(2, 4)), pl);
  insert_non_intersecting_curve(arr, Segment(Point(5, 0), Point(5, 5)));
  insert(arr, Segment(Point(1, 0), Point(5, 3)), pl);
  insert(arr, Segment(Point(0, 2), Point(6, 0)));
  insert(arr, Segment(Point(3, 0), Point(5, 5)), pl);
  print_arrangement_size(arr);

  // Perform a point-location query on the resulting arrangement and print
  // the boundary of the resulting face that contains the query point.
  Point         q(4, 1);
  CGAL::Object  obj = pl.locate(q);
  Arrangement::Face_const_handle   f;
  bool          success = CGAL::assign(f, obj);
  CGAL_assertion(success);
  std::cout << "The query point (" << q << ") is located in: ";
  print_face<Arrangement>(f);
  return 0;
}
```

3.4.2 Aggregate Insertion Functions

Given a set of n input curves, you can insert the curves in the set into an arrangement incrementally one by one. However, the *2D Arrangements* package also provides a couple of free (overloaded) functions that aggregately insert a range of curves into an arrangement.

- `insert_non_intersecting_curves(arr, begin, end)` inserts a set of x-monotone curves given by the range `[begin, end)` into an arrangement `arr`. The x-monotone curves should be pairwise disjoint in their interior and also interior-disjoint from all existing edges and vertices of `arr`.

- `insert(arr, begin, end)` operates on a range of x-monotone curves that may intersect one another.

- `insert(arr, begin, end)` inserts a set of general (not necessarily x-monotone and possibly intersecting one another) curves of type `Curve_2`, given by the range `[begin, end)`, into the arrangement `arr`.

We distinguish between two cases: (i) The given arrangement `arr` is empty (has only an unbounded face), so it must be constructed from scratch. (ii) The given arrangement `arr` is not empty.

In the first case, we sweep over the input curves, compute their intersection points, and construct the DCEL that represents their arrangement. This process is performed in $O((n + k) \log n)$ time, where k is the total number of intersection points. The running time is asymptotically better than the time needed for incremental insertion if the arrangement is relatively sparse (when k is $O(\frac{n^2}{\log n})$), but in practice it is recommended that this aggregate construction process be used even for dense arrangements, since the plane-sweep algorithm performs fewer geometric operations compared to the incremental insertion algorithms, and hence typically runs much faster in practice.

Another important advantage of the aggregate insertion functions is that they do not issue any point-location queries.[7] Thus, no point-location object needs to be attached to the arrangement. As explained in Section 3.1.1, there is a trade-off between construction time and query time in each of the point-location strategies, which affects the running times of the incremental insertion process. Naturally, this trade-off is absent in the case of aggregate insertion.

Example: The program below shows how to construct the same arrangement of the five line segments built incrementally in **ex_incremental_insertion.cpp**, shown in the previous section. Note that no point-location object needs to be constructed and attached to the arrangement.

```
// File: ex_aggregated_insertion.cpp

#include "arr_exact_construction_segments.h"
#include "arr_print.h"

int main()
{
  // Aggregately construct the arrangement of five line segments.
  Segment segments[] = {Segment(Point(1, 0), Point(2, 4)),
                        Segment(Point(5, 0), Point(5, 5)),
                        Segment(Point(1, 0), Point(5, 3)),
                        Segment(Point(0, 2), Point(6, 0)),
                        Segment(Point(3, 0), Point(5, 5))};
  Arrangement arr;
  insert(arr, segments, segments + sizeof(segments)/sizeof(Segment));
```

---

[7]The only queries they do issue are *one-dimensional* point-location queries of the y-coordinates of curve starting-points in the one-dimensional arrangement along the sweep line.

```
    print_arrangement_size(arr);
    return 0;
}
```

Next we handle the case where we have to insert a set of n curves into an existing arrangement. Let N denote the number of edges in the arrangement. If n is very small compared to N (in theory, we would say that if $n = o(\sqrt{N})$), we insert the curves one by one. For larger input sets, we use the aggregate insertion procedures.

Example: The program below aggregately constructs an arrangement of a set S_1 containing five line segments (drawn as solid lines in the figure to the right). Then, it inserts a single segment s (drawn as a dotted line) using the incremental insertion function. Finally, it adds a set S_2 with five more line segments (drawn as dashed lines) in an aggregate fashion. Notice that the line segments of S_1 are pairwise interior-disjoint, so **insert_non_intersecting_curves()** is safely used. S_2 also contain pairwise interior-disjoint segments, but as they intersect the existing arrangement, **insert()** must be used to insert them. Also note that the single segment s inserted incrementally partially overlaps an existing arrangement edge; the overlapped portion is drawn as a dash-dotted line.

// File: ex_global_insertion.cpp

```cpp
#include "arr_exact_construction_segments.h"
#include "arr_print.h"

int main()
{
  Segment      S1[] = {Segment(Point(1, 3), Point(4, 6)),
                       Segment(Point(1, 3), Point(6, 3)),
                       Segment(Point(1, 3), Point(4, 0)),
                       Segment(Point(4, 6), Point(6, 3)),
                       Segment(Point(4, 0), Point(6, 3))};
  Segment      s = Segment(Point(0, 3), Point(4, 3));
  Segment      S2[] = {Segment(Point(0, 5), Point(6, 6)),
                       Segment(Point(0, 4), Point(6, 5)),
                       Segment(Point(0, 2), Point(6, 1)),
                       Segment(Point(0, 1), Point(6, 0)),
                       Segment(Point(6, 1), Point(6, 5))};

  Arrangement  arr;
  insert_non_intersecting_curves(arr, S1, S1 + sizeof(S1)/sizeof(Segment));
  insert(arr, s);                                   // 1 incremental
  insert(arr, S2, S2 + sizeof(S2)/sizeof(Segment)); // 5 aggregate
  print_arrangement_size(arr);
  return 0;
}
```

We summarize the three levels of arrangement types based on curve monotonicity and intersection properties in the table below. The three levels, starting from the most restrictive, are (i) x-monotone curves that are pairwise disjoint in their interior, (ii) x-monotone curves (which are possibly intersecting one another), and (iii) general curves. We list the single-curve insertion functions.

Type of Curves	Geometry-Traits Concept	Insertion Function
x-monotone and pairwise disjoint	ArrangementBasicTraits_2 (or ArrangementLandmarkTraits_2)	`insert_non_intersecting_curve()`
x-monotone	ArrangementXMonotoneTraits_2	`insert()`
general	ArrangementTraits_2	`insert()`

The insertion function `insert()` is overloaded to (i) incrementally insert a single *x*-monotone curve, (ii) incrementally insert a single general curve, (iii) aggregately insert a set of *x*-monotone curves, and (iv) aggregately insert a set of general curves. The `insert_non_intersecting_curves()` function aggregately inserts a set of *x*-monotone pairwise interior-disjoint curves into an arrangement.

3.5 Removing Vertices and Edges

The free functions `remove_vertex()` and `remove_edge()` handle the removal of vertices and edges from an arrangement, respectively. The difference between these functions and the corresponding member functions of the `Arrangement_2` class template (see Section 2.2.2) has to do with the ability to merge two curves associated with adjacent edges to form a single curve associated with a single edge. An attempt to remove a vertex or an edge from an arrangement object `arr` using the free functions above requires that the traits class used to instantiate the arrangement type of `arr` models the concept ArrangementXMonotoneTraits_2, which refines the ArrangementBasicTraits_2 concept; see Section 5.1.

The function template `remove_vertex(arr, v)` removes the vertex *v* from the given arrangement `arr`, where *v* is either an isolated vertex or a *redundant* vertex. Namely, it has exactly two incident edges that are associated with two curves that can be merged to form a single *x*-monotone curve. If neither of the two cases applies, the function returns an indication that it has failed to remove the vertex.

The function template `remove_edge(arr, e)` removes the edge *e* from the arrangement by simply calling `arr.remove_edge(e)`; see Section 2.2.2. In addition, if either of the end vertices of *e* becomes isolated or redundant after the removal of the edge, it is removed as well.

Example: The program below demonstrates the usage of the free removal functions. It creates an arrangement of four line segments $s_1, \ldots, s_4$, as shown to the right. Then it removes the two horizontal edges induced by s_1 and s_2 (drawn as dashed lines) and clears all redundant vertices (these vertices are drawn as lightly shaded discs), such that the final arrangement consists of just two edges associated with the vertical line-segments s_3 and s_4.

```
// File: ex_global_removal.cpp

#include "arr_exact_construction_segments.h"
#include "arr_print.h"

int main()
{
  // Create an arrangement of four line segments.
  Arrangement       arr;

  Segment           s1(Point(1, 3), Point(5, 3));
  Halfedge_handle e1 = arr.insert_in_face_interior(s1, arr.unbounded_face());
  Segment           s2(Point(1, 4), Point(5, 4));
```

```
  Halfedge_handle e2 = arr.insert_in_face_interior(s2, arr.unbounded_face());
  insert(arr, Segment(Point(1, 1), Point(1, 6)));        // s3
  insert(arr, Segment(Point(5, 1), Point(5, 6)));        // s4

  std::cout << "The initial arrangement:" << std::endl;
  print_arrangement(arr);

//Remove e1 and its incident vertices using the member function remove_edge().
  Vertex_handle   v1 = e1->source(), v2 = e1->target();
  arr.remove_edge(e1);
  remove_vertex(arr, v1);
  remove_vertex(arr, v2);

  // Remove e2 using the free remove_edge() function.
  remove_edge(arr, e2);

  std::cout << "The final arrangement:" << std::endl;
  print_arrangement(arr);
  return 0;
}
```

3.6 Vertical Decomposition

As you have already seen, an arrangement face may have a fairly complicated structure; its outer boundary may be very large, and it may contain numerous holes. For many practical applications, it is more convenient to analyze the faces by decomposing each into a finite number of cells (preferably convex when dealing with arrangements of linear curves) of constant complexity. Vertical decomposition is a generic way to achieve such a simplification of the arrangement.

Given an arrangement, we consider each arrangement vertex v, and locate the feature lying vertically below the vertex and the feature lying vertically above it, or the unbounded face in case there is no such feature. (Such a feature is also the result of the vertical ray-shooting operation from the vertex v, as described in Section 3.1.2.) It is now possible to construct two vertical segments connecting v to the feature above it and to the feature below it, possibly extending the vertical segments to infinity. The collection of the vertical segments and rays computed for all arrangement vertices induces a subdivision of the arrangement into simple cells. There are two types of bounded two-dimensional cells, as follows:

- Cells whose outer boundaries comprise four edges, of which two originate from the original arrangement and the other two are vertical.

- Cells whose outer boundaries comprise three edges, of which two originate from the original arrangement and intersect at a common vertex and the remaining one is vertical.

In the case of an arrangement of line segments, two-dimensional cells of the former type are trapezoids (as they have a pair of parallel vertical edges), while two-dimensional cells of the latter type are triangles, which can be viewed as degenerate trapezoids. The unbounded cells can be similarly categorized into two types. Observe that the boundary of an unbounded cell contains only one original edge or none and it may contain only non-vertical edges. The resulting cells are therefore referred to as *pseudo trapezoids*. The figure 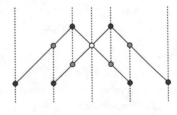 to the right depicts the vertical decomposition of an arrangement induced by four line segments. The decomposition consists of 14 pseudo trapezoids, out of which two are bounded triangles, two are bounded trapezoids, and 10 are unbounded trapezoids.

The decomposition of the arrangement into pseudo-trapezoidal cells has many useful applications. As mentioned in Section 3.1.1, it is possible to use the decomposition to construct an auxiliary search-structure over the pseudo trapezoids that answers point-location queries in an efficient manner; see, e.g., [45, Chapter 6] for the details. In Section 9.3.1 we plan a path for a robot moving amidst obstacles using a decomposed arrangement.

The function template `decompose()` accepts an arrangement $\mathcal{A}$ and outputs for each vertex v of $\mathcal{A}$ a pair of features—one that directly lies below v and another that directly lies above v. It is implemented as a plane-sweep algorithm employing a dedicated visitor (see Section 3.2), which aggregately computes the set of pairs of features. Let v be a vertex of $\mathcal{A}$. The feature above (respectively below) v may be one of the following:

- Another vertex u having the same x-coordinate as v.

- An arrangement edge associated with an x-monotone curve that contains v in its x-range.

- An unbounded face in case v is incident to an unbounded face, and there is no curve lying above (respectively below) it.

- An empty object, in case v is the lower (respectively upper) endpoint of a vertical edge in the arrangement.

──── *advanced* ────

In Chapter 4 you will learn that we also support arrangements induced by unbounded curves, referred to as *unbounded arrangements*. For such an arrangement we maintain an imaginary rectangle with left, right, bottom, and top boundaries, which bound the concrete vertices and edges of the arrangement. Halfedges that represent the boundaries of the rectangle are *fictitious*. The `decompose()` function template handles unbounded arrangements. If v is a vertex of an unbounded arrangement, and v is incident to an unbounded face and there is no curve lying above (respectively below) it, the feature above (respectively below) v returned by the function template `decompose()` is a fictitious edge.

──── *advanced* ────

3.6.1 Application: Decomposing an Arrangement of Line Segments

Given an arrangement of line segments, we can use the output of `decompose()` and construct a vertical edge emanating from v downward by connecting it to a vertex that lies below it or by splitting the curve below it, and connecting v to the split point. Similarly, we can construct the upward vertical segment emanating from v.

The function template `add_vertical_segment()` listed below accepts an arrangement object $\mathcal{A}$, a handle to a vertex v of $\mathcal{A}$, and a polymorphic object o of type `CGAL::Object` that wraps a handle to a feature of $\mathcal{A}$. It constructs a segment between the geometric embedding of v and the geometric embedding of the feature represented by o (in case o is not empty), and it inserts the newly constructed segment into $\mathcal{A}$. It returns a handle to one of the two halfedges that represents the new edge. If o represents a vertex, the construction of the new segment is trivial. If, however, o represents a non-fictitious edge e, the edge is split at the vertical projection of v onto the geometric embedding of e, and a segment between the geometric embedding of v and the split point is constructed. The function template returns a handle to one of the halfedges created by the insertion of the vertical segment, or an invalid halfedge-handle in case the arrangement remains intact. The function template is defined in the header file `vertical_decomposition.h`.

```
template <typename Arrangement, typename Kernel>
typename Arrangement::Halfedge_const_handle
add_vertical_segment(Arrangement& arr, typename Arrangement::Vertex_handle v,
                     CGAL::Object obj, Kernel& ker)
{
```

```
    typedef typename Arrangement::X_monotone_curve_2          Segment;

Segment                                                      seg;
typename Arrangement::Vertex_const_handle    vh;
typename Arrangement::Halfedge_const_handle  hh;
typename Arrangement::Face_const_handle       fh;
typename Arrangement::Vertex_handle           v2;

  if (CGAL::assign(vh, obj)) {
    // The given feature is a vertex.
    seg = Segment(v->point(), vh->point());
    v2 = arr.non_const_handle(vh);
  }
  else if (CGAL::assign(hh, obj)) {
    // The given feature is a halfedge. We ignore fictitious halfedges.
    if (hh->is_fictitious())
      return typename Arrangement::Halfedge_const_handle();

    // Check whether v lies in the interior of the x-range of the edge (in
    // which case this edge should be split).
    const typename Kernel::Compare_x_2 cmp_x = ker.compare_x_2_object();
    if (cmp_x(v->point(), hh->target()->point()) == CGAL::EQUAL) {
      // In case the target of the edge already has the same x-coordinate as
      // the vertex v, just connect these two vertices.
      seg = Segment(v->point(), hh->target()->point());
      v2 = arr.non_const_handle(hh->target());
    }
    else {
      // Compute the vertical projection of v onto the segment associated
      // with the halfedge. Split the edge and connect v with the split point.
      typedef typename Kernel::Line_2 Line;
      Line supp_line(hh->source()->point(), hh->target()->point());
      Line vert_line(v->point(), Point(v->point().x(), v->point().y() + 1));
      typename Arrangement::Point_2  point;
      CGAL::assign(point, ker.intersect_2_object()(supp_line, vert_line));
      seg = Segment(v->point(), point);
      arr.split_edge(arr.non_const_handle(hh),
                     Segment(hh->source()->point(), point),
                     Segment(point, hh->target()->point()));
      v2 = arr.non_const_handle(hh->target());
    }
  }
  // Ignore faces and empty objects.
  else return typename Arrangement::Halfedge_const_handle();

  // Add the vertical segment to the arrangement using its two end vertices.
  return arr.insert_at_vertices(seg, v, v2);
}
```

We do not care about unbounded faces in this particular application and the other application that uses this function template presented in Section 9.3.1. As a matter of fact, we do not even use unbounded arrangements in these applications. In light of this, and to simplify the code, when processing the output of the **decompose()** function template, we ignore fictitious edges. We do not even need to check for vertices that lie on the bottom or top boundaries, since such vertices cannot be present in the output set, as explained below.

─────── *advanced* ───────

A vertex u that lies on the top boundary of the imaginary rectangle represents an unbounded end of a vertical linear curve or an unbounded end of a curve with a vertical asymptote; see Section 4.1. Recall, that the `decompose()` function template returns an empty object in case v is the lower endpoint of a vertical edge. Thus, if u represents an unbounded end of a vertical linear curve, u cannot be returned. If the x-coordinate of a vertex v is equal to the x-coordinate of the vertical asymptote of some curve c, as depicted in the figure to the right, v is not even in the x-range of c; see Section 4.1.1 for the precise definition. The same argument holds for a vertex that lies on the bottom boundary. Thus, the vertex above (respectively below) any vertex cannot lie on the rectangle boundary.

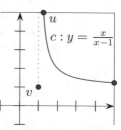

─────── *advanced* ───────

The function template `vertical_decomposition()` listed below constructs the vertical decomposition of a given arrangement.

```
template <typename Arrangement, typename Kernel>
void vertical_decomposition(Arrangement& arr, Kernel& ker)
{
  typedef std::pair<typename Arrangement::Vertex_const_handle,
                    std::pair<CGAL::Object, CGAL::Object> > Vd_entry;

  // For each vertex in the arrangment, locate the feature that lies
  // directly below it and the feature that lies directly above it.
  std::list<Vd_entry>         vd_list;
  CGAL::decompose(arr, std::back_inserter(vd_list));

  // Go over the vertices (given in ascending lexicographical xy-order),
  // and add segements to the feautres below and above it.
  const typename Kernel::Equal_2         equal = ker.equal_2_object();
  typename std::list<Vd_entry>::iterator  it, prev = vd_list.end();
  for (it = vd_list.begin(); it != vd_list.end(); ++it) {
    // If the feature above the previous vertex is not the current vertex,
    // add a vertical segment to the feature below the vertex.
    typename Arrangement::Vertex_const_handle v;
    if ((prev == vd_list.end()) ||
        !CGAL::assign(v, prev->second.second) ||
        !equal(v->point(), it->first->point()))
      add_vertical_segment(arr, arr.non_const_handle(it->first),
                           it->second.first, ker);
    // Add a vertical segment to the feature above the vertex.
    add_vertical_segment(arr, arr.non_const_handle(it->first),
                         it->second.second, ker);
    prev = it;
  }
}
```

The program below reads a set of line segments from an input file. It computes an axis-parallel iso-rectangle that contains the segments in its interior, and inserts the four segments that comprise its boundary into the input set. Then, the program constructs the arrangement of the line segments. The introduction of the iso-rectangle exempts us from the need to handle the unbounded set, as the faces adjacent to the bounding iso-rectangle represent the unbounded region. Finally, the program decomposes the arrangement into pseudo-trapezoidal cells. By

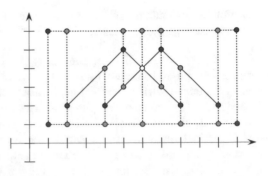

default the program reads the input file `segments.bat`. It consists of the description of four line segments drawn as solid lines in the figure to the right. The bounding rectangle is drawn as dashed lines, and the vertical segments inserted into the arrangement as part of the vertical decomposition are drawn as dotted lines. The final arrangement consists of 25 vertices, 38 edges, and 15 faces.

```cpp
// File: vertical_decomposition.cpp

#include <list>

#include "arr_exact_construction_segments.h"
#include "read_objects.h"
#include "arr_print.h"
#include "vertical_decomposition.h"
#include "bbox.h"

int main(int argc, char* argv[])
{
  // Get the name of the input file from the command line, or use the default
  // segments.dat file if no command-line parameters are given.
  const char* filename = (argc > 1) ? argv[1] : "segments.dat";
  std::list<Segment> segments;
  read_objects<Segment>(filename, std::back_inserter(segments));

  // Add four segments of an iso-rectangle that bounds all other segments.
  CGAL::Bbox_2 rect = bbox(segments.begin(), segments.end());
  const Number_type x_min = static_cast<int>(rect.xmin() + 0.5) - 1;
  const Number_type y_min = static_cast<int>(rect.ymin() + 0.5) - 1;
  const Number_type x_max = static_cast<int>(rect.xmax() + 0.5) + 1;
  const Number_type y_max = static_cast<int>(rect.ymax() + 0.5) + 1;

  segments.push_back(Segment(Point(x_min, y_min), Point(x_max, y_min)));
  segments.push_back(Segment(Point(x_max, y_min), Point(x_max, y_max)));
  segments.push_back(Segment(Point(x_max, y_max), Point(x_min, y_max)));
  segments.push_back(Segment(Point(x_min, y_max), Point(x_min, y_min)));

  Traits traits;
  Arrangement arr(&traits);
  insert(arr, segments.begin(), segments.end());
  print_arrangement_size(arr);

  // Add vertical edges that induce the vertical decomposition.
  Kernel* kernel = &traits;
  vertical_decomposition(arr, *kernel);
```

```
  print_arrangement_size(arr);
  return 0;
}
```

The program above uses the function template `bbox<Input_iterator>()` listed below. It accepts a range of geometric objects and computes the approximate iso-rectangle that bounds the geometric objects. The returned object is of type `CGAL::Bbox_2`. It can be queried for the minimum and maximum x- and y-coordinates of the rectangle in double-precision floating-point (`double`). When instantiated, the template parameter `Input_iterator` must be substituted with a valid input iterator the value type of which has a member function `bbox()`. The function template is defined in the header file `bbox.h`.

```
#include <CGAL/basic.h>

template <typename Input_iterator>
CGAL::Bbox_2 bbox(Input_iterator begin, Input_iterator end)
{
  CGAL_assertion(begin != end);
  CGAL::Bbox_2 rect = (*begin++).bbox();
  while (begin != end) rect = rect + (*begin++).bbox();
  return rect;
}
```

 Try: If you need to distinguish between the edges induced by original segments and the edges induced by vertical segments inserted as part of the vertical decomposition process, you have to alter the `vertical_decomposition()` function template or introduce a new function template, possibly overloading the existing one. Alter the `vertical_decomposition()` function template, so that it collects the set of the edges induced by vertical segments and returns them to the caller via an output iterator. Pass a null output iterator that simply discards the output in cases where obtaining the set of edges inserted as part of the vertical decomposition process is not needed. Here is one way to do it. First, define a model of the UnaryFunction concept, the function operator of which does nothing; here is one:

```
struct Nuller { template <typename T> void operator()(const T&) {} };
```

Then, construct the null output iterator with the help of the function output iterator adaptor provided by BOOST. You can pass it as an argument to an instance of the altered `vertical_decomposition()` function template as follows:

```
#include <boost/function_output_iterator.hpp>
  ⋮
vertical_decomposition(arr, *kernel,
                       boost::make_function_output_iterator(Nuller()));
```

3.7   Bibliographic Notes and Remarks

Point location in planar subdivision is a well-studied problem [199]. There are many efficient algorithms for it; we mention a few deterministic [135, 186] and randomized [164, 190] ones. The randomized algorithms are the basis of the so-called RIC point location in CGAL.

The walk-along-a-line approach to point location and its variants have mostly been discussed in the context of triangulations [38, 50, 51, 151, 162]. The landmark algorithm relies on a nearest-neighbor search-structure, and in particular we use *KD-trees* proposed by Bentley [17]. One can opt for using different implementations of KD-trees, such as CGAL's or those of the library for approximate nearest-neighbor search, ANN, by Mount and Arya.[8]

---

[8]http://www.cs.umd.edu/~mount/ANN/.

We conducted extensive experiments on the performance of various point-location strategies in practice. These are reported in Haran's M.Sc. thesis [110,111].

The fundamental sweep paradigm was suggested by Bentley and Ottmann [18], originally presented to find all the intersections of a given set of segments in the plane. It has since found a myriad of uses in computational geometry. The description of the algorithm in the textbook [45, Chapter 2] does not assume general position, and describes how to handle various degeneracies in the case of sweeping line segments. The plane-sweep paradigm nicely generalizes to curves rather than to segments. More degeneracies arise in the case of curves. We note that the CGAL *2D Arrangements* package handles all degeneracies, including two types that are typically ignored elsewhere: isolated points and overlapping curves.

The *zone* of a curve in an arrangement of curves is central to the study of arrangements, and it is discussed in almost all books and book chapters dedicated to arrangements; see Section 1.6 and, for example, [45, Chapter 8.3] for the computation of the zone of a line in an arrangement of lines. As the CGAL *2D Arrangements* package evolved we realized that constructing the zone of a curve in an arrangement is a key operation in a variety of algorithms; hence, it is presented in the book as an algorithmic framework parallel to that of the plane-sweep algorithm. The zone of a line in an arrangement $\mathcal{A}$ of n lines has complexity $O(n)$ and it can be computed in time $O(n)$ assuming $\mathcal{A}$ is represented, for example, by a DCEL.

Let $\mathcal{C}$ be a collection of n well-behaved curves (see Section 1.3.3), each pair of which intersects in at most s points. Let γ be another well-behaved curve intersecting each curve of $\mathcal{C}$ in at most some constant number of points. The complexity of the zone of γ in the arrangement $\mathcal{A}(\mathcal{C})$ is $O(\lambda_s(n))$ if the curves in $\mathcal{C}$ are all unbounded or $O(\lambda_{s+2}(n))$ if they are bounded, where $\lambda_s(n)$ is the maximum length of an (n, s)-Davenport-Schinzel sequence. The zone can be computed by an algorithm whose running time is close to the worst-case complexity of the computed zone. For more information on these combinatorial bounds and algorithms see [196].

The interchangeability of the different point-location algorithms in the library resembles the *strategy* design pattern as described in [83, Chapter 5].

A concrete algorithm that utilizes the plane sweep or the zone construction algorithmic frameworks is realized through a visitor, which resembles the *visitor* design-pattern [83, Chapter 5].[9]

3.8 Exercises

3.1 Write a program that reads the `world.in` data file on the book's website `http://acg.cs.tau.ac.il/cgal-arrangement-book`,[10] and constructs the map of the world using incremental insertion and aggregate insertion. Locate different cities of the world using different point-location strategies.

What is the most efficient construction method? What is the most efficient point-location strategy?

Do not forget to insert a bounding rectangle; otherwise, the continent of Antarctica and the Antarctica Ocean would share the same face of the induced arrangement.

3.2 The *2D Arrangements* package supports only vertical ray-shooting. Implement a program that supports approximate arbitrary ray-shooting in the following way. The program is given a direction $\vec{d}$ in the plane and a rational number ϵ as input. Denote the angle defined by the positive x-direction and $\vec{d}$ as α.

First, the program reads a set of line segments from an input file and computes the arrangement induced by the input segments. Then, it rotates the arrangement by an angle β' that approximates $\beta = -\alpha + \pi/2$ such that $|\sin\beta' - \sin\beta| \le \epsilon$ and $|\cos\beta' - \cos\beta| \le \epsilon$. Finally, it reads a sequence of input points in the plane. For each point p it applies vertical

---

[9]A visitor must provide a specific set of callback member-functions. Typically, visitors derive from a base class. We say *resemble* since our visitors do not necessarily derive from a base class, but merely implement the member functions.

[10]The data was taken from Gnuplot; see `http://www.gnuplot.info/`.

ray-shooting upwards on the rotated arrangement to obtain the first cell of the (original) arrangement encountered by a vector emanating from p in the $\vec{d}$ direction.

You may use the free function `rational_rotation_approximation()` provided by CGAL to obtain the desired $\sin\beta'$ and $\cos\beta'$. The implementation of this function is based on a method presented by Canny, Donald, and Ressler [33].

3.3 In a complete graph each pair of graph vertices is connected by an edge. The complete graph with n vertices is denoted by K_n and has $\binom{n}{2}$ undirected edges.

(a) Empirically compute the expected number of crossings in a straight-edge drawing of K_n for $n = 4, 5, \ldots, 21$, where the planar embedding of the vertices are points uniformly sampled at random within the unit circle.

(b) The rectilinear crossing number of a graph G with n vertices denoted by $\overline{\nu}(G)$ is the minimum number of crossings in a straight-edge drawing of G embedded in the plane. The hunt for crossing numbers of complete graphs was initiated by R. Guy in the 1960s. At the time this book was written only values for $n <= 17$ and for $n = 19$ and $n = 21$ had been found [6].

Obtain complete graphs that attain as small a number of rectilinear crossings as you can. Compute the number of crossings and compare them with the rectilinear crossing numbers (or with their upper bounds in cases where the tight bounds are not settled) for $n = 4, 5, \ldots, 21$, as listed in the table below.

n	3	4	5	6	7	8	9	10	11	12	13	14	15	16	17	18	19	20	21
$\overline{\nu}(K_n)$	0	0	1	3	9	19	36	62	102	153	229	324	447	603	798	1029	1318	[1652,1657]	2055

3.4 The expected complexity of the outer face in arrangements of line segments of a fixed length in the plane drawn uniformly at random within a square follows a phase transition, where it sharply drops as a function of the total number of segments [8]. Until the phase transition, the complexity of the outer face is almost linear in the number of segments n. The complexity of the outer face is roughly proportional to $\sqrt{n}$ after the phase transition.

Write a program that empirically computes the value n where the phase transition occurs for random arrangements defined as follows: Let S denote a unit square in the plane. Fix an integer n and a real positive length ℓ. Let $\mathcal{C}$ be a collection of n oriented line segments, all of length ℓ, where each $s \in \mathcal{C}$ is chosen independently by first choosing the location of an endpoint of s uniformly at random in S, and then choosing an orientation for s uniformly at random on the unit circle Γ. Thus, a segment in C may protrude through the boundary of S.

Chapter 4

Arrangements of Unbounded Curves

All the arrangements constructed and manipulated in previous chapters were induced only by line segments, which are, in particular, bounded curves. Such arrangements always have one unbounded face that contains all other arrangement features. In this chapter we explain how to construct arrangements of unbounded curves. For simplicity of exposition, we stay with linear objects and restrict our examples in this chapter to lines and rays. However, the discussion in this chapter, as well as the software described, apply more generally to arbitrary curves in the plane.

4.1 Representing Arrangements of Unbounded Curves

Given a set $\mathcal{C}$ of unbounded curves, a simple approach for representing the arrangement induced by $\mathcal{C}$ would be to clip the unbounded curves using an axis-parallel rectangle that contains all finite curve endpoints and intersection points among curves in $\mathcal{C}$. This process would result in a set of bounded curves (line segments if $\mathcal{C}$ contains lines or rays), and it would be straightforward to compute the arrangement induced by this set. However, we are interested in operating directly on the unbounded curves, without having to preprocess them. Moreover, if we are not given all the curves inducing the arrangement in advance, then the choice of a good bounding rectangle may change as more curves are introduced.

Instead of an explicit approach, we use an implicit bounding rectangle embedded in the DCEL structure. Figure 4.1 shows the arrangement of four lines that subdivide the plane into eight unbounded faces and two bounded ones. Notice that in this case portions of the unbounded faces now have outer boundaries (those portions inside the bounding rectangle), and the halfedges

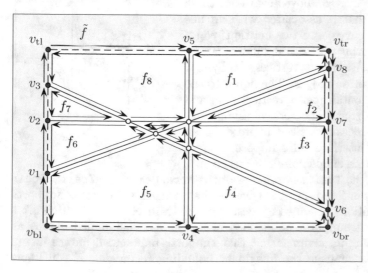

Fig. 4.1: A DCEL representing an arrangement of four lines. Halfedges are drawn as thin arrows. The vertices $v_1, \ldots, v_8$ lie at infinity and are not associated with valid points. The halfedges that connect them are fictitious and are not associated with concrete curves. The face denoted $\tilde{f}$ (lightly shaded) is the fictitious unbounded face, which lies outside the imaginary rectangle (dashed) that bounds the actual arrangement. The four fictitious vertices $v_{\mathrm{bl}}, v_{\mathrm{tl}}, v_{\mathrm{br}}$, and v_{tr} represent the four corners of the imaginary bounding rectangle.

E. Fogel et al., *CGAL Arrangements and Their Applications*, Geometry and Computing 7, DOI 10.1007/978-3-642-17283-0_4, © Springer-Verlag Berlin Heidelberg 2012

along these outer CCBs are drawn as arrows. The bounding rectangle is drawn with a dashed line. The vertices $v_1, \ldots, v_8$, which lie on the bounding rectangle, represent the unbounded ends of the four lines that approach infinity. The halfedges connecting them, which overlap with the bounding rectangle, are not associated with geometric curves. Thus, we refer to them as *fictitious*. Note that the outer CCBs of the unbounded faces contain fictitious halfedges. The twins of these halfedges form together one connected component that corresponds to the entire imaginary rectangle. It forms a single hole in the face $\tilde{f}$. We also refer to $\tilde{f}$ as *fictitious*, since it does not correspond to a real two-dimensional cell of the arrangement. Finally, there are four additional vertices denoted by $v_{\mathrm{bl}}, v_{\mathrm{tl}}, v_{\mathrm{br}},$ and v_{tr}, which coincide with the bottom-left, top-left, bottom-right, and top-right corners of the bounding rectangle, respectively. They do not lie on any curve, and are referred to as *fictitious* as well. These four vertices identify each of the fictitious edges as lying on the top, the bottom, the left, and the right edge of the imaginary bounding rectangle. The four fictitious vertices exist even when the arrangement is empty: In this case they are connected by four pairs of fictitious halfedges that define a single unbounded face (which represents the entire $\mathbb{R}^2$ plane) lying inside the imaginary bounding rectangle and a fictitious face lying outside.

In summary, there are four types of arrangement vertices, which differ from one another by their location with respect to the imaginary bounding rectangle as follows:

1. A vertex associated with a point in $\mathbb{R}^2$. Such a vertex always lies inside the bounding rectangle and has bounded coordinates.

2. A vertex that represents an unbounded end of an x-monotone curve that approaches $x = -\infty$ or $x = \infty$. In the case of a horizontal line or a curve with a horizontal asymptote, the y-coordinate of the curve end may be finite (see, for example, the vertices v_2 and v_7 in Figure 4.1), but in general the curve end also approaches $y = \pm\infty$; see for instance the vertices v_1, v_3, v_6, and v_8 in Figure 4.1. For convenience, we always take a "tall" enough bounding rectangle and treat such vertices as lying on either the left or the right rectangle edges (that is, if a curve approaches $x = -\infty$, its left end will be represented by a vertex on the left edge of the bounding rectangle, and if it approaches $x = \infty$, its right end will be represented by a vertex on the right edge).

3. A vertex that represents the unbounded end of a vertical linear curve (line or ray) or of a curve with a vertical asymptote (finite x-coordinate and an unbounded y-coordinate). Such a vertex always lies on one of the horizontal edges of the bounding rectangle (either the bottom one if $y = -\infty$, or the top one if $y = \infty$). The vertices v_4 and v_5 in Figure 4.1 are of this type.

4. The fictitious vertices that represent the four corners of the imaginary bounding rectangle.

A vertex at infinity of Type 2 or Type 3 above always has three incident edges—one concrete edge that is associated with an unbounded portion of an x-monotone curve, and two fictitious edges connecting the vertex to its neighboring vertices at infinity or the corners of the bounding rectangle. Fictitious vertices (of Type 4 above) have exactly two incident fictitious edges. The figure to the right depicts the portions of a horizontal line, a vertical line, and two rectangular hyperbolas with horizontal and vertical asymptotes confined to the imaginary bounding rectangle. See Chapter 5 for an explanation on how the traits-class interface helps impose the fact that a vertex at infinity is incident to a single non-fictitious edge.

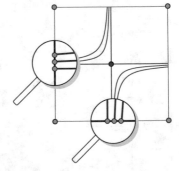

The traits class that substitutes the **Traits** parameter of the **Arrangement_2<Traits,Dcel>** class template when the latter is instantiated specifies whether the arrangement instance supports unbounded curves through the definitions of some tags nested in the traits class; see Section 5.1.1 for details. Every arrangement that supports unbounded curves supports bounded curves as well, but not vice versa. Maintaining an arrangement that supports unbounded curves incurs an overhead due to the necessity to manage the imaginary bounding rectangle. If you know beforehand that all input curves that induce a particular arrangement are bounded, define your arrangement accordingly. That is, use a traits class that does not support unbounded curves.

4.1.1 Basic Manipulation and Traversal Methods

The types `Vertex`, `Halfedge`, and `Face` nested in the `Arrangement_2` class template support the methods described below in addition to the ones listed in Section 2.2.1. Let v, e, and f be handles to a vertex of type `Vertex`, a halfedge of type `Halfedge`, and a face of type `Face`, respectively.

- The calls `v->parameter_space_in_x()` and `v->parameter_space_in_y()` determine the location of the geometric embedding of the vertex v. The call `v->parameter_space_in_x()` returns `CGAL::ARR_INTERIOR` if the x-coordinate associated with v is finite, `CGAL::ARR_LEFT_BOUNDARY` if v represents the end of a curve the x-coordinate of which approaches $-\infty$, and `CGAL::ARR_RIGHT_BOUNDARY` if v represents the end of a curve the x-coordinate of which approaches ∞. Similarly, the call `v->parameter_space_in_y()` returns `CGAL::ARR_INTERIOR` if the y-coordinate associated with v is finite and `CGAL::ARR_BOTTOM_BOUNDARY` or `CGAL::ARR_TOP_BOUNDARY` if v represents the end of a curve the y-coordinate of which approaches $-\infty$ or ∞, respectively.[1]

  The nested `Vertex` type also provides the Boolean predicate `is_at_open_boundary()`. If the call `v->is_at_open_boundary()` predicates evaluates to `false`, you can access the point associated with v. Otherwise, v is not associated with a `Point_2` object, as it represents the unbounded end of a curve that approaches infinity.

- The nested `Halfedge` type provides the Boolean predicate `is_fictitious()`. If the call `e->is_fictitious()` predicate evaluates to `false`, you can access the x-monotone curve associated with e. Otherwise, e is not associated with an `X_monotone_curve` object.

- The nested `Face` type provides the Boolean predicate `is_fictitious()`. The call `f->is_fictitious()` predicate evaluates to `true` only when f lies outside the bounding rectangle. The method `outer_ccb()` (see Section 2.2.1) must not be invoked for the fictitious face.[2] Note that a valid unbounded face (of an arrangement that supports unbounded curves) has a valid outer CCB, although the CCB comprises only fictitious halfedges if the arrangement is induced only by bounded curves.

Let $p_l = (x_l, y_l)$ and $p_r = (x_r, y_r)$ be the left and right endpoints of a bounded x-monotone curve c, respectively. We say that a point $p = (x_p, y_p)$ lies in the x-range of c if $x_l \le x_p \le x_r$. All the points in the figure to the right are in the x-range of the curve in the figure. Let $p_l = (x_l, y_l)$ be the left endpoint of an x-monotone curve c unbounded from the right. A point $p = (x_p, y_p)$ lies in the x-range of c if $x_l \le x_p$. Similarly, A point $p = (x_p, y_p)$ lies in the x-range of an x-monotone curve unbounded from the left if $x_p \le x_r$. Naturally, every 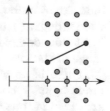 point $p \in \mathbb{R}^2$ is in the x-range of an x-monotone curve unbounded from the left and from the right. The template function `is_in_x_range()` listed below, and defined in the file `is_in_x_range.h`, checks whether a given point p is in the x-range of the curve associated with a given halfedge e. The function template also exemplifies how some of the above functions can be used.

```
template <typename Arrangement>
bool is_in_x_range(typename Arrangement::Halfedge_const_handle he,
                   const typename Arrangement::Point_2& p,
                   const typename Arrangement::Traits_2& traits)
{
```

---

[1] The term "parameter space" stems from a major extension the *2D Arrangements* package was going through at the time this book was written to support arrangements embedded on certain two-dimensional parametric surfaces in three (or higher) dimensions; see Section 11.1.

[2] Typically, the code is guarded against such malicious calls with adequate preconditions. However, these preconditions are suppressed when the code is compiled with maximum optimization.

```cpp
  typedef typename Arrangement::Traits_2                              Traits_2;
  typename Traits_2::Compare_x_2 cmp = traits.compare_x_2_object();

  // Compare p with the source vertex (which may lie at x = +/- oo).
  CGAL::Arr_parameter_space src_px = he->source()->parameter_space_in_x();
  CGAL::Comparison_result res_s =
    (src_px == CGAL::ARR_LEFT_BOUNDARY) ? CGAL::SMALLER :
    ((src_px == CGAL::ARR_RIGHT_BOUNDARY) ? CGAL::LARGER :
     cmp(he->source()->point(), p));
  if (res_s == CGAL::EQUAL) return true;

  // Compare p with the target vertex (which may lie at x = +/- oo).
  CGAL::Arr_parameter_space trg_px = he->target()->parameter_space_in_x();
  CGAL::Comparison_result res_t =
    (trg_px == CGAL::ARR_LEFT_BOUNDARY) ? CGAL::SMALLER :
    ((trg_px == CGAL::ARR_RIGHT_BOUNDARY) ? CGAL::LARGER :
     cmp(he->target()->point(), p));

  // p lies in the x-range of the halfedge iff its source and target lie
  // at opposite x-positions.
  return (res_s != res_t);
}
```

It is important to observe that the call `arr.number_of_vertices()` does not count the vertices at infinity in the arrangement `arr`. To find out this number you need to issue the call `arr.number_of_vertices_at_infinity()`. Similarly, `arr.number_of_edges()` does not count the fictitious edges (whose number is always `arr.number_of_vertices_at_infinity() + 4`) and `arr.number_of_faces()` does not count the fictitious faces. The vertex, halfedge, edge, and face iterators defined by the `Arrangement_2` class template only go over true features of the arrangement; namely, vertices at infinity and fictitious halfedges and fictitious faces are skipped. On the other hand, the `Ccb_halfedge_circulator` of the outer boundary of an unbounded face or the `Halfegde_around_vertex_circulator` of a vertex at infinity do traverse fictitious halfedges. While an arrangement induced by bounded curves has a single unbounded face, an arrangement induced by unbounded curves may have several unbounded faces. The calls `arr.unbounded_faces_begin()` and `arr.unbounded_faces_end()` return iterators of the type `Arrangement_2::Unbounded_face_iterator` (or its non-mutable, `const`, counterpart type) that define the range of the arrangement unbounded faces.

There is no way to directly obtain the fictitious vertices that represent the four corners of the imaginary bounding rectangle. However, you can obtain the fictitious face through the call `arr.fictitious_face()`, and then iterate over the boundary of its single hole, which represents the imaginary bounding rectangle. Recall that an arrangement of bounded curves does not have a fictitious face. In this case the call above returns a null handle.

Example: The example below exhibits the difference between an arrangement induced by bounded curves, which has a single unbounded face, and an arrangement induced by unbounded curves, which may have several unbounded faces. It also demonstrates the usage of the insertion function for pairwise interior-disjoint unbounded curves. In this case we use an instance of the traits class-template `Arr_linear_traits_2<Kernel>` to substitute the `Traits` parameter of the `Arrangement_2<Traits,Dcel>` class template when instantiated. This traits class-template is capable of representing line segments as well as unbounded linear curves, namely, lines and rays. Observe that objects of the `X_monotone_curve_2` type nested in this traits class-template are constructible from objects of types `Line_2`, `Ray_2`, and `Segment_2` also nested in the traits class-template. These three types and the `Point_2` type are defined by the traits class-template `Arr_linear_traits_2<Kernel>` to be the corresponding types of the kernel used to instantiate the traits class-template; see Section 5.2.3.

The type definitions used by the example below, as well as by other examples that use the **Arr_linear_traits_2**, are listed next. These types are defined in the header file **Arr_linear.h**.

```
#include <CGAL/Exact_predicates_exact_constructions_kernel.h>
#include <CGAL/Arr_linear_traits_2.h>
#include <CGAL/Arrangement_2.h>

typedef CGAL::Exact_predicates_exact_constructions_kernel  Kernel;
typedef Kernel::FT                                         Number_type;

typedef CGAL::Arr_linear_traits_2<Kernel>                  Traits;
typedef Traits::Point_2                                    Point;
typedef Traits::Segment_2                                  Segment;
typedef Traits::Ray_2                                      Ray;
typedef Traits::Line_2                                     Line;
typedef Traits::X_monotone_curve_2                         X_monotone_curve;

typedef CGAL::Arrangement_2<Traits>                        Arrangement;
typedef Arrangement::Vertex_handle                         Vertex_handle;
typedef Arrangement::Halfedge_handle                       Halfedge_handle;
typedef Arrangement::Face_handle                           Face_handle;
```

The first three curves, c_1, c_2, and c_3, are inserted using the specialized insertion functions for x-monotone curves whose location in the arrangement is known. Notice that inserting an unbounded curve in the interior of an unbounded face, or from an existing vertex that represents an bounded end of the curve, may cause an unbounded face to split. (This is never the case when inserting a bounded curve—compare with Section 2.2.2.) Three additional rays are then inserted incrementally using the insertion function for x-monotone curves, the interior of which is disjoint from all arrangement features; see the illustration in the figure to the right. Finally, the program prints the size of the arrangement using the template function **print_unbounded_arrangement_size()** defined in header file **arr_print.h**. (Its listing is omitted here.) The program also traverses the outer boundaries of its six unbounded faces and prints the curves along these boundaries. The header file **arr_linear.h**, which contains the definitions used by the example program, is listed immediately after the listing of the main function.

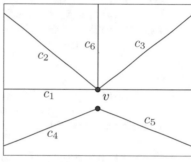

```
// File: ex_unbounded_non_intersecting.cpp

#include "arr_linear.h"
#include "arr_print.h"

int main()
{
  Arrangement arr;

  //Insert a line in the (currently single) unbounded face of the arrangement,
  // then split it into two at (0,0). Assign v to be the split point.
  X_monotone_curve c1 = Line(Point(-1, 0), Point(1, 0));
  Halfedge_handle  e1 = arr.insert_in_face_interior(c1, arr.unbounded_face());
  X_monotone_curve c1_left = Ray(Point(0, 0), Point(-1, 0));
  X_monotone_curve c1_right = Ray(Point(0, 0), Point(1, 0));
  e1 = arr.split_edge(e1, c1_left, c1_right);
  Vertex_handle v = e1->target();
```

```
CGAL_assertion(! v->is_at_open_boundary());

// Add two more rays using the specialized insertion functions.
arr.insert_from_right_vertex(Ray(Point(0, 0), Point(-1, 1)), v); // c2
arr.insert_from_left_vertex(Ray(Point(0, 0), Point(1, 1)), v);   // c3

// Insert three more interior-disjoint rays, c4, c5, and c6.
insert_non_intersecting_curve(arr, Ray(Point(0, -1), Point(-2, -2)));
insert_non_intersecting_curve(arr, Ray(Point(0, -1), Point(2, -2)));
insert_non_intersecting_curve(arr, Ray(Point(0, 0), Point(0, 1)));

print_unbounded_arrangement_size(arr);

// Print the outer CCBs of the unbounded faces.
int k = 1;
Arrangement::Unbounded_face_const_iterator it;
for (it = arr.unbounded_faces_begin(); it != arr.unbounded_faces_end(); ++it)
{
  std::cout << "Face no. " << k++ << "(" << it->number_of_outer_ccbs()
            << "," << it->number_of_inner_ccbs() << ")" << ": ";
  Arrangement::Ccb_halfedge_const_circulator first, curr;
  curr = first = it->outer_ccb();
  if (! curr->source()->is_at_open_boundary())
    std::cout << "(" << curr->source()->point() << ")";

  do {
    Arrangement::Halfedge_const_handle he = curr;
    if (! he->is_fictitious()) std::cout << "   [" << he->curve() << "]   ";
    else std::cout << "   [ ... ]   ";

    if (! he->target()->is_at_open_boundary())
      std::cout << "(" << he->target()->point() << ")";
  } while (++curr != first);
  std::cout << std::endl;
}
return 0;
}
```

4.1.2   Free Functions

All the free functions that operate on arrangements of bounded curves (see Chapter 3) can also be applied to arrangements of unbounded curves. For example, consider a container of linear curves that has to be inserted into an arrangement object, the type of which is an instance of the `Arrangement_2<Traits,Dcel>` class template, where the `Traits` parameter is substituted with the traits class that handles linear curves; see Section 5.2. You can do it incrementally; namely, insert the curves one by one as follows:

```
Curve_container::const_iterator  it;
for (it = curves.begin(); it != curves.end(); ++it) insert(arr, *it);
```

Alternatively, the curves can be inserted aggregately using a single call as follows:

```
insert(arr, curves.begin(), curves.end());
```

It is also possible to issue point-location queries and vertical ray-shooting queries (see also Section 3.1) on arrangements of lines, where the only restriction is that the query point has finite coordinates. Note that all the point-location strategies mentioned above, except the trapezoidal map strategy, are capable of handling arrangements of unbounded curves.

We are now ready to introduce a nontrivial application of unbounded arrangements: Finding the minimum-area triangle defined by a set of points in the plane. Before that we discuss a general tool for transforming problems on point sets into problems on arrangements.

4.2   Point-Line Duality

One of the reasons why arrangements are so useful is that there are various (fairly easy) ways to transform problems on collections of objects into problems on arrangements of curves and surfaces. We present here a simple instance of such a transform, namely the *point-line duality*. Given points and lines in the *primal* plane, we transform them into lines and points, respectively, in the *dual* plane. Given an object o in the primal plane, we denote its dual by o^*. Using this transform we can turn a problem stated for planar sets of points (a.k.a. *point configurations*) into a problem on arrangements of lines.

Even for this basic duality, there are several different useful transforms. Most of the transforms preserve incidence; namely, a point p lies on a line ℓ in the primal plane iff the line p^* contains the point ℓ^* in the dual plane. We describe here a specific transform: A point p with coordinates (a, b) is transformed into the line p^* whose equation is $y = ax - b$. A non-vertical line ℓ whose equation is $y = cx + d$ is transformed into the point ℓ^* whose coordinates are $(c, -d)$. It is trivial to verify that

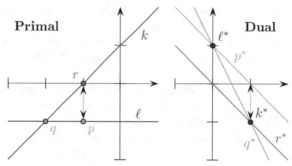

this transform is incidence preserving. In addition, this specific transform has a nice property useful for solving the problem at hand: It preserves the vertical distance between a point and a line. Namely, the distance between a point p and its vertical projection onto the line ℓ in the primal plane is equal to the distance between the point ℓ^* and its vertical projection onto the line p^* in the dual plane, as depicted in the figure above. In fact this transform preserves the *signed* vertical distance between points and lines, which in turn implies that it preserves the above/below relation between points and lines. That is, a point p is below the line ℓ in the primal plane iff the point ℓ^* lies below the line p^* in the dual plane.

The call `primal_point<Traits_2>(l)` constructs the primal point given its dual line l (with the precondition that l is not a vertical line). The call `dual_line<Traits_2>(p)` constructs the dual line given its primal point p. The template parameter `Traits_2` in both cases must be substituted with a model of a concept that defines the nested types `Point_2` and `X_monotone_curve_2`, which are used to represent a point and a line, respectively. The function templates `primal_point()` and `dual_line()` are used by the application presented in the following section. These function templates are listed below, and defined in the header file `dual_plane.h`.

```
template <typename Traits> typename Traits::Point_2
primal_point(const typename Traits::X_monotone_curve_2& cv)
{
  // If the supporting dual line of the linear curve is a*x + b*y + c = 0,
  // the primal point is (-a/b, c/b).
  const typename Traits::Line_2&  line = cv.supporting_line();
  CGAL_assertion(CGAL::sign(line.b()) != CGAL::ZERO);
  return typename Traits::Point_2(-line.a() / line.b(), line.c() / line.b());
}


template <typename Traits> typename Traits::X_monotone_curve_2
dual_line(const typename Traits::Point_2& p)
```

```
{
    // The line dual to the point (p_x, p_y) is y = p_x*x - p_y
    // (or p_x*x - y - p_y = 0).
    typename Traits::Line_2 line(p.x(), -1, -p.y());
    return typename Traits::X_monotone_curve_2(line);
}
```

The code implicitly requires that (i) the **Point_2** type nested in the traits class includes the member functions **x()** and **y()**, which return the x and y Cartesian coordinates of the point, respectively, and (ii) an object of **Point_2** type be constructible from the x and y coordinates. Similar requirements hold for the **Line_2** type, also nested in the traits class. Here, (i) the **Line_2** type must include the member functions **a()**, **b()**, and **c()**, which return the line coefficients, and (ii) an object of the **Line_2** type must be constructible from the a, b, and c coefficients.

4.2.1 Application: Minimum-Area Triangle

The minimum-area triangle problem: *Given a set $P = \{p_1, p_2, \ldots, p_n\}$ of n points in the plane, find three distinct points $p_i, p_j, p_k \in P$ such that the area of the triangle $\triangle p_i p_j p_k$ is minimal among all other triangles defined by three distinct points in the set.*

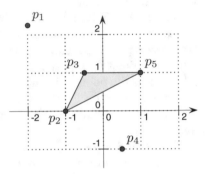

The area of the shaded triangle in the figure to the right is minimal among all the ten triangles defined by the five points. This problem can be naively solved in $O(n^3)$ time by going over all possible triplets of points in P. However, we can do much better if we use the arrangement of lines dual to the points in P. We need to locate three points that define the minimum-area triangle.

We use the duality presented above and dualize the points in P to lines; following the convention presented there, we denote this set of lines by P^*. Note that a line in P^* cannot be vertical. The intersection point ℓ_{ij}^* between the lines p_i^* and p_j^*, which are dual to the points $p_i \in P$ and $p_j \in P$, respectively, is the dual of the line ℓ_{ij} that connects p_i and p_j in the primal plane.

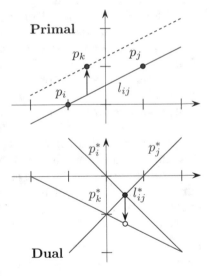

Consider a pair of points $p_i, p_j \in P$. Which is the other point p_k from P that defines the minimum-area triangle with p_i and p_j? Obviously, this is the point closest to the line ℓ_{ij}. It is an easy exercise to show that the closest point to ℓ_{ij} is also the closest point to ℓ_{ij} in its *vertical* distance. Recall that the duality transform that we are using preserves the vertical distance between a point and a line. Therefore, the quest for the closest point p_k to the line ℓ_{ij} translates in the dual plane to the line immediately above or immediately below the point ℓ_{ij}^*—one of them, or in a degenerate case both, will determine the other corner of the minimum-area triangle having p_i and p_j as corners.[3]

There are at most $\binom{n}{2} = O(n^2)$ vertices in the arrangement of the dual lines $p_1^*, \ldots, p_n^*$, which correspond to intersection points between the dual lines. Observe that if three (or more) lines intersect at a common vertex, at least three points are collinear in P. In this case, the minimum-area triangle has zero area and we are done. Otherwise, all arrangement vertices are of degree 4.

The special case of three (or more) collinear points that lie on a vertical line must be handled separately. We can sort the original points by their x-coordinates and go over the sorted list to

---

[3]Here we use the term *corner* for the vertices of a triangle, so as not to confuse these vertices with the vertices of the arrangement of dual lines.

detect three points with the same x-coordinate. This independent procedure does not impair the asymptotic complexity of the entire algorithm, but we can do slightly better if we operate on the arrangement in the dual plane, which we construct anyway.

Let p_1, p_2, and p_3 be such points ordered in increasing y-coordinates. The lines dual to these points, p_1^*, p_2^*, and p_3^*, are parallel. Loosely speaking they all intersect at infinity. Let e_1, e_2, and e_3 be the edges that correspond to the dual lines. Evidently, the fictitious vertices on the right (or left) boundary side of the imaginary bounding rectangle incident to e_1 and e_2, respectively, are neighbors, and so are the fictitious vertices incident to e_2 and e_3; see the figure to the right for an illustration. The template function `find_parallel_lines(arr, cv1, cv2, cv3)` finds three such parallel lines, if they exist, and returns **true**. If they do not exist, it returns **false**. The function template is listed below and defined in the header file `find_parallel_lines.h`. First, we obtain the fictitious face through the call `arr.fictitious_face()`. Then, we obtain the single hole in the fictitious face that represents the imaginary bounding rectangle. Finally, we iterate over the (fictitious) halfedges of the connected component of the boundary of the hole, skipping halfedges incident to the corner vertices, while looking for two adjacent halfedges h_1 and h_2 that are also adjacent to e_1, e_2, and e_3, as depicted in the figure above.

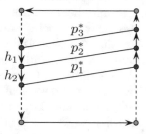

```
template <typename Arrangement, typename Kernel>
bool find_parallel_lines(const Arrangement& arr,
                         typename Arrangement::X_monotone_curve_2& cv1,
                         typename Arrangement::X_monotone_curve_2& cv2,
                         typename Arrangement::X_monotone_curve_2& cv3,
                         Kernel& kernel)
{
  typename Kernel::Are_parallel_2 are_parallel =
    kernel.are_parallel_2_object();
  typename Arrangement::Face_const_handle uf = arr.fictitious_face();
  typename Arrangement::Hole_const_iterator hit = uf->holes_begin();
  typename Arrangement::Ccb_halfedge_const_circulator first = *hit;
  typename Arrangement::Ccb_halfedge_const_circulator next = first;
  do {
    typename Arrangement::Ccb_halfedge_const_circulator curr = next++;
    if ((curr->source()->degree() == 2) || (curr->target()->degree() == 2) ||
        (next->target()->degree() == 2))
      continue;

    CGAL::Arr_parameter_space ps1 = curr->source()->parameter_space_in_x();
    CGAL::Arr_parameter_space ps2 = curr->target()->parameter_space_in_x();
    CGAL::Arr_parameter_space ps3 = next->target()->parameter_space_in_x();
    if ((ps1 != ps2) || (ps2 != ps3)) continue;

    cv1 = next->twin()->prev()->curve();
    cv2 = next->twin()->next()->curve();
    cv3 = curr->twin()->next()->curve();
    if (are_parallel(cv1.supporting_line(), cv2.supporting_line()) &&
        are_parallel(cv2.supporting_line(), cv3.supporting_line()))
      return true;
  } while (next != first);

  return false;
}
```

To complete the description of the algorithm, we explain how to locate the closest line that lies above or below a given vertex v in an efficient manner. Assume v corresponds to the point of intersection between the dual lines p_i^* and p_j^*. It has four incident halfedges, and it lies on the boundary of four different faces of the arrangement. As there are no vertical lines in the arrangement, one face lies above the vertex, one below it, and one face lies to the right of v, and another to its left. We are interested in the two faces that lie above and below v (the shaded faces in the figure to the right). We locate these faces by traversing the halfedges incident to v. We look for

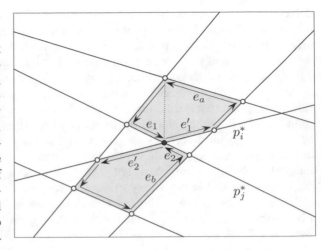

any halfedge e that has the same direction (left to right, or right to left) as its successor halfedge e' in the chain of halfedges that winds along the boundary of the face incident to both e and e'. This face lies above or below v. (These are the edge e_1 and its successor e_1' in the figure above, which are incident to the face above v, and e_2 and its successor e_2', which are incident to the face below v. The two other halfedges incident to v, namely the twin halfedges of e_1' and e_2', are irrelevant in this case.) We simply have to find the first halfedge along the CCB between e' and e whose associated segment contains the x-coordinate of the point associated with v in its x-range. As all the faces in an arrangement of lines are convex, there is either one such edge (see the face below v in the figure above) or two in a degenerate case, where the two edges have a common vertex that lies vertically above or below v (see the face above v in the figure above). If there are two, they induce two triangles of the same area. Naturally, we skip fictitious halfedges, which indicate that there is no dual line above (or below) v. Suppose we have reached a halfedge above (or below) v whose supporting line is p_k^*. We compute the area of the triangle defined by the primal points $\triangle p_i p_j p_k$ and check whether it is the minimal area so far.

You should be ready now to follow the complete implementation of the minimum-area triangle algorithm. It uses the arrangement machinery introduced so far. Notice how the various traversal functions are applied to the constructed arrangement. Also notice the usage of the kernel functors (see Section 1.4.4) to perform primitive operations on geometric objects, such as comparing the x-coordinates of two points, or computing the (signed) area of a triangle defined by three points.

It is a nontrivial fact that this algorithm running-time is $O(n^2)$. The arrangement $\mathcal{A}(P^*)$ can be constructed in $O(n^2)$ by standard incremental insertion; see Section 3.4.1. (The same running time does not necessarily apply to other types of well-behaved curves; see Section 1.3.3.) It may seem more surprising that even traversing the faces above and below each and every vertex of the arrangement takes total time $O(n^2)$. The bounds in both stages follow from the fact that the complexity of the *zone* (see Section 3.2) of a line in an arrangement of k lines is $O(k)$.

Below, however, we use the aggregate insertion method to construct the arrangement, which takes $O(n^2 \log n)$ time in this case, as it uses the plane-sweep algorithm; see Section 3.4.2. Recall that while carrying out the plane-sweep algorithm to insert the dual lines, we maintain a one-dimensional arrangement along the vertical sweep line called the *status structure*, which records the intersection of the sweep line with other lines. We process events associated with vertices sorted in increasing xy-lexicographic order. When we process an event associated with a vertex v, we can look for the features that lie immediately above and immediately below v using the status structure. This, however, requires a dedicated plane-sweep visitor (see Section 3.2), which is not implemented.

Try: While the asymptotic complexity of the aggregate insertion is greater than the asymptotic complexity of the incremental insertion it exhibits better performance in practice. Compare the running time (in practice) of these two alternatives. To carry out the incremental construction, the call to the method **insert()** in the program below should be substituted with a loop where the curves are inserted one by one.

We can simplify the code by using the **decompose()** function template, which outputs for each vertex v the pair of features that directly lie above and below v; see Section 3.6. Now, replace the **for**-loop compound statement that iterates over all vertices in a quest for the minimum-area triangle, with a call to an instance of the **decompose()** function template followed by the appropriate code that processes the output of the function. Observe that the asymptotic complexity of the new code is greater than the asymptotic complexity of the replaced code, as the function template **decompose()** is implemented as a plane-sweep algorithm. Compare the running times of both versions.

```cpp
// File: min_area_triangle.cpp

#include <CGAL/basic.h>

#include "arr_linear.h"
#include "arr_print.h"
#include "read_objects.h"
#include "is_in_x_range.h"
#include "dual_plane.h"
#include "find_parallel_lines.h"

typedef Arrangement::Halfedge_around_vertex_const_circulator
  Halfedge_around_vertex_const_circulator;

int main(int argc, char* argv[])
{
    // Get the name of the input file from the command line, or use the default
    // points.dat file if no command-line parameters are given.
    const char* filename = (argc > 1) ? argv[1] : "points.dat";
    std::list<Point> points;
    read_objects<Point>(filename, std::back_inserter(points));

    std::list<X_monotone_curve> dual_lines;
    std::list<Point>::const_iterator it;
    for (it = points.begin(); it != points.end(); ++it)
      dual_lines.push_back(dual_line<Traits>(*it));

    // Aggregately construct the arrangement of dual lines.
    Traits traits;
    Arrangement arr(&traits);
    insert(arr, dual_lines.begin(), dual_lines.end());
    print_unbounded_arrangement_size(arr);

    //Detect 3 parallel lines, as they are the dual of 3 primal collinear points.
    X_monotone_curve cv1, cv2, cv3;
    Kernel                                          ker;
    if (find_parallel_lines(arr, cv1, cv2, cv3, ker)) {
        std::cout << "The minimum-area triangle is "
                << "(" << primal_point<Traits>(cv1) << ")"
                << "(" << primal_point<Traits>(cv2) << ")"
```

```cpp
                  << "(" << primal_point<Traits>(cv3) << ")"
                  << ", with area = 0" << std::endl;
      return 0;
  }


  // Go over all arrangement vertices and look for the vertex (intersection
  // of two lines) and the halfedge that induce the minimum-area triangle.
  Kernel::Compute_area_2  area = ker.compute_area_2_object();
  Point                   min_p1, min_p2, min_p3;
  Number_type             min_area;
  bool                    found = false;

  Arrangement::Vertex_const_iterator    vit;
  for (vit = arr.vertices_begin(); vit != arr.vertices_end(); ++vit) {
    //Each vertex is incident to at least 2 dual lines; get their dual points.
    Halfedge_around_vertex_const_circulator  circ = vit->incident_halfedges();
    Point p1 = primal_point<Traits>(circ++->curve());
    Point p2 = primal_point<Traits>(circ->curve());

    // If the vertex degree is greater than 4, it is incident to three lines,
    // whose primal points are collinear.
    if (vit->degree() > 4) {
      std::cout << "The minimum-area triangle is (" << p1 << ") (" << p2
                << ") (" << primal_point<Traits>((++circ)->curve())
                << "), with area = 0" << std::endl;
      return 0;
    }

    // Locate the halfedges that lie above and below the vertex.
    Halfedge_around_vertex_const_circulator  first;
    circ = first = vit->incident_halfedges();
    do {
      // Check whether the current halfedge is incident to a face that lies
      // above (or below) the current vertex. If so, look for a non-fictitious
      //halfedge along the face boundary that lies above (or below) the vertex.
      if (circ->direction() == circ->next()->direction()) {
        // The halfedge circ and its next halfedge are incident to the current
        // vertex, so there is no need to check them.
        Arrangement::Halfedge_const_handle  he;
        for (he = circ->next()->next(); he != circ; he = he->next()) {
          if (he->is_fictitious() ||            // skip ficititious edges
              !is_in_x_range<Arrangement>(he, vit->point(), traits))
            continue;

          // Compute the area induced by the three points and compare it
          // to the minimal area so far.
          Point p3 = primal_point<Traits>(he->curve());
          Number_type curr_area = CGAL::abs(area(p1, p2, p3));
          if (!found || CGAL::compare(curr_area, min_area) == CGAL::SMALLER) {
            min_p1 = p1;
            min_p2 = p2;
            min_p3 = p3;
            min_area = curr_area;
            found = true;
```

```
        }
      break;
    }
  }
  } while (++circ != first);
} // End loop on the arrangement vertices.

  CGAL_assertion(found);
  std::cout << "The_minimum–area_triangle_is_(" << min_p1 << ")_(" << min_p2
        << ")_(" << min_p3 << "),_with_area_=_" << min_area << std::endl;
  return 0;
}
```

4.2.2   A Note on the Input Precision

Throughout the book we follow the exact geometric-computing approach to guarantee robustness and consistency of the implementation of geometric algorithms. There is one stage, however, where we can give up exactness without risking the robustness of our implementation: When we read off floating-point numbers from the input. More often than not, floating-point input is approximate to begin with, and we can rely on this to convert the input numbers into compact rational numbers. We emphasize that the discussion below applies only to applications that do not assume that the input must be read exactly. If exactness is of cardinal importance for your application, then you should be aware of such considerations.

Consider, for example, the extractor ($>>$) operator for the `Point_2` type implicitly used in the code that solves the **minimum-area triangle problem** above.[4] It expects two values—the x and y Cartesian coordinates of the point. It is capable of reading integers or floating-point numbers. In any case they are converted to multi-precision rational numbers, and from that point on all arithmetic operations are carried out in an exact manner. If it is not assumed that the input should be read exactly, because, for example, the input comprises floating-point numbers that only approximate real numbers, it is safe to truncate each input number to a near-rational number with shorter bit lengths for the numerator and for the denominator, which in turn can expedite the computation drastically. (It may introduce errors with respect to the input, which is often imprecise to start with.) For example, the rational number 1/10 can be represented by two integers with relatively small bit lengths. However, it cannot be represented as a binary fraction. Thus, when stored as a double-precision floating-point (`double`) it is approximated as 3602879701896397/36028797018963968—a number that when represented as a fraction of two integers, namely a rational, requires two integers with much larger bit lengths. The code excerpt listed below allows you to some extent to trade-off between exactness and performance. The function template `numerical_cast()` defined in the header file `numerical_cast.h` casts a floating-point number to a multi-precision rational number, where the numerator and the denominator are represented by integers in unlimited precision (restricted to memory limitations). Observe that before the numerator is passed to the constructor of the rational number it is statically cast to the `int` storage type. This imposes a restriction explained in the next paragraph. Two specialized implementations are listed for two target types of rational numbers—one for `Gmpq` and one for `Exact_predicates_exact_constructions_kernel::FT`, which is a filtered `Gmpq`.

```
#include <CGAL/Exact_predicates_exact_constructions_kernel.h>
#include <CGAL/Exact_predicates_inexact_constructions_kernel.h>

// Convert a double–precision value to a rational number.
template <int denom>
CGAL::Gmpq numerical_cast_imp(const double& val, CGAL::Gmpq)
```

---

[4] CGAL kernels provide extractors for the geometric primitive types they define.

```cpp
{ return CGAL::Gmpq(static_cast<int>(val*denom + 0.5), denom); }

template <int denom>
CGAL::Exact_predicates_exact_constructions_kernel::FT
numerical_cast_imp(const double& val,
                   CGAL::Exact_predicates_exact_constructions_kernel::FT)
{
  CGAL::Gmpq tmp = numerical_cast_imp<denom>(val, CGAL::Gmpq());
  return CGAL::Exact_predicates_exact_constructions_kernel::FT(tmp);
}

template <int denom>
CGAL::Exact_predicates_inexact_constructions_kernel::FT
numerical_cast_imp(const double& val,
                   CGAL::Exact_predicates_inexact_constructions_kernel::FT)
{ return val; }

template <typename Target, int denom, typename Source>
Target numerical_cast(const Source& val)
{ return numerical_cast_imp<denom>(val, Target()); }
```

The `denom` template parameter controls the trade-off between exactness and performance. It multiplies the numerator and is used as the denominator. Substitute this parameter with a number that, on the one hand, is not too large, so that it does not increase the value of the numerator along with its bit length, or, even worse, beyond the capacity of the `int` storage type, which is used to temporarily store the numerator, and, on the other hand, is not too small, so that the introduced error is contained within a tolerable bound. For example, if it is known that the fraction of every input (floating-point) number consists of at most three decimal digits, a denominator of 1000 will eliminate the fraction, and seemingly will not introduce any errors at all. This is not entirely true, as the many floating-point numbers, and in particular their fractions, cannot be represented in decimal notation. (The decimal printed fraction, for example, is in many cases an approximation of the number in memory.) However, this conversion is harmless for most practical purposes. Note that a denominator of 1000 will produce reasonable results only if the absolute value of every input number does not exceed the largest integer that can be represented as an `int` divided by 1000 (e.g., $2^{63}/1000$ on 64-bit machines). If this condition is not met, all bets are off, as an overflow will occur.

The following code excerpt lists an extractor for the `Point_2` type. It can be used to override the extractor provided by CGAL kernels as default.

```cpp
std::istream& operator>>(std::istream& is, Point_2& point)
{
  double x, y;
  is >> x >> y;
  point = Point_2(numerical_cast<Number_type, 1000> (x),
                  numerical_cast<Number_type, 1000> (y));
  return is;
}
```

 Try: Develop a program that reads a large set of line segments, constructs their arrangement, and measures the running time of the program. The coordinates of the endpoints of the input line-segments are given in floating-point numbers. Compare the running times when using the default point-extractor and the approximate point-extractor listed above. Test the effect of different denominator-values on the running time.

4.3   Bibliographic Notes and Remarks

The extension of CGAL's *2D Arrangements* package to handle unbounded curves is described in [21, 22].

Duality transforms are presented in several books, including [45], [53], and [171]. The solution to the **minimum-area triangle problem** was given by Chazelle et al. [36]. In $\mathbb{R}^d$ we can similarly ask for the minimum-volume simplex defined by d out of n given points. The solution of the planar problem is extended to any fixed dimension based on a point-hyperplane duality [53]. Moreover, Edelsbrunner gives several applications of arrangements to problems defined on point sets. de Berg et al. [45, Chapter 8] give an application for computing the so-called discrepancy of points and halfplanes.[5]

Finding the specific set of n points within a fixed-area shape (e.g., a disc or a square) that maximizes the area of the minimum-area triangle is called *Heilbronn's triangle problem* [40].

There are more general transforms that handle not only points and hyperplanes in $\mathbb{R}^d$, namely, points and $(d-1)$-dimensional flats, but also arbitrary dimensional flats (often called k-flats). A useful transform maps lines in three-dimensional space into points and hyperplanes in five-dimensional space through the usage of Plücker coordinates. The end result is that problems stated on lines in 3-space are eventually solved by looking at the resulting arrangements of hyperplanes in 5-space; see, for example, the survey [179]. It may seem odd at first that we prefer to work in 5-space than in lower-dimensional spaces. However, from a computational point of view arrangements of hyperplanes in 5-space are more wieldy than collections of lines in 3-space.

Finally, you may wonder if one can solve the **minimum-area triangle problem** in $o(n^2)$ time, namely faster than $O(n^2)$ time. It turns out that this problem is one in a family of problems referred to as 3-sum hard problems [82] that are closely related to one another, and for which no solution faster than quadratic is known. Furthermore, in a certain restricted computational model a tight lower bound $\Omega(n^2)$ is known [61].

4.4   Exercises

4.1 Enhance the `add_vertical_segment()` function template (see Section 3.6.1) to split fictitious edges, and hence produce a complete vertical decomposition of unbounded arrangements.

4.2 We call an arrangement the faces of which are all convex a *convex arrangement*. Develop a function template called `is_convex()` that determines whether a given input arrangement of linear curves is convex.

4.3 Show that the point-line duality transform presented in the beginning of Section 4.2 preserves the vertical distance between a point and a line; that is, the distance between a point p and its vertical projection (parallel to the y-axis) onto the line ℓ is the same as the distance between the point ℓ^* and its vertical projection onto the line p^*.

4.4 [**Project**] Given a finite set P of points in the plane, a k-set of P is a subset of points in P of size k that can be separated from the remaining points in P by a line. Implement an efficient algorithm that, given a set P of n points in the plane, computes the number of k-sets of P for $k = 1, 2, \ldots, \lfloor \frac{n}{2} \rfloor$. **Hint:** Rephrase the statement "k points of P lie below (or above) a line ℓ" in the dual plane, and solve the problem there.

---

[5] A half-plane is a planar region consisting of all points on one side of a line, and no points on the other side.

Chapter 5

Arrangement-Traits Classes

It is time for you to make an acquaintance with the arrangement *traits* classes, mentioned a number of times in the previous chapters. A traits class encapsulates the definitions of the geometric entities and the implementation of the geometric predicates and constructions that handle these geometric entities, used by the `Arrangement_2` class and by its peripheral modules. The identified minimal requirements imposed by the various algorithms that apply to arrangements are organized in a hierarchy of refined geometry-traits concepts. The requirements listed by each concept include only the utterly essential types and operations needed to implement specific algorithms. This modular structuring yields controllable parts that can be produced, maintained, and utilized with less effort. For each operation, all the preconditions that its operands must satisfy are specified as well, as these may simplify the implementation of models of these concepts even further. Each traits class models one or more concepts. This chapter contains a detailed description of the concepts in the refinement hierarchy and the various traits classes that model these concepts.

All the algebra required for constructing and manipulating arrangements is concentrated in the traits classes. The knowledge required to devise a good traits class is very different from the knowledge required for the development of the rest of the package or for using the package. It has less to do with computational geometry and it involves mainly algebra and numerical computation. This way, traits classes for new types of curves can be developed with little knowledge of algorithms and data structures in computational geometry. In this chapter we discuss how to use existing traits classes, but we also explain the concepts these traits classes model—a starting point for every novice developer of such classes.

This chapter is roughly made of three parts. The first part describes the refinement hierarchy of the arrangement-traits concepts. The second part reviews various models of these concepts. These are traits classes included in the public distribution of CGAL, which handle different curve families, such as line segments, polylines, conic arcs, Bézier curves, and algebraic curves. The last part introduces decorators for geometric traits classes distributed with CGAL. A decorator of a traits class attaches auxiliary data to the geometric objects handled by the original traits class, thereby extending it.

5.1 The Hierarchy of Traits-Class Concepts

A hierarchy of related concepts can be viewed as a directed acyclic graph, where a node of the graph represents a concept and an arc represents a refinement relation. An arc directed from concept A to concept B indicates that concept B refines concept A. A rather large directed acyclic graph is required to capture the entire hierarchy of the geometry traits-class concepts. In the following subsections we review individual clusters and describe the relations between them. The figure to the right depicts the central cluster.

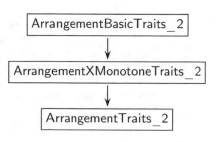

E. Fogel et al., *CGAL Arrangements and Their Applications*, Geometry and Computing 7, DOI 10.1007/978-3-642-17283-0_5, © Springer-Verlag Berlin Heidelberg 2012

5.1.1   The Basic Concept

A model of the basic concept ArrangementBasicTraits_2 needs to define the nested types **Point_2** and **X_monotone_curve_2**, where objects of the former type are the geometric mappings of arrangement vertices, and objects of the latter type are the geometric mappings of edges. In addition, it has to support the set of predicates listed below. This basic set of predicates is sufficient for constructing arrangements of bounded x-monotone curves that are pairwise disjoint in their interiors and points that do not lie in the interior of the curves. The basic set of predicates is also sufficient for answering vertical ray-shooting queries and point-location queries with a small exception: Locating a point using the landmark strategy requires a traits class that models the refined concept ArrangementLandmarkTraits_2; see Section 5.1.4.

Before we continue with the description of the basic-concept requirements, we introduce some definitions and notations used globally in this chapter. The x- and y-coordinates of a point p are denoted by x_p and y_p, respectively. An x-monotone curve is a parametric curve with parametric equations $x = X(t)$ and $y = Y(t)$. More precisely, it is a continuous function $C : I \to \mathbb{R}^2$, where I is an open, half-open, or closed interval with endpoints 0 and 1, and C is bijective. $0 \leq t_1 < t_2 \leq 1$ implies that $C(t_1)$ is lexicographically smaller than $C(t_2)$. Also, if $X(t_1) = X(t_2) = q, 0 \leq t_1 < t_2 \leq 1$, for some $q \in \mathbb{R}^2$, then $X(t) = q, \forall t \in I$; such a curve is called vertical. If $0 \notin I$, either $\lim_{t \to 0+} X(t) = -\infty$ or $X(0)$ exists and $\lim_{t \to 0+} Y(t) = \pm\infty$. Similarly, if $1 \notin I$, either $\lim_{t \to 1-} X(t) = \infty$ or $X(1)$ exists and $\lim_{t \to 1-} Y(t) = \pm\infty$. We say that a curve C has two *ends*, the 0-end and the 1-end. Let $d \in \{0, 1\}$ be an index that specifies a curve end. If $d \in I$, the d-end has a geometric interpretation; it is a point in $\mathbb{R}^2$. If $d \notin I$, the d-end has no geometric interpretation. In this case the d-end of the curve is open, and in fact unbounded.[1]

The three-valued functions $\mathrm{cmp}_x()$ and $\mathrm{cmp}_y()$ accept pairs of points in the plane and compare them by their x-coordinates and by their y-coordinates, respectively. These functions and all comparison predicates evaluate to **CGAL::SMALLER**, **CGAL::EQUAL**, or **CGAL::LARGER**.

Compare_x_2: Compare the x-coordinates of two points p_1 and p_2; return $\mathrm{cmp}_x(p_1, p_2)$.

Compare_xy_2: Compare two points lexicographically, first by their x-coordinates, and if they are equal, by their y-coordinates. Formally, if $\mathrm{cmp}_x(p_1, p_2)$ does not evaluate to **CGAL::EQUAL**, return $\mathrm{cmp}_x(p_1, p_2)$. Otherwise, return $\mathrm{cmp}_y(p_1, p_2)$. This predicate is used to maintain the order of event points during the preprocessing step of the plane-sweep procedure.

Equal_2: There are two overloaded Boolean operators that test equality. One returns **true** iff two given points represent the same geometric point in the plane, and the second returns **true** iff the graphs of two given x-monotone curves are the same.[2]

Construct_min_vertex_2: Return the lexicographically smallest endpoint of an x-monotone curve (the 0-end). A precondition ensures that the curve is closed at this end.

Construct_max_vertex_2: Return the lexicographically largest endpoint of an x-monotone curve (the 1-end). A precondition ensures that the curve is closed at this end.

Is_vertical_2: Determine whether an x-monotone curve is vertical.

Compare_y_at_x_2: Given an x-monotone curve $C = (X(t), Y(t))$ and a point p such that p lies in the x-range[3] of C, determine whether p is above, below, or lies on C. Formally, if C is vertical, determine whether $y_p < Y(0)$, $Y(0) \leq y_p \leq Y(1)$, or $Y(1) < y_p$. Otherwise, there must be a unique $t' \in I$, such that $X(t') = x_p$. In this case return $\mathrm{cmp}_y(p, C(t'))$. This predicate is used to insert a new curve with a minimal end at p into the status structure during the execution of the plane-sweep procedure.

---

[1] We use the term "open" instead of "unbounded" because we intend to enhance the code to handle open curves that are still bounded.

[2] The predicate that tests equality between graphs of two x-monotone curves is used only for testing as part of the CGAL test suite.

[3] We say that a point p lies in the x-range of a curve $C = (X(t), Y(t))$, if $X(0) \leq x_p \leq X(1)$.

Compare_y_at_x_right_2: Given two x-monotone curves C_1 and C_2 that share a common mini-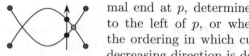 mal end at p, determine whether C_1 lies above or below C_2 immediately to the right of p, or whether the two curves overlap there. Determining the ordering in which curves emanate from a point p in lexicographically increasing direction is done by comparing the curves infinitesimally to the right of p. Formally, return $\mathrm{cmp}_y(C_1(\epsilon_1), C_2(\epsilon_2))$, where $\epsilon_1, \epsilon_2 > 0$ are infinitesimally small and $x_{C_1(\epsilon_1)} = x_{C_2(\epsilon_2)}$. This predicate is used to insert new curves into the status structure when the minimal end lies on an existing curve in the status structure.

Compare_y_at_x_left_2: Given two x-monotone curves C_1 and C_2 that share a common maxi-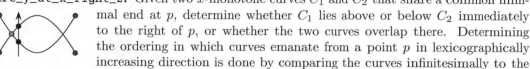 mal end at p, determine whether C_1 lies above or below C_2 immediately to the left of p, or whether the two curves overlap there. Determining the ordering in which curves emanate from a point p in lexicographically decreasing direction is done by comparing the curves infinitesimally to the left of p. Formally, return $\mathrm{cmp}_y(C_1(1 - \epsilon_1), C_2(1 - \epsilon_2))$, where $\epsilon_1, \epsilon_2 > 0$ are infinitesimally small and $x_{C_1(1-\epsilon_1)} = x_{C_2(1-\epsilon_2)}$. This predicate is used by some point-location strategies and by the zone-computation algorithm. This is an *optional* requirement with ramifications in case it is not fulfilled; see Section 5.1.6.

Every model of the ArrangementBasicTraits_2 needs to define a nested type named **Has_left_category**, which determines whether the traits class supports the optional predicate **Compare_y_at_x_left_2** described above. If the **Has_left_category** type nested in a model of the basic concept is defined as **Tag_true**,[4] the model must support the predicate. If the type is defined as **Tag_false**, we resort to an internal version, which is based just on the reduced set of provided operations; see Section 5.1.6 for details about the technique used to implement the selection. The internal version might be less efficient, but it exempts the traits developer from providing an (efficient) implementation of this predicate—a task that turns out to be nontrivial in some cases.

The curves of an arrangement are embedded in an rectangular two-dimensional area called the parameter space; see Section 11.1. The parameter space is defined as $X \times Y$, where X and Y are open, half-open, or closed intervals with endpoints in the compactified real line $\mathbb{R} \cup \{-\infty, +\infty\}$. Let b_l, b_r, b_b, and b_t denote the endpoints of X and Y, respectively. We refer to these values as the left, right, bottom, and top sides of the boundary of the parameter space. If the parameter space is, for example, the entire compactified plane, which is currently the only option supported by the package, $b_l = b_b = -\infty$ and $b_r = b_t = +\infty$. The basic set of predicates is sufficient for constructing arrangements of x-monotone curves that do not reach or approach the boundary of the parameter space. The nature of the input curves, whether they are expected to reach or approach the left, right, bottom, or top side of the boundary of the parameter space, must be conveyed by the traits class. This is done through the definition of four additional nested types, namely, **Left_side_category**, **Right_side_category**, **Bottom_side_category**, and **Top_side_category**. Each such type must be convertible to the type **Arr_oblivious_side_tag** for the class to be a model of the concept ArrangementBasicTraits_2.

5.1.2 Supporting Intersections

Constructing an arrangement induced by x-monotone curves that may intersect in their interior requires operations that are not listed by the ArrangementBasicTraits_2 concept. The additional operations are listed by the concept ArrangementXMonotoneTraits_2, which refines the basic arrangement-traits concept described above. While models of the ArrangementXMonotoneTraits_2 concept still handle only x-monotone curves, the curves are not restricted to be disjoint in their interiors. Such a model must be capable of computing points of intersection between x-monotone curves, splitting curves at these intersection points to obtain pairs of interior-disjoint subcurves, and optionally merging pairs of subcurves. A point of intersection between two curves is also represented by the **Point_2** type. A model of the refined concept must define an additional type

---

[4] In principle, the category type may only be convertible to the tag type, but in practice the category is typically explicitly defined as the tag.

called `Multiplicity`. An object of this type indicates the multiplicity of the intersection point of two curves, also referred to as the *intersection number*.[5] Loosely speaking, if two curves intersect at a point p but have different tangents (first derivatives) at p, p is of multiplicity 1. If the curve tangents are equal but their curvatures (second derivatives) are not, p is of multiplicity 2, and so on. The multiplicity of points of intersection between line segments is always 1, and the multiplicity of a point of intersection between two polylines is 1 if the intersection point is interior to the corresponding two line segments of the polylines, and undefined (coded as 0) otherwise. A model of the refined concept thus has to support the following additional operations:

`Split_2:` Split a given x-monotone curve C at a given point p, which lies in the interior of C, into two interior-disjoint subcurves.

`Are_mergeable_2:` Determine whether two x-monotone curves C_1 and C_2 that share a common endpoint can be merged into a single x-monotone curve representable by the traits class.[6]

`Merge_2:` Merge two mergeable curves C_1 and C_2 into a single curve C.

`Intersection_2:` Compute all intersection points and overlapping sections of two given x-mono-
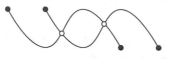
tone curves, C_1 and C_2. If possible, compute also the multiplicity of each intersection point. Providing the multiplicity of an intersection point is not required, but it can expedite the arrangement construction. Typically, the multiplicity is a byproduct of the intersection computation. However, if it is not available, undefined multiplicity (coded as 0) is assumed. Having the multiplicity of intersection points while constructing arrangements enables the exploitation of the geometric knowledge intersection points may have. In particular, costly calls to other traits-class functions can be avoided, as in many cases the order of incident curves to the right of a common intersection point can be deduced from their order to the left of the point and the intersection multiplicity. For example, two curves swap their relative order at intersection points of odd multiplicities. More details can be found in the Bibliographic Notes and Remarks section at the end of the chapter.

Using a model of the ArrangementXMonotoneTraits_2 concept, it is possible to construct arrangements induced by x-monotone curves (and points) that may intersect one another. The two last operations listed above, regarding the merging of curves, are optional, and should be provided only if the type `Has_merge_category` nested in a model of the ArrangementXMonotoneTraits_2 concept is defined as `Tag_true`. Otherwise, it is not possible to merge x-monotone curve and redundant vertices may be left in the arrangement due to the removal of edges; see more details and examples in Section 3.5.

──────── *advanced* ────────

We invite you to take a small detour and delve into the world of traits development. We define the exact requirement for the `Intersect_2` operation, and we use this definition to point out the correct qualification of traits methods and traits-functor methods. Let `Traits_2` be a model of the ArrangementXMonotoneTraits_2 concept. Each required operation entails (i) the definition of a type nested in `Traits_2`, which models some concept that refines a Functor concept variant, and (ii) a `const`-qualified method of `Traits_2` that returns an object of the nested type. The requirement for the `Intersect_2` operation is no exception.

The type `Intersect_2` nested in `Traits_2` models the concept Intersect_2. This concept requires the expression

op(c1, c2, oi)

────────────────────────

[5] See, e.g., `http://en.wikipedia.org/wiki/Intersection_number` for more information about intersection numbers.

[6] On the face of it this seems a difficult predicate to implement. In practice we use very simple tests to decide whether two curves are mergeable: We check whether their underlying curves are identical and whether they do not bend to form a non-x-monotone curve. The two curves must share a common endpoint as a precondition.

to be valid, where **op** is an object of type `Traits_2::Intersect_2`, **c1** and **c2** are objects of type `Traits_2::X_monotone_curve_2`, **oi** is an object of type `Output_iterator`, and the expression return type is convertible to `Output_iterator`. The expression computes the intersections of c_1 and c_2 and inserts the results into an ascending lexicographic xy-order sequence associated with the output iterator **oi**. The value type of `Output_iterator` is `CGAL::Object`, where each object of this type is polymorphic and wraps either an object of type `std::pair<Traits_2::Point_2,Traits_2::Multiplicity>`, which represents a point of intersection with its multiplicity (in case the multiplicity is undefined or unknown, it should be set to 0), or an object of type `Traits_2::X_monotone_curve_2`, which represents an overlapping subcurve of c_1 and c_2. The expression return value is a past-the-end iterator for the output sequence. In addition, the ArrangementXMonotoneTraits_2 concept requires the expression

```
traits.intersect_2_object()
```

to be valid, where **traits** is of type `Traits_2` (a model of the ArrangementXMonotoneTraits_2 concept), and the expression return type is convertible to `Traits_2::Intersect_2`.

The `intersect_2_object()` member function of `Traits_2` and the function operator of the Traits_2::Intersect_2 must be `const` qualified. This variant may seem problematic at first glance for traits models that maintain state to expedite expensive operations, such as computing the points of intersection between high-degree algebraic curves; see Section 5.4. Typically, traits classes have no data members that need to be dynamically updated. For data members that need to be occasionally modified C++ presents a simple solution—they must be qualified as `mutable`.

— advanced —

5.1.3 Supporting Arbitrary Curves

The concept ArrangementTraits_2 refines the ArrangementXMonotoneTraits_2 concept. A model of the refined concept must define an additional type that is used to represent general, not necessarily x-monotone and not necessarily connected, curves, named **Curve_2**. It also has to support the subdivision of a curve of that type into a set of continuous x-monotone curves and isolated points. For example, the curve $C : (x^2 + y^2)(x^2 + y^2 - 1) = 0$ comprises the unit circle (the locus of all points for which $x^2 + y^2 = 1$) and the origin (the singular point $(0,0)$). C should therefore be subdivided into 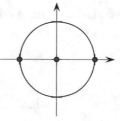 two circular arcs, the upper part and the lower part of the unit circle, and a single isolated point. In particular a model of the refined concept has to support the following additional operation:

Make_x_monotone_2: Divide a given general curve of type `Curve_2` into continuous weakly x-monotone curves and isolated points.

 Note that a model of the refined concept `ArrangementTraits_2` is required only when using the free `insert()` function (see Section 3.4), which accepts an object of type `Curve_2` or a range of objects of that type. In all other cases it suffices to use a model of the ArrangementXMonotoneTraits_2 concept.

— advanced —

We define the exact requirement for the `Make_x_monotone_2` operation, and we use this definition to clarify the difference between the types `X_monotone_curve_2` and `Curve_2`. Let `Traits_2` be a model of the ArrangementTraits_2 concept. The type `Make_x_monotone_2` nested in `Traits_2` must model the concept MakeXMonotone_2. This concept requires the expression

```
op(c, oi)
```

to be valid, where **op** is an object of type `Traits_2::Make_x_monotone_2`, **c** is an object of type `Traits_2::Curve_2`, **oi** is an object of type `Output_iterator`, and the expression evaluates to a type convertible to `Output_iterator`. The operation coded by the expression subdivides the curve c into x-monotone subcurves and isolated points, and inserts the results into a container associated with the output iterator **oi**. The value type of **Output_iterator** is **CGAL::Object**, where each object of this type wraps either a **Traits_2::X_monotone_curve_2** object or a **Traits_2::Point_2** object. The expression evaluates to a past-the-end iterator for the output sequence. In addition, the ArrangementTraits_2 concept requires the expression

```
traits.make_x_monotone_2_object()
```

to be valid, where **traits** is of type **Traits_2** (a model of the ArrangementTraits_2 concept), and the expression return type is convertible to **Traits_2::Make_x_monotone_2**. (This expression is used to obtain an object of type **Traits_2::Make_x_monotone_2**, which is not necessarily default constructible.)

Consider a model of the concept ArrangementTraits_2 that handles arbitrary curves, which are always x-monotone, such as a traits class that handles linear curves. The nested types **Curve_2** and **X_monotone_curve_2** can be defined as equivalent types. Moreover, defining them as equivalent types is advantageous, as it enables a generic simple implementation of the function operator of the **Make_x_monotone_2** nested type:

```
template <typename Output_iterator>
Output_iterator operator()(const Curve_2& c, Output_iterator oi) const
{ return *oi++ = make_object(c);    // wrap the input curve. }
```

On the other hand, consider a model of the ArrangementTraits_2 concept that handles arbitrary curves, which may be not x-monotone. The **Curve_2** and **X_monotone_curve_2** nested types must be defined as different types to allow proper dispatching of the free functions that accept such curves, e.g., **insert()**; see Section 3.4.

└──────── *advanced* ────────

5.1.4   The Landmark Concept

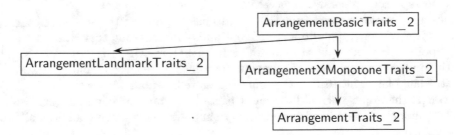

Fig. 5.1: The traits-concept hierarchy for arrangements induced by bounded curves.

The type of an arrangement associated with the landmark point-location (see Section 3.1.1) must be an instance of the **Arrangement_2<Traits,Dcel>** class template, where the **Traits** parameter is substituted with a model of the concept ArrangementLandmarkTraits_2. (Naturally, it can also model either the ArrangementXMonotoneTraits_2 concept or the ArrangementTraits_2 concept.) A model the ArrangementLandmarkTraits_2 concept must define a fixed-precision number type named **Approximate_number_type** (typically the double-precision floating-point **double**) and support the following additional operations:

Approximate_2: Given a point p, approximate the x- and y-coordinates of p using a not necessarily multi-precision number type. We use this operation for approximate computations—there are certain operations performed during the search for the location of the query point that need not be exact, and that can be performed faster when carried out, for example, using a fixed-precision number type.

Construct_x_monotone_curve_2: Given two points p_1 and p_2, construct an x-monotone curve connecting p_1 and p_2.

As you will see, most traits classes model the ArrangementTraits_2 concept, and some also model the ArrangementLandmarkTraits_2 concept.

5.1.5 Supporting Unbounded Curves

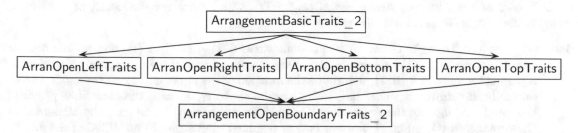

Fig. 5.2: The traits-concept hierarchy for arrangements induced by unbounded curves.

An arrangement that supports unbounded x-monotone curves maintains an implicit bounding rectangle in the DCEL structure; see Chapter 4. The unbounded ends of vertical rays, vertical lines, and curves with vertical asymptotes are represented by vertices that lie on the bottom or top sides of this bounding rectangle. These vertices are not associated with points, but are associated with (finite) x-coordinates. The unbounded ends of all other curves are represented by vertices that lie on the left or right sides of this bounding rectangle. These vertices are not associated with points either. Edges connect these vertices and the four vertices that represents the corners of this bounding rectangle to form the rectangle. Several predicates are required to handle x-monotone curves that approach infinity and thus approach the boundary of the parameter space. These predicates are sufficient to handle not only curves embedded in an unbounded parameter space, but also curves embedded in a bounded parameter space with open boundaries. When the parameter space is, for example, bounded, it is the exact geometric embedding of the implicit bounding rectangle.

The additional requirements are organized in four different concepts, namely, ArrangementOpenLeftTraits, ArrangementOpenRightTraits, ArrangementOpenBottomTraits, and ArrangementOpenTopTraits; see Figure 5.2. Every traits class for arrangements models a subset of the four concepts above. (A traits class that models the empty subset supports only bounded curves.)

Let b_l and b_r denote the x-coordinates of the left and right boundaries of the parameter space, respectively. Let b_b and b_t denote the y-coordinates of the bottom and top boundaries of the parameter space, respectively. If the parameter space is, for example, $\mathbb{R}^2$, $b_l = b_b = -\infty$ and $b_r = b_t = +\infty$. If some curves inserted into an arrangement object are expected to approach the left boundary of the parameter space, namely, there exists $d \in \{0,1\}$ such that $\lim_{t \to d} X(t) = b_l$ holds for at least one input curve $C(t) = (X(t), Y(t))$, the arrangement template must be instantiated with a model of the ArrangementOpenLeftTraits concept. Similarly, if some curves inserted into the arrangement are expected to approach the right ($\lim_{t \to d} X(t) = b_r$), the bottom ($\lim_{t \to d} Y(t) = b_b$), or the top ($\lim_{t \to d} Y(t) = b_t$) boundary of the parameter space, the arrangement template must be instantiated with a model of the corresponding concept.

If the arrangement template is instantiated with a traits class that models at least one of the four concepts above, the DCEL is initialized to handle unbounded curves as described in Chapter 4. The arrangement template instantiated with a traits class that models all the four concepts above can handle x-monotone curves that are unbounded in any direction. In this case the traits class models the combined concept ArrangementOpenBoundaryTraits_2; see Figure 5.2.

The type `Left_side_category` nested in a model of the concept ArrangementOpenLeftTraits must be convertible to `Arr_open_side_tag`.[7] Similarly, the types `Right_side_category`, `Bottom_side_category`, and `Top_side_category` nested in models of the concepts ArrangementOpen-

---

[7]The tags `Arr_oblivious_side_tag` and `Arr_open_side_tag` are only two of a larger number of options for the side categories included in a major extension the code was going through at the time this book was written; see Section 11.1.

RightTraits, ArrangementOpenBottomTraits, and ArrangementOpenTopTraits, respectively, must be convertible to **Arr_open_side_tag**. For example, the **Arr_rational_function_traits_2** traits model supports unbounded curves; see Section 5.4.3. Thus, all four nested types are defined as **Arr_open_side_tag**. Adversely, all four types nested in the **Arr_segment_traits_2** traits model (see Section 5.2.1) are defined as **Arr_oblivious_side_tag**, as segments are always bounded.

A model of the concepts ArrangementOpenLeftTraits and ArrangementOpenRightTraits must provide the additional predicates listed below.

Parameter_space_in_x_2: Given an x-monotone curve $C(t) = (X(t), Y(t))$ and an enumerator i that specifies either the minimum end or the maximum end of the curve, and thus maps to a parameter value $d \in \{0, 1\}$, determine the location of the curve end along the x-dimension. Formally, determine whether $\lim_{t \to d} X(t)$ evaluates to b_l, in case the class that provided this predicate models the ArrangementOpenLeftTraits concept, b_r, in case the class models the ArrangementOpenRightTraits concept, or a value in between. Return **CGAL::ARR_LEFT_BOUNDARY**, **CGAL::ARR_RIGHT_BOUNDARY**, or **CGAL::ARR_INTERIOR**, accordingly.

Compare_y_near_boundary_2: Given two x-monotone curves C_1 and C_2 and an enumerator i

that specifies either the minimum ends or the maximum ends of the two curves, compare the y-coordinates of the curves near their respective ends. That is, compare the y-coordinates of the vertical projection of a point p onto C_1 and onto C_2. If i specifies the minimum ends, the curves must approach the left boundary-side. In this case p is located far to the left, such that the result is invariant under a translation of p farther to the left. If i specifies the maximum ends, the curves must approach the right boundary-side. In that case p is located far to the right in a similar manner.

Models of the concepts ArrangementOpenLeftTraits and ArrangementOpenRightTraits differ in the preconditions the two predicates above have. If the class that provides those predicates does not model the ArrangementOpenRightTraits concept, then the enumerator i cannot specify the maximum end. If the class does not model the ArrangementOpenLeftTraits concept, then the enumerator i cannot specify the maximum end. Recall that preconditions are runtime characteristics.

A model of the ArrangementOpenBottomTraits and ArrangementOpenTopTraits concepts must provide the following additional predicates:

Parameter_space_in_y_2: Given an x-monotone curve $C(t) = (X(t), Y(t))$ and an enumerator that specifies either the minimum end or the maximum end of the curve, and thus maps to a parameter value $d \in \{0, 1\}$, determine the location of the curve end along the y-dimension. Formally, determine whether $\lim_{t \to d} Y(t)$ evaluates to b_b, in case the class that provided this predicate models the ArrangementOpenBottomTraits concept, b_t, in case the class models the ArrangementOpenTopTraits concept, or a value in between. Return **CGAL::ARR_BOTTOM_BOUNDARY**, **CGAL::ARR_INTERIOR**, or **CGAL::ARR_TOP_BOUNDARY**, accordingly.

Compare_x_at_limit_2: This predicate is overloaded with two versions as follows:

(i) Given a point p, an x-monotone curve $C(t) = (X(t), Y(t))$, and an enumerator that specifies either the minimum end or the maximum end of the curve, and thus maps to a parameter value $d \in \{0, 1\}$, compare x_p and $\lim_{t \to d} X(t)$. If the parameter space is unbounded, a precondition ensures that C has a vertical asymptote at its d-end; that is $\lim_{t \to d} X(t)$ is finite. (ii) Given two x-monotone curves $C_1(t) = (X_1(t), Y_1(t))$ and $C_2(t) = (X_2(t), Y_2(t))$ and two enumerators that specify either the minimum ends or the maximum ends of the curves, and thus map to parameter values $d_1 \in \{0, 1\}$ and $d_2 \in \{0, 1\}$ for C_1 and for C_2, respectively, compare $\lim_{t \to d_1} X_1(t)$ and $\lim_{t \to d_2} X_2(t)$. If the parameter space is unbounded, a precondition ensures that C_1 and C_2 have vertical asymptotes at their respective ends; that is, $\lim_{t \to d_1} X_1(t)$ and $\lim_{t \to d_2} X_2(t)$ are finite.

`Compare_x_near_limit_2:` Given two x-monotone curves C_1 and C_2 and an enumerator i that specifies either the minimum ends or the maximum ends of the two curves, compare the x-coordinates of the curves near their respective ends. That is, compare the x-coordinates of the horizontal projection of a point p onto C_1 and onto C_2. A precondition ensures that both curves approach the same boundary-side, either the bottom or the top, at their respective ends. If both curves approach the bottom boundary-side, p is located far to the bottom, such that the result is invariant under a translation of p farther toward the bottom. If both curves approach the top boundary-side, p is located far to the top in a similar manner. Another precondition ensures that the x-coordinates of the limits of the curves at their respective ends are equal. That is, the predicate `Compare_x_at_limit_2` applied to C_1, C_2, (at the end marked by the enumerator i) evaluates to `CGAL::EQUAL`.

5.1.6 The Traits Adaptor

The traits adaptor class-template implements various geometric operations using the operations supplied by a model of the geometry-traits concept as basic building blocks. It decreases the effort required to develop traits models (as the developer needs only to implement the functions listed in the concepts), and at the same time extends the usability of the traits models, adapting them for other purposes. The traits adaptor has a single template parameter that must be substituted with a model of the ArrangementBasicTraits_2 concept. The adaptor inherits from the substituted traits-model, centralizing all geometric operations. In cases where the efficiency of methods is crucial, a traits developer has the freedom to override some of the methods the adaptor implements with optimized ones.

For example, in order to determine whether a point p is in the x-range of an x-monotone curve C, the adaptor simply compares p with the endpoints of C. It checks whether p lies to the right of the left endpoint and to the left of the right endpoint.

Another example is the `Compare_y_at_x_left_2` predicate. If this predicate is not provided by the substituted traits-class, an internal version implemented by the adaptor is used; see Section 5.1.1. The internal version locates a new reference point to the left of the original point, and checks what happens to its right. In the example depicted in the figure to the right, the position of the left end point of c_1, namely q, relative to c_2 is sought using the predicate `Compare_y_at_x_2`.

———— *advanced* ————

The geometry-traits adaptor class uses a *tag-dispatching* mechanism to select the appropriate implementation of a geometry-traits class operation (or an empty implementation in case the operation is not needed and thus never invoked). Tag dispatching is a technique that uses function overloading to dispatch a function *at compile time*, based on properties of the types of the arguments the function accepts. This mechanism enables developers to implement their traits class with a reduced or alternative set of operations. The adaptor dispatches the appropriate implementation based on the tags listed above that every traits class must define, i.e., `Has_left_category`, `Has_merge_category`, `Left_side_category`, `Right_side_category`, `Bottom_side_category`, and `Top_side_category`.

———— *advanced* ————

5.2 Traits Classes for Line Segments and Linear Objects

There are two distinct traits classes that handle line segments. One caches information in the curve records (see Section 5.2.1), while the other retains the minimal amount of data (see Section 5.2.2). Operations on arrangements instantiated with the former traits class consume more space, but they are more efficient for dense arrangements (namely, arrangements induced by line segments with a large number of intersections). Another model handles not only (bounded) line segments, but also rays and lines; see Section 5.2.3.

5.2.1 The Caching Segment-Traits Class

An instance of the `Arr_segment_traits_2<Kernel>` class template used in most example programs so far is instantiated by substituting the `Kernel` template parameter with a geometric kernel that must conform to the CGAL Kernel concept; see Section 1.4.4 for more details about this concept. This traits class defines its point type to be the `Kernel::Point_2` type. However, neither the `Curve_2` nor the `X_monotone_curve_2` nested types of the traits are defined as the `Kernel::Segment_2` type. A kernel segment is represented by its two endpoints, and these may have a large bit size representation when the segment is the result of several split operations in comparison with the representation of the original-segment endpoints.

The figure to the right demonstrates the cascading effect of segment intersection with exponential growth of the bit lengths of the intersection-point coordinates. The large bit size representation may significantly slow down the various traits-class operations involving such segments.[8]

In contrast, the `Arr_segment_traits_2` class template represents a segment using its supporting line in addition to the two endpoints. Most computations are performed on the supporting line, which never changes as the segment is split. The `Arr_segment_traits_2` class template also caches some additional information with each segment to speed up various predicates, i.e., two

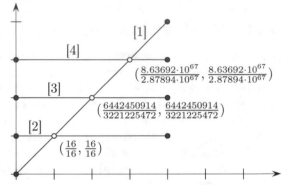

Fig. 5.3: An arrangement of four line segments inserted in the order indicated in brackets.

Boolean flags indicating (i) whether the segment is vertical and (ii) whether the segment target-point is lexicographically larger than its source. An `X_monotone_curve_2` object is still constructible from two endpoints or from a kernel segment and converted to a `Kernel::Segment_2` object. Moreover, an `X_monotone_curve_2` instance can also be cast to a `Kernel::Segment_2` object. The two types are thus convertible to one another.

An instance of the `Arr_segment_traits_2<Kernel>` class template can be very efficient for constructing arrangements induced by line segments with a large number of intersections. Efficiency is affected by the substituted geometric kernel. Using `Cartesian<Gmpq>` as the kernel type is in general not a bad choice; the coordinates of the segment endpoints are represented as multi-precision rational-numbers, and this ensures the correctness of all computations regardless of the input. Computations on multi-precision number types, such as `Gmpq`, typically take longer than computations on machine-precision floating-point. However, in almost all cases it is possible to expedite the computation using numerical filtering; see Section 1.3.2. If the input set of line segments do not have degeneracies; namely, no two segments in the set share a common end-point, and no three segments intersect at a common point, or at least, degeneracies exist but their number is relatively small, then filtered computation incurs only negligible overhead compared to floating-point arithmetic, which is error-prone. Indeed, in almost all examples and applications given in this book, a predefined filtered kernel is used to instantiate the line-segment traits class; see Section 1.4.4.

5.2.2 The Non-Caching Segment-Traits Class

The *2D Arrangements* package offers an alternative traits class that handles line segments, namely the `Arr_non_caching_segment_traits_2<Kernel>` class template. This class template and the `Arr_segment_traits_2<Kernel>` class template are both parameterized by a geometric kernel and

---

[8]A straightforward solution would be to repeatedly normalize the results of all computations. However, our experience shows that indiscriminate normalization considerably slows down the arrangement construction.

model the concepts ArrangementTraits_2 and ArrangementLandmarkTraits_2.[9] The class template `Arr_non_caching_segment_traits_2<Kernel>` derives from the instance `Arr_non_caching_segment_basic_traits_2<Kernel>`, which models the ArrangementLandmarkTraits_2 traits concept but not the refined ArrangementXMonotoneTraits_2 concept.[10] Like the `Arr_segment_traits_2` class template it derives from the `Kernel` type. Unlike the `Arr_segment_traits_2` class template it defines its point and segment types as `Kernel::Point_2` and `Kernel::Segment_2`, respectively, and most of its defined operations are delegations of the corresponding operations of the `Kernel` type. For example, the functor `Compare_xy_2` is defined as `Kernel::Compare_xy_2`. The remaining operations are defined in terms of just a few other kernel operations. For example, the `Compare_y_at_x_right_2` predicate is defined in terms of the `Kernel::Compare_slope_2` predicate (ignoring preconditions for the sake of clarity); see Section 5.1.1 for the description of this predicate. The class template `Arr_non_caching_segment_basic_traits_2<Kernel>` is slightly less efficient than the `Arr_segment_traits_2` class template for constructing arrangements of pairwise interior-disjoint line-segments in many cases, as it does not exploit caching at all. Nevertheless, you may choose to use this traits class, as it consumes less memory. For arrangements of line segments that do intersect you may use the class template `Arr_non_caching_segment_traits_2<Kernel>`. However, the performance difference in favor of the `Arr_segment_traits_2` class template is much larger, especially when the number of intersections is large.

Example: The example below reads an input file containing a set of line segments that are pairwise disjoint in their interior. As the segments do not intersect, no new points are constructed. Therefore, it is safe to instantiate the `Arr_non_caching_segment_traits_basic_2<Kernel>` class template and substitute its `Kernel` parameter with a kernel that does not support exact constructions, such as the `Exact_predicates_inexact_constructions_kernel` predefined kernel. Note that the `insert_non_intersecting_curves()` function is used to construct the arrangement. By default, the example opens and reads the content of the input file `Europe.dat`, located in the example folder, which contains more than 3,000 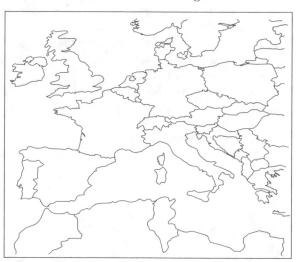 line segments with floating-point coordinates describing an old map of Europe, as shown in the figure above. The example uses an instance of the `print_arrangement_size()` function template listed on Page 21. It also uses an instance of the function template `read_objects()` explained in Section 4.2.1.

// File: ex_non_caching_segments.cpp

```
#include <list>
#include <iostream>
#include <boost/timer.hpp>

#include <CGAL/Exact_predicates_inexact_constructions_kernel.h>
#include <CGAL/Arr_non_caching_segment_basic_traits_2.h>
#include <CGAL/Arrangement_2.h>
#include <CGAL/Timer.h>
```

---

[9] They also model the refined concept ArrangementDirectionalXMonotoneTraits_2 defined in Chapter 8; see also Section 5.6.

[10] It does, however, model the refined concept ArrangementDirectionalBasicTraits_2 defined in Section 8.2.1.

```
#include "arr_print.h"
#include "read_objects.h"

typedef CGAL::Exact_predicates_inexact_constructions_kernel   Kernel;
typedef Kernel::FT                                            Number_type;
typedef CGAL::Arr_non_caching_segment_basic_traits_2<Kernel>  Traits;
typedef Traits::Point_2                                       Point;
typedef Traits::X_monotone_curve_2                            Segment;
typedef CGAL::Arrangement_2<Traits>                           Arrangement;

int main(int argc, char* argv[])
{
  // Get the name of the input file from the command line, or use the default
  // Europe.dat file if no command-line parameters are given.
  const char* filename = (argc > 1) ? argv[1] : "Europe.dat";
  std::list<Segment> segments;
  read_objects<Segment>(filename, std::back_inserter(segments));

  // Construct the arrangement by aggregately inserting all segments.
  Arrangement arr;
  std::cout << "Performing aggregated insertion of "
            << segments.size() << " segments." << std::endl;
  boost::timer timer;
  insert_non_intersecting_curves(arr, segments.begin(), segments.end());
  double secs = timer.elapsed();

  print_arrangement_size(arr);
  std::cout << "Construction took " << secs << " seconds." << std::endl;
  return 0;
}
```

5.2.3 The Linear-Traits Class

The `Arr_linear_traits_2<Kernel>` class template used in the previous chapter for demonstrating the construction of arrangements of unbounded curves is capable of handling bounded and unbounded linear objects, namely, lines, rays, and line segments. It models the concepts ArrangementTraits_2, ArrangementLandmarkTraits_2, and ArrangementOpenBoundaryTraits_2. It is parameterized by a geometric kernel and its nested type `Point_2` is defined to be the kernel-point type. The `Curve_2` (and `X_monotone_curve_2`) nested types are constructible from a `Kernel::Line_2`, a `Kernel::Ray_2`, or a `Kernel::Segment_2` object. Given a linear-curve object c, you can use the calls `c.is_line()`, `c.is_ray()`, and `c.is_segment()` to find out whether it has a source point or a target point. Based on the curve type, you can access its endpoints using the methods `c.source()` (for rays and for line segments) and `c.target()` (for segments only). It is also possible to cast a curve into a `Kernel::Line_2`, a `Kernel::Ray_2`, or a `Kernel::Segment_2` type, using the methods `c.line()`, `c.ray()`, and `c.segment()`, respectively.

Note that, like the default line-segment traits class, the linear-curve traits class uses caching techniques to speed up its predicate evaluations and object constructions.

5.3 The Polyline-Traits Class

Polylines are continuous piecewise linear curves. Polylines are of particular interest, as they can be used to approximate more complex curves in the plane. At the same time they are easier to handle in comparison to higher-degree algebraic curves, as rational arithmetic is sufficient to carry out computations on polylines, and to construct arrangements of polylines in an exact and robust manner.

The `Arr_polyline_traits_2<SegmentTraits>` class template handles polylines. It models the concepts ArrangementTraits_2 and ArrangementLandmarkTraits_2 depending on the type that substitutes the `SegmentTraits` parameter when the `Arr_polyline_traits_2>` class template is instantiated. The substituted type must be a geometry-traits class that handles segments and models the concept ArrangementXMonotoneTraits_2. If it also models the ArrangementTraits_2 concept, the instantiated `Arr_polyline_traits_2<SegmentTraits>` type models the concept ArrangementTraits_2 as well. The same holds for the ArrangementLandmarkTraits_2 concept. The curves handled by each operation defined by the type that substitutes the `SegmentTraits` parameter are restricted to line segments. (This precondition cannot be enforced by the compiler.) An instance of the polyline class-template inherits its point type from the substituted line-segment traits class, and defines the new type `Curve_2`, which is used to represent polylines. A curve of this type is stored as a vector of `SegmentTraits::X_monotone_curve_2` objects (namely line segments). Polyline objects are constructible from a range of points, where two succeeding points in the range represent the endpoints of a segment of the polyline. The nested `X_monotone_curve_2` type inherits from the type `Curve_2`. The points in an x-monotone polyline curve are always stored in lexicographically increasing order of their coordinates. The polyline traits class also supports the traversal over the range of defining points, whose first and past-the-end iterators can be obtained through the methods `c.begin()` and `c.end()` of a polyline c.

The polyline-traits class does not perform any geometric operations directly. Instead, it solely relies on the functionality of the substituted segment-traits class. For example, when we need to determine the position of a point with respect to an x-monotone polyline, we use binary search to locate the relevant segment that contains the point in its x-range. Then, we compute the position of the point with respect to this segment. Thus, operations on x-monotone polylines of size m typically take $O(\log m)$ time.

You are free to choose the underlying line-segment traits class. Your decision could be based, for example, on the number of expected intersection points (see discussion above in Section 5.2.1). Moreover, it is possible to substitute the `SegmentTraits` template parameter with a traits class that handles segments with some additional data attached to each individual segment; see Section 5.5. This makes it possible to associate different data objects with the different segments that compose a polyline.

———————— *advanced* ————————

Consider, for example, the `Equal_2` predicate that returns **true** iff the graphs of two given x-monotone polylines are the same; see Section 5.1.1. The implementation of this predicate for polylines follows the algorithm below presented in pseudo-code, where, first_segment(c) and last_segment(c) return the first and last segments of the polyline c, respectively; next_segment (c, s) returns the segment of the polyline c following the segment s; minpoint(s) and maxpoint(s) return the lexicographically minimum and maximum endpoints of a segment s, respectively; and ison(p, s) determines whether the point p lies on a segment s. These functions are trivially implemented with the operations of the traits class other than the `Equal_2` predicate and the access functions of the polyline class. The correctness of the algorithm is based on the assumption that the subcurve that goes through two points is unique, namely, that the subcurve is linear.

Are the graphs of the polylines c_1 and c_2 equal?

$s_1 \leftarrow \text{first\_segment}(c_1)$
$s_2 \leftarrow \text{first\_segment}(c_2)$
if ($\text{minpoint}(s_1) \neq \text{minpoint}(s_1)$) **return false**
while ($s_1 \neq \text{last\_segment}(c_1) \wedge s_2 \neq \text{last\_segment}(c_2)$)
   if ($\text{maxpoint}(s_1) = \text{maxpoint}(s_2)$)
      $s_1 \leftarrow \text{next\_segment}(c_1, s_1)$
      $s_2 \leftarrow \text{next\_segment}(c_2, s_2)$
   else if ($\text{maxpoint}(s_1) < \text{maxpoint}(s_2)$)
      if $\neg\text{ison}(\text{maxpoint}(s_1), s_2)$ **return false**
      $s_1 \leftarrow \text{next\_segment}(c_1, s_1)$
   else
      assert($\text{maxpoint}(s_2) < \text{maxpoint}(s_1)$)
      if $\neg\text{ison}(\text{maxpoint}(s_2), s_1)$ **return false**
      $s_2 \leftarrow \text{next\_segment}(c_2, s_2)$
if ($s_1 \neq \text{last\_segment}(c_1) \vee s_2 \neq \text{last\_segment}(c_2)$) **return false**
if ($\text{maxpoint}(s_1) \neq \text{maxpoint}(s_2)$) **return false**
return true

——— *advanced* ———

Example: The program listed below constructs an arrangement of three polylines, π_1, π_2, and π_3, as depicted in the figure to the right. In this example, each polyline is constructed from points stored in a different container, i.e., array, list, and vector. Points defining the polylines are not necessarily associated with arrangement vertices. The arrangement vertices are either the extreme points of each x-monotone polyline (drawn as dark discs) or the intersection points between two polylines (drawn as rings). Observe that the polyline π_2 is split into three x-monotone polylines, and that the two curves π_1 and π_3 have *two* overlapping sections—an impossible scenario in arrangements of line segments.

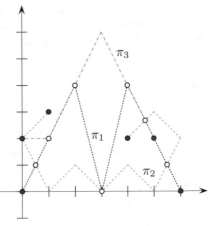

// File: ex_polylines.cpp

#include $<\text{list}>$
#include $<\text{vector}>$

#include `"arr_polylines.h"`
#include `"arr_print.h"`

int main()
{
  // Construct three polylines and compute their arrangement.
  Point pts1[] =
    {Point(0, 0), Point(2, 4), Point(3, 0), Point(4, 4), Point(6, 0)};

  std::list$<\text{Point}>$ pts2;
  pts2.push_back(Point(1, 3));   pts2.push_back(Point(0, 2));
  pts2.push_back(Point(1, 0));   pts2.push_back(Point(2, 1));
  pts2.push_back(Point(3, 0));   pts2.push_back(Point(4, 1));
  pts2.push_back(Point(5, 0));   pts2.push_back(Point(6, 2));
  pts2.push_back(Point(5, 3));   pts2.push_back(Point(4, 2));

```
std::vector<Point>   pts3(4);
pts3[0] = Point(0, 2);              pts3[1] = Point(1, 2);
pts3[2] = Point(3, 6);              pts3[3] = Point(5, 2);

Arrangement          arr;
insert(arr, Polyline(&pts1[0], &pts1[5]));       // pi1
insert(arr, Polyline(pts2.begin(), pts2.end())); // pi2
insert(arr, Polyline(pts3.begin(), pts3.end())); // pi3

print_arrangement(arr);
return 0;
}
```

The common types used by the example programs involving arrangements of polylines are listed below, and defined in the header file **arr_polylines.h**. As we do in the case of segments and linear objects, we use the predefined CGAL kernel to instantiate our traits class. However, in this case we have yet another layer of template instantiation, namely the instantiation of the polyline-traits class-template with an instance of a line-segment traits class-template. Naturally, we use the **Arr_segment_traits_2** class, instantiated with the predefined filtered kernel of CGAL.

```
#include <CGAL/Exact_predicates_exact_constructions_kernel.h>
#include <CGAL/Arr_segment_traits_2.h>
#include <CGAL/Arr_polyline_traits_2.h>
#include <CGAL/Arrangement_2.h>

typedef CGAL::Exact_predicates_exact_constructions_kernel   Kernel;
typedef Kernel::FT                                          Number_type;

typedef CGAL::Arr_segment_traits_2<Kernel>                 Segment_traits;
typedef CGAL::Arr_polyline_traits_2<Segment_traits>        Traits;
typedef Traits::Point_2                                    Point;
typedef Traits::Curve_2                                    Polyline;
typedef CGAL::Arrangement_2<Traits>                        Arrangement;
```

5.4   Traits Classes for Algebraic Curves

A curve in our context is typically (but not necessarily) defined as the zero set of a bivariate nonzero polynomial with rational (or, equivalently, integral) coefficients. We call such polynomials and the curves they define *algebraic*. When dealing with linear curves (e.g., line segments and polylines), having rational coefficients guarantees that all intersection points also have rational coordinates, such that the arrangement of such curves can be constructed and maintained using only rational arithmetic. The *2D Arrangements* package also offers geometry traits-classes that handle algebraic curves defined by algebraic polynomials of degree higher than 1. Unfortunately, the coordinates of the intersection points constructed by these traits models are in general algebraic numbers[11] of degree higher than 1. It is therefore clear that we have to use number types different from plain rational to represent point coordinates and be able to apply arithmetic operations on them.

Several types of algebraic curves are handled by more than one traits model. Section 5.2 introduces a few different traits models that handle line segments. This duplication becomes more evident with the introduction of traits classes that handle algebraic curves. The different traits models have different properties. In some cases they were developed by different authors at different times exploiting different tools that were available at the time they were developed. As

---

[11] A number is called *algebraic*, if it is the root of a univariate algebraic polynomial with rational coefficients.

a general rule, you should always use the minimal traits model that still satisfies your needs, as the most dedicated model is most likely to be the most efficient.

5.4.1 A Traits Class for Circular Arcs and Line Segments

Arrangements induced by circular arcs and line segments are very useful and frequently arise in applications, where curves of interleaved line segments and circular arcs are used to model the boundaries of complex shapes. Such curves can fit the original boundary more tightly and more compactly than, for example, a simple polyline; see Section 5.7 for more details and references. We present two applications of arrangements induced by line segments and circular arcs in the following chapters, namely, (i) computing the union of general polygons the boundaries of which comprise line segments and circular arcs (see Section 8.3), and (ii) coordinating two disc robots amidst polygonal obstacles; see Section 9.3.2.

Besides the importance of arrangements of circular arcs and line segments it turns out that it is possible to implement efficient traits models that handle curves restricted to circular arcs and line segments. Rational numbers cannot represent the coordinates of intersection points that may arise in such arrangements in an exact manner. Thus, algebraic numbers must be used. However, the performance impact that (general) algebraic numbers incur can be reduced by the use of an efficient type of exact algebraic numbers called *square root extension* that uses rational arithmetic in an almost straightforward fashion.

Square root numbers have the form $\alpha + \beta\sqrt{\gamma}$, where α, β, and γ are rational numbers.[12] The rational number γ is referred to as the extension. Each subset that has a particular extension is closed under arithmetic operations and order relations; hence, it is a valid algebraic structure.[13]

The class template `CGAL::Sqrt_extension<NT,Root>` implements the square root extension type. Instances of this template represent square root numbers. It is equipped with the implementation of a set of arithmetic operations and order relations that exploit identical extensions of operands. It also provides the ability to compare two numbers with different extensions $\gamma_1 \neq \gamma_2$ efficiently. Each operation is implemented using only few rational arithmetic operations. Here, `NT` is the type of α and β, and `Root` is the type of γ.

The running times of the arithmetic operations and order relations provided by the square root extension when the identical-extension condition is met are comparable to the running times of the corresponding arithmetic operations of rational-number types. The running times of order relations when the condition is not met are only slightly larger. In practice, using number types that represent (arbitrary) algebraic numbers increases the running time of the application significantly.

We call circles whose center coordinates and squared radii are rational numbers *rational circles*. The equation of such a circle, that is, $(x - x_0)^2 + (y - y_0)^2 = r^2$, where (x_0, y_0) and r denote the circle center and its radius, respectively, has rational coefficients. The coordinates of the points of intersection between two such circles are therefore solutions of quadratic equations with rational coefficients, in other words, algebraic numbers of degree 2 or simply square root numbers. The same applies to intersection points between such a rational circle and a line, or a line segment, with rational coefficients (a line whose equation is $ax + by + c = 0$, where a, b, and c are rational).

The *2D Arrangements* package offers a traits class-template called `Arr_circle_segment_traits_2<Kernel>` that exclusively handles line segments, circular arcs, and whole circles and models the concept ArrangementTraits_2 (but it does not model the ArrangementLandmarkTraits_2 concept).[14] It exploits efficient computations with square root numbers, which makes it attractive for arrangements induced by line segments, circular arcs, and whole circles. When the traits class-template is instantiated, the `Kernel` template parameter must be substituted with a geometric kernel that models the Kernel concept; see Section 1.4.4 for more details about this concept.

---

[12]Square root numbers are also called one-root numbers.

[13]The term algebraic structure refers to the set closed under one or more operations satisfying some axioms; see, e.g., http://en.wikipedia.org/wiki/Algebraic_structure. For example, all numbers that can be expressed as $\alpha + \beta\sqrt{K}$, where α and β are rational and K is a rational constant, compose an algebraic structure.

[14]It also models the refined concept ArrangementDirectionalXMonotoneTraits_2 defined in Chapter 8; see, in addition, Section 5.6.

Observe that the nested type **Point_2** defined by the traits class, whose coordinates are typically algebraic numbers of degree 2, is *not* the same as the **Kernel::Point_2** type, which is only capable of representing a point with rational coordinates. The coordinates of a point are represented using the number type **CoordNT**, nested in the traits class-template.

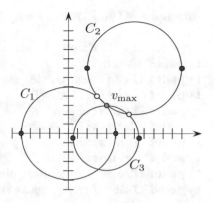

Example: In the example below an arrangement of three full circles is constructed, as shown in the figure to the right. Each circle is split into two x-monotone circular arcs, the endpoints of which are drawn as full discs; rings mark vertices that correspond to intersection points. Once the arrangement is constructed, we locate the vertex of maximal degree in the arrangement. The geometric mapping of this vertex, denoted by $v_{\max}$, is the point $(4, 3)$, as all three circles intersect at this point and the associated vertex has six incident edges.

```
// File: ex_circles.cpp

#include "arr_circular.h"

int main()
{
  Arrangement  arr;

  // Create a circle centered at the origin with radius 5 (C1).
  insert(arr, Curve(Circle(Rational_point(0, 0), Number_type(25))));

  // Create a circle centered at (7,7) with radius 5 (C2).
  insert(arr, Curve(Circle(Rational_point(7, 7), Number_type(25))));

  // Create a circle centered at (4,-0.5) with radius 3.5 (= 7/2) (C3).
  Rational_point c3 = Rational_point(4, Number_type(-1) / Number_type(2));
  insert(arr, Curve(Circle(c3, Number_type(49) / Number_type(4))));

  // Locate the vertex with maximal degree.
  Arrangement::Vertex_const_iterator  vit = arr.vertices_begin();
  Arrangement::Vertex_const_handle    v_max = vit;
  for (++vit; vit != arr.vertices_end(); ++vit)
    if (vit->degree() > v_max->degree()) v_max = vit;

  std::cout << "The vertex with maximal degree in the arrangement is: "
            << "v_max = (" << v_max->point() << ") "
            << "with degree " << v_max->degree() << "." << std::endl;
  return 0;
}
```

The types common to the example programs that use the **Arr_circle_segment_traits_2** class template are listed below and defined in the header file **arr_circular.h**. Even though algebraic numbers are required to represent coordinates of points where the inducing curves are circles or circular arcs, such as the curves handled by the **Arr_circle_segment_traits_2** class template, an (exact) rational kernel suffices, and a filtered one improves the performance further.

```
#include <CGAL/Exact_predicates_exact_constructions_kernel.h>
#include <CGAL/Arr_circle_segment_traits_2.h>
```

#include <CGAL/Arrangement_2.h>

typedef CGAL::Exact_predicates_exact_constructions_kernel	Kernel;
typedef Kernel::FT	Number_type;
typedef CGAL::Arr_circle_segment_traits_2<Kernel>	Traits;
typedef Traits::CoordNT	CoordNT;
typedef Traits::Point_2	Point;
typedef Traits::Curve_2	Curve;
typedef Traits::Rational_point_2	Rational_point;
typedef Traits::Rational_segment_2	Segment;
typedef Traits::Rational_circle_2	Circle;
typedef CGAL::Arrangement_2<Traits>	Arrangement;

The `Arr_circle_segment_traits_2<Kernel>` class template uses the `CGAL::Sqrt_extension` class template to represent numbers of the square root extension type.[15] The impact of substituting `Kernel` with a filtered kernel as opposed to a non-filtered kernel when the traits class-template is instantiated can be substantial despite the limited use of kernel functors.

Try: Measure the impact of a filtered kernel. Develop a simple program that constructs an arrangement of m line segments and n circular arcs uniformly distributed in the unit square at random. Use the program to compare the performance of two instances of the `Arr_circle_segment_traits_2<Kernel>` class template—one is obtained by substituting the `Kernel` parameter with the `Exact_predicates_exact_construction_kernel` type and the other is by substituting it with the non-filtered kernel type `Cartesian<Gmpq>`.

The `Curve_2` type nested in `Arr_circle_segment_traits_2` can be used to represent circles, circular arcs, or line segments. We now describe and demonstrate a variety of ways in which the curves of this type can be constructed. A curve object is constructible from a `Kernel::Circle_2` object or from a `Kernel::Segment_2` object. A circular arc is typically defined by the supporting circle and two endpoints, where the endpoints are objects of type `Point_2`, with rational or irrational coordinates. The orientation of the arc is determined by the orientation of the supporting circle. Similarly, we also support the construction of line segments given their supporting line (of type `Kernel::Line_2`) and two endpoints, which may have irrational coordinates (unlike the `Kernel::Segment_2` type).

Note that the `Kernel::Circle_2` type is used to represent a circle whose *squared radius* is rational, where the radius itself may be irrational. However, if the radius is known to be rational, its use is recommended for efficiency. It is therefore also possible to construct a circle, or a circular arc specifying the circle center (a `Kernel::Point_2` object), its rational radius (of type `Kernel::FT`), and its orientation. Finally, we also support the construction of a circular arc that is defined by two endpoints and an arbitrary interior point that lies on the arc in between its endpoints. In this case, all three points are required to have rational coordinates; namely, they are all given as `Kernel::Point_2` objects.

---

[15] At the time this book was written the `Arr_circle_segment_traits_2` class template still utilized its own implementation of the square root extension number type, as it had been developed before the `CGAL::Sqrt_extension` class template became available. The latter was expected to replace the former soon after.

Example: The example below demonstrates the usage of the various construction methods for circular arcs and line segments. The resulting arrangement is depicted in the figure to the right. Note the usage of the constructor of `CoordNT(alpha, beta, gamma)`, which creates a degree-2 algebraic number whose value is $\alpha + \beta\sqrt{\gamma}$.

```cpp
// File: ex_circular_arc.cpp

#include "arr_circular.h"
#include "arr_print.h"

int main()
{
    std::list<Curve>  curves;

    // Create a circle (C1) centered at the origin with squared radius 2.
    curves.push_back(Curve(Circle(Rational_point(0, 0), Number_type(2))));

    // Create a circle (C2) centered at (2, 3) with radius 3/2. Note that
    // as the radius is rational we use a different curve constructor.
    Number_type three_halves = Number_type(3) / Number_type(2);
    curves.push_back(Curve(Rational_point(2, 3), three_halves));

    // Create a segment (C3) of the line (y = x) with rational endpoints.
    Segment s3 = Segment(Rational_point(-2, -2), Rational_point(2, 2));
    curves.push_back(Curve(s3));

    // Create a line segment (C4) with the same supporting line (y = x), but
    // having one endpoint with irrational coefficients.
    CoordNT            sqrt_15 = CoordNT(0, 1, 15);  // = sqrt(15)
    curves.push_back(Curve(s3.supporting_line(),
                     Point(3, 3), Point(sqrt_15, sqrt_15)));

//Create a circular arc (C5) that is the upper half of the circle centered at
//(1, 1) with squared radius 3. Create the circle with clockwise orientation,
    // so the arc is directed from (1 - sqrt(3), 1) to (1 + sqrt(3), 1).
    Rational_point c5 = Rational_point(1, 1);
    Circle            circ5 = Circle(c5, 3, CGAL::CLOCKWISE);
    CoordNT           one_minus_sqrt_3 = CoordNT(1, -1, 3);
    CoordNT           one_plus_sqrt_3 = CoordNT(1, 1, 3);
    Point             s5 = Point(one_minus_sqrt_3, CoordNT(1));
    Point             t5 = Point(one_plus_sqrt_3, CoordNT(1));
    curves.push_back(Curve(circ5, s5, t5));

    // Create an arc (C6) of the unit circle, directed clockwise from
    // (-1/2, sqrt(3)/2) to (1/2, sqrt(3)/2).
    // The supporting circle is oriented accordingly.
    Rational_point c6 = Rational_point(0, 0);
    Number_type       half = Number_type(1) / Number_type(2);
    CoordNT           sqrt_3_div_2 = CoordNT(Number_type(0), half, 3);
    Point             s6 = Point(-half, sqrt_3_div_2);
```

```
Point              t6 = Point(half, sqrt_3_div_2);
curves.push_back(Curve(c6, 1, CGAL::CLOCKWISE, s6, t6));

// Create a circular arc (C7) defined by two endpoints and a midpoint,
// all having rational coordinates. This arc is the upper right
// quarter of a circle centered at the origin with radius 5.
Rational_point s7 = Rational_point(0, 5);
Rational_point mid7 = Rational_point(3, 4);
Rational_point t7 = Rational_point(5, 0);
curves.push_back(Curve(s7, mid7, t7));

// Construct the arrangement of the curves and print its size.
Arrangement  arr;
insert(arr, curves.begin(), curves.end());
print_arrangement_size(arr);

return 0;
}
```

It is also possible to construct x-monotone curves, which represent x-monotone circular arcs or line segments, using similar constructors (full circles are not x-monotone).

5.4.2   A Traits Class for Conic Arcs

A *conic curve* is an algebraic curve of degree 2. Namely, it is the locus of all points (x, y) satisfying the equation $\kappa : rx^2 + sy^2 + txy + ux + vy + w = 0$, where the six coefficients $\langle r, s, t, u, v, w \rangle$ completely characterize the curve. The sign of the expression $\Delta_\kappa = 4rs - t^2$ determines the type of the curve.

- If $\Delta_\kappa > 0$, the curve is an *ellipse*. A circle is a special case of an ellipse where $r = s$ and $t = 0$.

- If $\Delta_\kappa = 0$, the curve is a *parabola*—an unbounded conic curve with a single connected branch. When $r = s = t = 0$ we have a line, which can be considered as a degenerate parabola.

- If $\Delta_\kappa < 0$, the curve is a *hyperbola*. That is, it comprises two disconnected unbounded branches.

The `Arr_conic_traits_2<RatKernel, AlgKernel, NtTraits>` class template is capable of handling only bounded arcs of conic curves, referred to as *conic arcs*. A conic arc a may be either (i) a full ellipse, or (ii) defined by the tuple $\langle \kappa, p_s, p_t, o \rangle$, where κ is a conic curve and p_s and p_t are two points on κ (namely, $\kappa(p_s) = \kappa(p_t) = 0$) that define the *source* and *target* of the arc, respectively. The arc is formed by traversing κ from the source to the target in the orientation specified by o, which is typically `CGAL::CLOCKWISE` or `CGAL::COUNTERCLOCKWISE`, but may also be `CGAL::COLLINEAR` in the case of degenerate conic-curves, namely, lines or pairs of lines.

We always assume that the conic coefficients $\langle r, s, t, u, v, w \rangle$ are rational. The coordinates of points of intersection between two conic curves with rational coefficients are in general algebraic numbers of degree 4. Namely, they are roots of algebraic polynomials of degree 4. In addition, conic arcs may not necessarily be x-monotone, and must be split at points where the tangent to the arc is vertical. In the general case such points typically have coordinates that are algebraic numbers of degree 2. Note that as arrangement vertices induced by intersection points and points with vertical tangents are likely to have algebraic coordinates, we also allow the original endpoints of the input arcs p_s and p_t to have algebraic coordinates.

The `Arr_conic_traits_2<RatKernel, AlgKernel, NtTraits>` class template is designed for efficient handling of arrangements of bounded conic arcs. The template has three parameters, defined as follows:

- The **RatKernel** template parameter must be substituted with a geometric kernel whose field type is an exact rational type. It is used to define basic geometric entities (e.g., a line segment or a circle) with rational coefficients. Typically, we use one of the standard CGAL kernels instantiated with the number type **NtTraits::Rational** (see below).

- The **AlgKernel** template parameter must be substituted with a geometric kernel whose field type represents an exact algebraic number. It is used to define points with algebraic coordinates. Typically, we use one of the standard CGAL kernels instantiated with the number type **NtTraits::Algebraic** (see below).

- The **NtTraits** template parameter must be substituted with a type that encapsulates all the numerical operations needed for performing the geometric computation carried out by the geometric traits-class. It defines the **Integer**, **Rational**, and **Algebraic** number types, and supports several operations on these types, such as conversion between number types, solving quadratic equations, and extraction of real roots of polynomials with integral coefficients. The use of the **CORE_algebraic_number_traits** class, which is included in the *2D Arrangements* package, is highly recommended. The traits class-template relies on the multi-precision number types implemented in the CORE library and performs exact computations on the number types it defines.

The instantiation of the conic traits class-template is slightly more complicated than the instantiation of the traits classes you have encountered so far. This instantiation is exemplified in the header file **arr_conics.h**. Note how we first define the rational and algebraic kernels using the number types given by the CORE number type traits-class, then use them to define the conic traits class-template. Also note the types defined by the rational kernels, which we need for conveniently constructing conic arcs.

```
#include <CGAL/Cartesian.h>
#include <CGAL/CORE_algebraic_number_traits.h>
#include <CGAL/Arr_conic_traits_2.h>
#include <CGAL/Arrangement_2.h>

typedef CGAL::CORE_algebraic_number_traits        Nt_traits;
typedef Nt_traits::Rational                        Rational;
typedef CGAL::Cartesian<Rational>                  Rat_kernel;
typedef Rat_kernel::Point_2                        Rat_point;
typedef Rat_kernel::Segment_2                      Rat_segment;
typedef Rat_kernel::Circle_2                       Rat_circle;
typedef Nt_traits::Algebraic                       Algebraic;
typedef CGAL::Cartesian<Algebraic>                 Alg_kernel;

typedef CGAL::Arr_conic_traits_2<Rat_kernel, Alg_kernel, Nt_traits>
                                                   Traits;
typedef Traits::Point_2                            Point;
typedef Traits::Curve_2                            Conic_arc;
typedef Traits::X_monotone_curve_2                 X_monotone_conic_arc;
typedef CGAL::Arrangement_2<Traits>               Arrangement;
```

The **Arr_conic_traits_2** models the ArrangementTraits_2 and ArrangementLandmarkTraits_2 concepts.[16] Its **Point_2** type is derived from **AlgKernel::Point_2**, while the **Curve_2** type is used to represent a bounded, not necessarily *x*-monotone, conic arc. The **X_monotone_curve_2** type is derived from **Curve_2**, but its constructors are used only by the traits class. Users should therefore construct only **Curve_2** objects and insert them into the arrangement using the **insert()** function.

---

[16]It also models the refined concept ArrangementDirectionalXMonotoneTraits_2 defined in Chapter 8.

Conic arcs are constructible from full ellipses or by specifying a supporting curve, two endpoints, and an orientation. However, several constructors of **Curve_2** are available to allow for some special cases, such as line segments or circular arcs. The **Curve_2** and the derived **X_monotone_curve_2** classes also support basic access functions such as **source()**, **target()**, and **orientation()**.

Example: The example below demonstrates the usage of the various constructors for conic arcs. The resulting arrangement is depicted in the figure to the right. Especially noteworthy are the constructor of a circular arc that accepts three points, the constructor of a conic arc that accepts five points, and the constructor that allows specifying approximate endpoints, where the exact endpoints are given explicitly as intersections of the supporting conic with two other conic curves. The approximate endpoints are used to select the specific exact endpoints out of all intersection points of the pair of curves (the supporting conic curve and the auxiliary conic curve). Also note that as the preconditions required by some of these constructors are rather complicated (see the Reference Manual for the details), a precondition violation does not cause the program to terminate—instead, an *invalid* arc is created. We can verify the validity of an arc a by using the **a.is_valid()** method. Naturally, inserting invalid arcs into an arrangement is not allowed, so the validity of an arc should be checked once it is constructed.

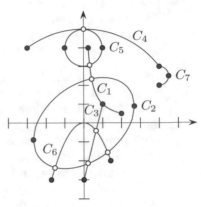

```
// File: ex_conic_arcs.cpp

#include "arr_conics.h"
#include "arr_print.h"

int main()
{
  Arrangement  arr;

  // Insert a hyperbolic arc (C1), supported by the hyperbola y = 1/x
  // (or: xy - 1 = 0) with the endpoints (1/4, 4) and (2, 1/2).
  // The arc is counterclockwise oriented.
  insert(arr, Conic_arc(0, 0, 1, 0, 0, -1, CGAL::COUNTERCLOCKWISE,
                  Point(Rational(1,4), 4), Point(2, Rational(1,2))));

  // Insert a full ellipse (C2), which is (x/4)^2 + (y/2)^2 = 0 rotated by
  // phi = 36.87 degrees (such that sin(phi) = 0.6, cos(phi) = 0.8),
  // yielding: 58x^2 + 72y^2 - 48xy - 360 = 0.
  insert(arr, Conic_arc (58, 72, -48, 0, 0, -360));

  // Insert the segment (C3) (1, 1) — (0, -3).
  insert(arr, Conic_arc(Rat_segment(Rat_point(1, 1), Rat_point(0, -3))));

  // Insert a circular arc (C4) supported by the circle x^2 + y^2 = 5^2,
  // with (-3, 4) and (4, 3) as its endpoints. We want the arc to be
  // clockwise-oriented, so it passes through (0, 5) as well.
  Conic_arc  c4(Rat_point(-3, 4), Rat_point(0, 5), Rat_point(4, 3));
  CGAL_assertion(c4.is_valid());
  insert(arr, c4);

  // Insert a full unit circle (C5) that is centered at (0, 4).
```

```
insert(arr, Conic_arc(Rat_circle(Rat_point(0,4), 1)));

// Insert a parabolic arc (C6) supported by the parabola y = -x^2
// with endpoints (-sqrt(3),-3) (~(-1.73,-3)) and (sqrt(2),-2) (~(1.41,-2)).
// Since the x-coordinates of the endpoints cannot be acccurately represented,
// we specify them as the intersections of the parabola with the lines
// y = -3 and y = -2, respectively. The arc is clockwise-oriented.
Conic_arc   c6 =
  Conic_arc(1, 0, 0, 0, 1, 0,          // The parabola.
            CGAL::CLOCKWISE,
            Point(-1.73, -3),          // Approximation of the source.
            0, 0, 0, 0, 1, 3,          // The line: y = -3.
            Point(1.41, -2),           // Approximation of the target.
            0, 0, 0, 0, 1, 2);         // The line: y = -2.
CGAL_assertion(c6.is_valid());
insert(arr, c6);

// Insert the right half of the circle centered at (4, 2.5) whose radius
// is 1/2 (therefore its squared radius is 1/4) (C7).
Rat_circle   circ7(Rat_point(4, Rational(5,2)), Rational(1,4));
insert(arr, Conic_arc(circ7, CGAL::CLOCKWISE, Point(4, 3), Point(4, 2)));
print_arrangement_size(arr);

  return 0;
}
```

5.4.3   A Traits Class for Arcs of Rational Functions

A *rational function* is given by the equation $y = \frac{P(x)}{Q(x)}$, where P and Q are polynomials of arbitrary degree. In particular, if $Q(x) = 1$, then y is simply a polynomial function. A bounded *rational arc* is defined by the graph of a rational function over some interval $[x_{min}, x_{max}]$, where Q does not have any real roots in this interval (thus, the arc does not contain any singularities). However, we may consider functions defined over unbounded intervals, namely, over $(-\infty, x_{max}]$, $[x_{min}, \infty)$, or $(-\infty, \infty)$. Rational functions, and polynomial functions in particular, are not only interesting in their own right, they are also useful for approximating more complicated curves and for interpolation.

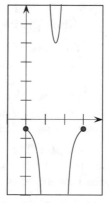

Using the `Arr_rational_function_traits_2` class template it is possible to construct and maintain arrangements induced by rational arcs. Every instance of the `Arr_rational_function_traits_2` class template models the concepts ArrangementTraits_2 and ArrangementOpenBoundaryTraits_2 (but it does not model the ArrangementLandmarkTraits_2 concept).[17]

A rational arc is always x-monotone in the mathematical sense. However, it is not necessarily continuous, as it may have singularities. An arc that has singularities must be split into continuous portions before being inserted into the arrangement. Consider, for example, the rational arc given by the equation $y = \frac{1}{(x-1)(2-x)}$ defined over the interval $[0,3]$, as depicted in the figure to the right. This arc has two singularities, at $x = 1$ and at $x = 2$. It is split into three continuous portions defined over the intervals $[0,1)$, $(1,2)$, and $(2,3]$ by the traits operation `Make_x_monotone_2`. Arbitrary rational functions are represented by the nested type `Curve_2` and continuous portions of rational functions are represented by the nested type `X_monotone_curve_2`. Constructors for both types are provided by the traits in form of functors.

---

[17]It also models the refined concept ArrangementDirectionalXMonotoneTraits_2 defined in Chapter 8.

When the `Arr_rational_function_traits_2<AlgebraicKernel_d_1>` class template is instantiated, the template parameter must be substituted with a model of the AlgebraicKernel_d_1 concept. Models of this concept, such as the `Algebraic_kernel_d_1<Coefficient>` class template provided by the CGAL package *Algebraic Foundations*, are meant to support algebraic functionalities on univariate polynomials of arbitrary degree. See the documentation of the concept AlgebraicKernel_d_1 for more information. A rational function is then represented as the quotient of two polynomials, P and Q, of type `Polynomial_1` nested in every model of the concept AlgebraicKernel_d_1 and in particular in the algebraic-kernel type that substitutes the template parameter `AlgebraicKernel_d_1` when the traits class-template is instantiated. Such a rational function is constructible from a single polynomial P (with $Q(x) = 1$), or from two polynomials P and Q. The type of the polynomial coefficients, namely `Coefficient`, and the type of the interval bounds, namely `Bound`, are also nested in the algebraic-kernel type. If an instance of the `Algebraic_kernel_d_1<Coefficient>` class template is used, for example, as the algebraic-kernel type, the type that substitutes its template parameter is defined as the `Coefficient` type. This type cannot be algebraic. Moreover, it is recommended that this type is not made rational either, since using rational (as opposed to integral) coefficients does not extend the range of the rational arcs and is typically less efficient.[18] The `Bound` type, however, can be algebraic. A point of type `Point_2` nested in the `Arr_rational_function_traits_2` class template is represented by a rational function and its x-coordinate, which is derived from the type `Algebraic_real_1` nested in the algebraic-kernel type. An explicit representation by the nested type `Algebraic_real_1` of the y-coordinate is only computed upon request, as it can be a rather costly operation.

The aforementioned types, `Polynomial_1`, `Coefficient`, `Bound`, and `Algebraic_real_1`, are conveniently nested in the `Arr_rational_function_traits_2` class template among the others and obtained from there in the type definitions used in the examples given in this section and listed below. These types are defined in the header file `arr_rat_functions.h`.

```
#include <CGAL/basic.h>
#include <CGAL/CORE_BigInt.h>
#include <CGAL/Algebraic_kernel_d_1.h>
#include <CGAL/Arr_rational_function_traits_2.h>
#include <CGAL/Arrangement_2.h>


typedef CORE::BigInt                                    Number_type;
typedef CGAL::Algebraic_kernel_d_1<Number_type>         AK1;
typedef CGAL::Arr_rational_function_traits_2<AK1>       Traits;

typedef Traits::Polynomial_1                            Polynomial;
typedef Traits::Algebraic_real_1                        Alg_real;
typedef Traits::Bound                                   Bound;
```

The constructed rational functions are cached by the traits class. The cache is local to each traits class object. It is therefore necessary to construct curves using only the constructor objects provided by member functions of the traits class. Moreover, a curve must only be used by the traits-class object that was used to construct it. The cache is automatically cleaned up from time to time. The amortized clean up costs are constant. In addition, there is also a separate member function that cleans up the cache upon request.

Example: The example below demonstrates the construction of an arrangement induced by the bounded rational arcs depicted in Figure 5.4a. It uses constructors both for polynomial arcs and for rational arcs.

```
// File: ex_rational_functions.cpp

#include "arr_rat_functions.h"
```

---

[18] The `Algebraic_kernel_d_1` class template uses the types provided by the *Polynomial* package of CGAL to define its nested `Polynomial_1` type and conveniently expose it to the user.

```cpp
#include "arr_print.h"

int main()
{
  CGAL::set_pretty_mode(std::cout);                    // for nice printouts.

  // Define a traits class object and a constructor for rational functions.
  Traits traits;
  Traits::Construct_x_monotone_curve_2 construct =
    traits.construct_x_monotone_curve_2_object();

  // Define a polynomial representing x.
  Polynomial x = CGAL::shift(Polynomial(1), 1);

  // Define a container storing all arcs.
  std::vector<Traits::X_monotone_curve_2>  arcs;

  // Create an arc (C1) supported by the polynomial y = x^4 - 6x^2 + 8,
  // defined over the interval [-2.1, 2.1].
  Polynomial P1 = x*x*x*x - 6*x*x + 8;
  Alg_real l(Bound(-2.1)), r(Bound(2.1));
  arcs.push_back(construct(P1, l, r));

  // Create an arc (C2) supported by the function y = x / (1 + x^2),
  // defined over the interval [-3, 3].
  Polynomial P2 = x;
  Polynomial Q2 = 1+x*x;
  arcs.push_back(construct(P2, Q2, Alg_real(-3), Alg_real(3)));

  // Create an arc (C3) supported by the parbola y = 8 - x^2,
  // defined over the interval [-2, 3].
  Polynomial P3 = 8 - x*x;
  arcs.push_back(construct(P3, Alg_real(-2), Alg_real(3)));

  // Create an arc (C4) supported by the line y = -2x,
  // defined over the interval [-3, 0].
  Polynomial P4 = -2*x;
  arcs.push_back(construct(P4, Alg_real(-3), Alg_real(0)));

  // Construct the arrangement of the four arcs.
  Arrangement arr(&traits);
  insert(arr, arcs.begin(), arcs.end());
  print_arrangement(arr);

  return 0;
}
```

Example: The example below demonstrates the construction of an arrangement of six rational arcs, four unbounded arcs, and two bounded arcs, as depicted in Figure 5.4b. Note the usage of the various constructors. Also observe that the hyperbolas $y = \pm\frac{1}{x}$ and $y = \pm\frac{1}{2x}$ never intersect, although they have common vertical and horizontal asymptotes, so very "thin" unbounded faces are created between them.

// File: ex_unbounded_rational_functions.cpp

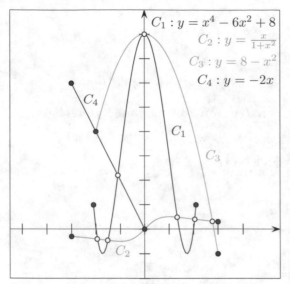
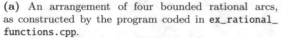

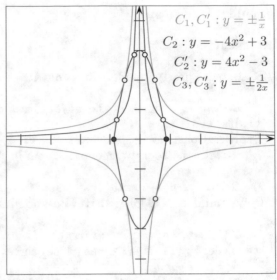

(a) An arrangement of four bounded rational arcs, as constructed by the program coded in `ex_rational_functions.cpp`.

(b) An arrangement of six unbounded arcs of rational functions, as constructed by the program coded in `ex_unbounded_rational_functions.cpp`.

Fig. 5.4: Arrangements of rational functions.

```
#include "arr_rat_functions.h"
#include "arr_print.h"

int main()
{
  CGAL::set_pretty_mode(std::cout);              // for nice printouts.

  // Define a traits class object and a constructor for rational functions.
  AK1 ak1;
  Traits traits(&ak1);
  Traits::Construct_curve_2 construct = traits.construct_curve_2_object();

  // Define a polynomial representing x.
  Polynomial x = CGAL::shift(Polynomial(1), 1);

  // Define a container storing all arcs.
  std::vector<Traits::Curve_2>  arcs;

//Create the arcs (C1, C'1) of the rational functions (y = 1 / x, y = -1 / x).
  Polynomial P1(1);
  Polynomial minusP1(-P1);
  Polynomial Q1 = x;
  arcs.push_back(construct(P1, Q1));
  arcs.push_back(construct(minusP1, Q1));

  // Create the bounded segments (C2, C'2) of the parabolas (y = -4*x^2 + 3)
  // and (y = 4*x^2 - 3), defined over [-sqrt(3)/2, sqrt(3)/2].
  Polynomial P2 = -4*x*x+3;
  Polynomial minusP2 = -P2;
```

```
std :: vector<std :: pair<Alg_real, int> > roots;
ak1.solve_1_object()(P2, std::back_inserter(roots));//[-sqrt(3)/2, sqrt(3)/2]
arcs.push_back(construct(P2, roots[0].first, roots[1].first));
arcs.push_back(construct(minusP2, roots[0].first, roots[1].first));

// Create the arcs (C3, C'3) of (i) the rational function (y = 1 / 2*x) for
// x > 0, and (ii) the rational function (y = -1 / 2*x) for x < 0.
Polynomial P3(1);
Polynomial minusP3(-P3);
Polynomial Q3 = 2*x;
arcs.push_back(construct(P3, Q3, Alg_real(0), true));
arcs.push_back(construct(minusP3, Q3, Alg_real(0), false));

// Construct the arrangement of the six arcs and print its size.
Arrangement  arr(&traits);
insert(arr, arcs.begin(), arcs.end());
print_unbounded_arrangement_size(arr);

  return 0;
}
```

The curve constructors have an additional advantage. They conveniently enable the provision of two polynomials that define a rational arc using rational coefficients. For example, let P and Q denote two polynomials with integral coefficients that define a rational arc at interest, and let P' and Q' denote two polynomials with rational coefficients that define the same rational arc; that is, the quotients P/Q and P'/Q' are identical. You can construct the rational arc providing the coefficients of P' and Q' to the constructor. In this case the constructor normalizes the coefficients and generates the desired polynomials P and Q. To this end, the curve constructors of both types, namely **Curve_2** and **X_monotone_curve_2**, have operators that accept ranges of polynomial coefficients as well as polynomials. The coefficients in a given range must be in the order of the degrees of the corresponding variables starting from the constant term.

5.4.4   A Traits Class for Planar Bézier Curves

A planar *Bézier curve* B is a parametric curve defined by a sequence of *control points* $p_0, \ldots, p_n$ as follows:

$$B(t) = (X(t), Y(t)) = \sum_{k=0}^{n} p_k \cdot \frac{n!}{k!(n-k)!} \cdot t^k (1-t)^{n-k} \ ,$$

where $t \in [0, 1]$. The degree of the curve is therefore n; namely, $X(t)$ and $Y(t)$ are polynomials of degree n. Bézier curves have numerous applications in computer graphics and solid modelling. They are used, for example, in free-form sketches and for defining the true-type fonts.

Using the **Arr_Bezier_curve_traits_2<RatKernel, AlgKernel, NtTraits>** class template you can construct and maintain arrangements induced by Bézier curves (including self-intersecting Bézier curves). The curves are given by rational control points (a sequence of objects of the **RatKernel::Point_2** type). The template parameters are the same ones used by the **Arr_conic_traits_2** class template. Here, the use of the **CORE_algebraic_number_traits** class is also recommended with Cartesian kernels instantiated with the **Rational** and **Algebraic** number types defined by this class. The examples given in this book use the type definitions listed below. These types are defined in the header file **arr_Bezier.h**.

```
#include <CGAL/Cartesian.h>
#include <CGAL/CORE_algebraic_number_traits.h>
#include <CGAL/Arr_Bezier_curve_traits_2.h>
```

```
#include <CGAL/Arrangement_2.h>

typedef CGAL::CORE_algebraic_number_traits          Nt_traits;
typedef Nt_traits::Rational                         NT;
typedef Nt_traits::Rational                         Rational;
typedef Nt_traits::Algebraic                        Algebraic;
typedef CGAL::Cartesian<Rational>                   Rat_kernel;
typedef CGAL::Cartesian<Algebraic>                  Alg_kernel;
typedef Rat_kernel::Point_2                         Rat_point;
typedef CGAL::Arr_Bezier_curve_traits_2<Rat_kernel, Alg_kernel, Nt_traits>
                                                    Traits;
typedef Traits::Curve_2                             Bezier_curve;
typedef CGAL::Arrangement_2<Traits>                 Arrangement;
```

As mentioned above, we assume that the coordinates of all control points that define a Bézier curve are rational numbers, so both $X(t)$ and $Y(t)$ are polynomials with rational coefficients. The intersection points between curves are, however, algebraic numbers, and their exact computation is time-consuming. The traits class therefore contains a layer of geometric filtering that performs all computations in an approximate manner whenever possible. Thus, it resorts to exact computations only when the approximate computation fails to produce an unambiguous result. Most arrangement vertices are therefore associated with approximated points. You cannot obtain the coordinates of such points as algebraic numbers. However, access to the approximate coordinates is possible; see the Reference Manual for the exact interface of the Point_2, Curve_2, and X_monotone_curve_2 types defined by the traits class.

Every instance of the Arr_Bezier_curve_traits_2 class template models the concept ArrangementTraits_2 (but it does not model the ArrangementLandmarkTraits_2 concept).[19]

Example: The example below reads a set of Bézier curves from an input file, where each file is specified by an integer stating its number of control points, followed by the sequence of control points given as integer or rational coordinates. By default, the program uses the Bezier.dat file, which contains ten curves of degree 5 each; their resulting arrangement is depicted in the figure to the right.

```
// File: ex_Bezier_curves.cpp

#include "arr_Bezier.h"
#include "arr_print.h"
#include "read_objects.h"

int main(int argc, char* argv[])
{
    // Get the name of the input file from the command line, or use the default
    // Bezier.dat file if no command-line parameters are given.
    const char* filename = (argc > 1) ? argv[1] : "Bezier.dat";

    // Read the Bezier curves.
    std::list<Bezier_curve>  curves;
    read_objects<Bezier_curve>(filename, std::back_inserter(curves));
```

---

[19] It also models the refined concept ArrangementDirectionalXMonotoneTraits_2 defined in Chapter 8.

```
    // Construct the arrangement.
    Arrangement    arr;
    insert(arr, curves.begin(), curves.end());
    print_arrangement_size(arr);

    return 0;
}
```

5.4.5 A Traits Class for Planar Algebraic Curves of Arbitrary Degree

The *2D Arrangements* package of Version 3.7 included for the first time a traits class that supports planar algebraic curves of arbitrary degree. The traits class, namely `Arr_algebraic_segment_traits_2`, is based on the `Algebraic_kernel_d_1` class template, which models the algebraic-kernel concept AlgebraicKernel_d_1; see Section 5.4.3. The traits class handles (i) algebraic curves and (ii) continuous (weakly) x-monotone segments of algebraic curves, which, however, are not necessarily maximal. (A formal definition is given below.) Observe that non-x-monotone segments are not supported. Still, it is the traits class that supports the most general type of curves among the traits classes included in the package.

Recall that an algebraic curve C in the plane is defined as the (real) zero set of a bivariate polynomial $f(x, y)$. The curve is uniquely defined by f, although several polynomials may define the same curve. We call f a *defining polynomial* of C.

A formal definition of (weakly) x-monotone segments of algebraic curves follows. A point p on a curve $C_f \subset \mathbb{R}^2$ (with f its defining polynomial) is called *semi-regular* if, locally around p, C_f can be written as a function graph of some continuous function in x or in y. (We also say that p is parameterizable in x or y, respectively.) The only two cases of non-semi-regular points are isolated points and self-intersections. A *segment* of a curve in this context is a closed and continuous point set such that each interior point is semi-regular. It follows that a segment is either vertical or a closed connected point set, with all interior points parameterizable in x.

Every instance of the `Arr_algebraic_segment_traits_2<Coefficient>` class template models the ArrangementTraits_2 and ArrangementOpenBoundaryTraits_2 concepts (but it does not model the ArrangementLandmarkTraits_2 concept). The template parameter `Coefficient` determines the type of the scalar coefficients of the polynomial. Currently supported integral number types are `Gmpz`, `leda::integer`, and `CORE::BigInt`. This is reflected in the statements included in the header file **integer_type.h**, the listings of which are omitted here. This header file is used by the two example programs listed in this section. The template parameter `Coefficient` can be substituted in addition with an instance of the `Sqrt_extension<NT,Root>` class template, where the template parameters `NT` and `Root` are substituted in turn with one of the integral number types above. Finally, the template parameter `Coefficient` can be substituted also with a rational number type, where the type of the numerator and denominator is one of the types above.

The type `Curve_2` nested in the `Arr_algebraic_segment_traits_2` class template defines an algebraic curve. An object of this type can be constructed by the `Construct_curve_2` functor also nested in the class template. Its function operator accepts as an argument an object of type `Polynomial_2`, nested as well in the traits class-template. The type `Polynomial_2` models the concept Polynomial_d. An object of the nested type `Polynomial_2` represents a bivariate polynomial. It can be constructed in a few convenient ways, some are exemplified by the programs listed below. Consult the reference guide for the complete set of options.

Example: The example below computes the arrangement depicted in Figure 5.5a. The arrangement is induced by four algebraic curves, C_1, C_2, C_3, and C_4, of degrees 1, 2, 3, and 6, respectively. For each curve the defining polynomial is constructed first. Then, the algebraic curve is constructed using the `Construct_curve_2` functor. Finally, the curve is inserted into the arrangement.

```
// File: ex_algebraic_curves.cpp

#include <iostream>
```

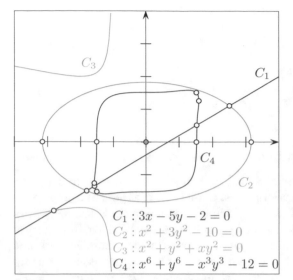

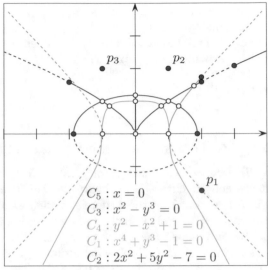

$C_1 : 3x - 5y - 2 = 0$
$C_2 : x^2 + 3y^2 - 10 = 0$
$C_3 : x^2 + y^2 + xy^2 = 0$
$C_4 : x^6 + y^6 - x^3y^3 - 12 = 0$

$C_5 : x = 0$
$C_3 : x^2 - y^3 = 0$
$C_4 : y^2 - x^2 + 1 = 0$
$C_1 : x^4 + y^3 - 1 = 0$
$C_2 : 2x^2 + 5y^2 - 7 = 0$

(a) An arrangement of four algebraic curves, as constructed by the program coded in ex_algebraic_curves.cpp.

(b) An arrangement of five algebraic segments and three isolated points, as constructed by the program coded in algebraic_segments.cpp. The supporting curves are drawn as dashed lines.

Fig. 5.5: Arrangements of algebraic curves.

```
#include <CGAL/basic.h>
#include <CGAL/Gmpz.h>
#include <CGAL/Arrangement_2.h>
#include <CGAL/Arr_algebraic_segment_traits_2.h>

#include "integer_type.h"
#include "arr_print.h"

typedef CGAL::Arr_algebraic_segment_traits_2<Integer>  Traits;
typedef CGAL::Arrangement_2<Traits>                    Arrangement;
typedef Traits::Curve_2                                Curve;
typedef Traits::Polynomial_2                           Polynomial;

int main()
{
  CGAL::set_pretty_mode(std::cout);                 // for nice printouts.

  Traits         traits;
  Arrangement arr(&traits);

  // Functor to create a curve from a Polynomial.
  Traits::Construct_curve_2 construct_cv = traits.construct_curve_2_object();

  Polynomial x = CGAL::shift(Polynomial(1), 1, 0);
  Polynomial y = CGAL::shift(Polynomial(1), 1, 1);

  // Construct an unbounded line (C1) with the equation 3x - 5y - 2 = 0.
  Polynomial f1 = 3*x - 5*y - 2;
  Curve cv1 = construct_cv(f1);
```

```
std::cout << "Inserting curve " << f1 << std::endl;
CGAL::insert(arr, cv1);

// Construct the ellipse (C2) with the equation x^2 + 3*y^2 - 10 = 0.
Polynomial f2 = CGAL::ipower(x, 2) + 3*CGAL::ipower(y, 2) - 10;
Curve cv2 = construct_cv(f2);
std::cout << "Inserting curve " << f2 << std::endl;
CGAL::insert(arr, cv2);

// Construct a cubic curve (C3) with the equation x^2 + y^2 + xy^2 = 0,
// with isolated point at (0,0) and vertical asymptote at x = 1.
Polynomial f3 = CGAL::ipower(x, 2) + CGAL::ipower(y, 2) +
  x*CGAL::ipower(y, 2);
Curve cv3 = construct_cv(f3);
std::cout << "Inserting curve " << f3 << std::endl;
CGAL::insert(arr, cv3);

// Construct a curve of degree 6 (C4) with the equation
// x^6 + y^6 - x^3y^3 - 12 = 0.
Polynomial f4 = CGAL::ipower(x, 6) + CGAL::ipower(y, 6) -
  CGAL::ipower(x, 3)*CGAL::ipower(y, 3) - 12;
Curve cv4 = construct_cv(f4);
std::cout << "Inserting curve " << f4 << std::endl;
CGAL::insert(arr, cv4);

print_arrangement_size(arr);               // print the arrangement size
return 0;
}
```

The `Arr_algebraic_segment_traits_2` class template carries state to expedite some of its computations. Thus, it is essential to have only one copy of the traits object during the life time of a program that utilizes this traits class. To this end, the example above uses the constructor of the `Arrangement_2` data structure that accepts the traits object as input. Carrying state is not a unique property of the `Arr_algebraic_segment_traits_2` class template; it is common to many traits classes, especially to traits classes that handle algebraic curves. Therefore, as a general rule, if your application requires direct access to a traits object, define it locally, and pass it to the constructor of the `Arrangement_2` data structure to avoid the construction of a duplicate traits-class object.

A weakly x-monotone segment of an algebraic curve is represented by the `X_monotone_curve_2` type nested in the traits class-template. You can construct such segments in two ways as follows: (i) using the `Make_x_monotone_2` functor or (ii) using the `Construct_x_monotone_segment_2` functor. Both functors are nested in the traits class-template. The former is required by the concept ArrangementTraits_2 our traits class models; see Section 5.1.3. The latter enables the construction of individual segments. The `X_monotone_curve_2` type represents weakly x-monotone segments of a curve; however, for technical reasons, segments may need to be further subdivided into several sub-segments, called terminal segments. Therefore, `Construct_x_monotone_segment_2` constructs a sequence of `X_monotone_curve_2` objects, whose union represents the desired weakly x-monotone segment. The function operator of the `Construct_x_monotone_segment_2` functor accepts as arguments the underlying algebraic curve, the leftmost point of the segment, the rightmost point of the segment, and an output iterator associated with a container of output terminal segments. The function operator is overloaded. In addition to the variant above, there is one that accepts the underlying algebraic curve, a single point p, an enumerator that delimits the segment, and an output iterator. It returns the maximal x-monotone segment that contains p. The enumerator specifies whether p is interior to the returned segment, its left endpoint, or its right endpoint. The third variant accepts only two delimiting points and an output iterator. It constructs line segments.

———— *advanced* ————

The subdivision into terminal segments is due to the internal representation of x-monotone segments, which is based on a vertical decomposition. We assume the defining polynomial f of the curve C to be *square-free*. That is, it contains no divisor g^2 of total degree greater than zero. We define a *(complex) critical point* $p \in \mathbb{C}^2$ by $f(p) = 0 = \frac{\partial f}{\partial y}(p)$. An x-coordinate $\alpha \in \mathbb{R}$ is *critical*, either if some critical point has x-coordinate α, or if the leading coefficient of f, considered as a polynomial in y, vanishes. In particular, vertical lines and isolated points of C can only take place at critical x-coordinates. Between two consecutive critical x-coordinates the curve decomposes into a finite number of x-monotone segments (the same holds on the left of the leftmost and on the right of the rightmost critical x-coordinate). The type `X_monotone_curve_2` is only capable of representing such segments (and sub-segments of them). Formally, a terminal segment is either a vertical line-segment or a segment of an x-monotone curve whose x-range does not contain critical points in its interior. The figure below depicts a quartic curve and its decomposition into terminal segments. Notice that six vertices split the curve into x-monotone segments, and four additional vertices further split the corresponding x-monotone segments into terminal segments.

———— *advanced* ————

The type `Algebraic_real_1` must be defined by any model of the AlgebraicKernel_d_1 concept. The traits class-template `Arr_algebraic_segment_traits_2` exploits an instance of the `Algebraic_kernel_d_1` class template, which models the concept AlgebraicKernel_d_1. The exploited instance is nested in the traits class-template. You can use this model to create algebraic numbers as roots of univariate polynomials, and process them, for instance, compare them or approximate them to any precision. See the documentation of the concept AlgebraicKernel_d_1 for more information. Coordinates of points are represented by the type `Algebraic_real_1` nested in the traits class-template. This type is defined as the corresponding type nested in the instance of `Algebraic_kernel_d_1`.

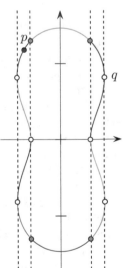

You can construct an object of type `Point_2` using the `Construct_point_2` functor nested in the traits class-template. Its function operator is overloaded with a couple of variants that accepts the x and y coordinates of the point. Their types must be either `Algebraic_real_1` or `Coefficient`. Another efficient variant accepts a triple $\langle x', C, i \rangle$, which identifies the ith point (counted from the bottom) in the fiber of C at the x-coordinate x'.[20] In the example depicted in the figure to the right, if x_p denotes the x-coordinate of p, and C represents the algebraic curve, then p could be represented by $\langle x_p, C, 3 \rangle$. If x_q is the x-coordinate of q, then $\langle x_q, C, 1 \rangle$ is a valid representation of q. Points are presented internally using the triple described above. The y-coordinates of points are not explicitly stored. Although the y-coordinate of a point represented by an object of the nested type `Algebraic_real_1` can be obtained, we advise caution with that option, since computing an explicit representation of the y-coordinate can be rather expensive.

 Try: The curve depicted in the figure to the right belongs to a family of curves called Ovals of Cassini. A curve in this family is the locus of points such that for each point the product of its distances from two fixed points, a distance $2a$ apart, is a constant b^2. In the figure above $a = 1$ and $b^2 = \sqrt{4/3}$. The polynomial $(y^2 + x^2 + 1)^2 - 4y^2 = 4/3$ defines the curve. Write a program that constructs an arrangement induced by this curve. Verify that the resulting arrangement consists of 2 faces, 10 edges, and 10 vertices.

---

[20]The fiber of a curve C at some x-coordinate x' is the set of all points on C with x-coordinate x'. Formally, for a curve C and $x' \in \mathbb{R}$, the fiber of C at x' is $C \cap \{(x', b) \mid b \in \mathbb{R}\}$.

Example: The program listed below exemplifies the method to construct points and the various methods to construct algebraic segments. The computed arrangement is depicted in Figure 5.5b.

```cpp
// File: ex_algebraic_segments.cpp

#include <iostream>

#include <CGAL/basic.h>
#include <CGAL/Gmpz.h>
#include <CGAL/Arrangement_2.h>
#include <CGAL/Arr_algebraic_segment_traits_2.h>

#include "integer_type.h"
#include "arr_print.h"

typedef CGAL::Arr_algebraic_segment_traits_2<Integer> Traits;
typedef CGAL::Arrangement_2<Traits>                    Arrangement;
typedef Traits::Curve_2                                Curve;
typedef Traits::Polynomial_2                           Polynomial;
typedef Traits::Algebraic_real_1                       Algebraic_real;
typedef Traits::X_monotone_curve_2                     X_monotone_curve;
typedef Traits::Point_2                                Point;

int main()
{
  Traits   traits;

  Traits::Make_x_monotone_2 make_xmon = traits.make_x_monotone_2_object();
  Traits::Construct_curve_2 ctr_cv = traits.construct_curve_2_object();
  Traits::Construct_point_2 ctr_pt = traits.construct_point_2_object();
  Traits::Construct_x_monotone_segment_2 construct_xseg =
    traits.construct_x_monotone_segment_2_object();

  Polynomial x = CGAL::shift(Polynomial(1), 1, 0);
  Polynomial y = CGAL::shift(Polynomial(1), 1, 1);

  // Construct a curve (C1) with the equation x^4+y^3-1=0.
  Curve cv1 = ctr_cv(CGAL::ipower(x, 4) + CGAL::ipower(y, 3) - 1);
  // Construct all x-monotone segments using the Make_x_mononotone functor.
  std::vector<CGAL::Object> pre_segs;
  make_xmon(cv1, std::back_inserter(pre_segs));
  // Cast all CGAL::Objects into X_monotone_segment
  // (the vector might also contain Point objects for isolated points,
  // but not in this case).
  std::vector<X_monotone_curve> segs;
  for (size_t i = 0; i < pre_segs.size(); ++i) {
    X_monotone_curve curr;
    bool check = CGAL::assign(curr, pre_segs[i]);
    CGAL_assertion(check);
    segs.push_back(curr);
  }
  // Construct an ellipse (C2) with the equation 2*x^2+5*y^2-7=0.
  Curve cv2 = ctr_cv(2*CGAL::ipower(x,2)+5*CGAL::ipower(y,2)-7);
```

```
// Construct point on the upper arc (counting of arc numbers starts with 0).
Point p11 = ctr_pt(Algebraic_real(0), cv2, 1);

construct_xseg(cv2, p11, Traits::POINT_IN_INTERIOR,
               std::back_inserter(segs));

// Construct a vertical cusp (C3) with the equation x^2−y^3=0.
Curve cv3 = ctr_cv(CGAL::ipower(x, 2)−CGAL::ipower(y, 3));

// Construct a segment containing the cusp point.
// This adds two X_monotone_curve objects to the vector,
// because the cusp is a critical point.
Point p21 = ctr_pt(Algebraic_real(−2), cv3, 0);
Point p22 = ctr_pt(Algebraic_real(2), cv3, 0);
construct_xseg(cv3 ,p21, p22, std::back_inserter(segs));

// Construct an unbounded curve, starting at x=3.
Point p23 = ctr_pt(Algebraic_real(3), cv3, 0);
construct_xseg(cv3, p23, Traits::MIN_ENDPOINT, std::back_inserter(segs));

// Construct another conic (C4) with the equation y^2−x^2+1=0.
Curve cv4 = ctr_cv(CGAL::ipower(y,2)−CGAL::ipower(x,2)+1);

Point p31 = ctr_pt(Algebraic_real(2), cv4, 1);
construct_xseg(cv4, p31, Traits::MAX_ENDPOINT, std::back_inserter(segs));

// Construct a vertical segment (C5).
Curve cv5 = ctr_cv(x);
Point v1 = ctr_pt(Algebraic_real(0), cv3, 0);
Point v2 = ctr_pt(Algebraic_real(0), cv2, 1);
construct_xseg(cv5, v1, v2, std::back_inserter(segs));

Arrangement arr(&traits);
CGAL::insert(arr, segs.begin(), segs.end());

// Add some isolated points (must be wrapped into CGAL::Object).
std::vector<CGAL::Object> isolated_pts;
// p1
isolated_pts.push_back(CGAL::make_object(ctr_pt(Algebraic_real(2), cv4,
                                                0)));
// p2
isolated_pts.push_back(CGAL::make_object(ctr_pt(Integer(1), Integer(2))));
// p3
isolated_pts.push_back(CGAL::make_object(ctr_pt(Algebraic_real(−1),
                                                Algebraic_real(2))));
CGAL::insert(arr, isolated_pts.begin(), isolated_pts.end());

print_arrangement_size(arr);                    // print the arrangement size
return 0;
}
```

5.5   Traits-Class Decorators

Assume that we are given two sets of polyline curves that represent rivers and roads, respectively. We can construct their arrangement and compute their intersection points. However, we cannot tell the difference between an intersection of two roads (namely a *junction*) and an intersection of a road and a river (namely a *bridge*) without storing the type of each curve (*road* or *river* in our case), and probably also its name, in an accessible place, such that this information can be retrieved later on. If an efficient retrieval operation is desired, it seems that we cannot use the `Arr_polyline_traits_2` class template as is, and have to write a new traits class from scratch— or at least write a new traits class that inherits from `Arr_polyline_traits_2` and extends its curve type with the desired information. The CGAL *2D Arrangements* package contains traits-class *decorators* that do exactly that for your convenience.

Geometric traits-class decorators allow you to attach auxiliary data to the geometric objects (curves and points). The data is automatically manipulated by the decorators and distributed to the constructed geometric entities. Additional information can alternatively be maintained by extending the vertex, halfedge, or face types provided by the DCEL class and used by the arrangement; see Section 6.2 for details. In many cases, however, it is convenient to attach the data to the curve itself, exploiting the automatic proliferation of the additional data fields from each curve to all its induced subcurves. Moreover, as two halfedges are associated with a single curve, storing the data once in the curve record either saves space or avoids an indirect access from one halfedge to its twin.

The *2D Arrangements* package includes a traits-class decorator used to attach a data field to curves and to x-monotone curves. It is a class template named `Arr_curve_data_traits_2<BaseTraits, XMonotoneCurveData, Merge, CurveData, Convert>` parameterized by a base-traits class, which must be substituted with one of the geometric traits models described in the previous sections or with a user-defined traits model, when the decorator is instantiated. The curve-data decorator derives from the base-traits class, and in particular inherits its `Point_2` type. The remaining nested types are defined as follows:

- `Curve_2` is derived from the basic type `BaseTraits::Curve_2`, extending it by an extra field of type `CurveData`.

- `X_monotone_curve_2` is derived from the basic type `BaseTraits::X_monotone_curve_2`, extending it by an extra field of type `XMonotoneCurveData`, which must model the concepts CopyConstructible, Comparable,[21] and DefaultConstructible. The latter ensures that every instance of the class template `Arr_curve_data_traits_2` obtained by substituting the `BaseTraits` template parameter with a model of the ArrangementLandmarkTraits_2 concept models the ArrangementLandmarkTraits_2 concept as well.

Note that the nested types `Curve_2` and `X_monotone_curve_2` are not necessarily the same, even if the `BaseTraits::Curve_2` and `BaseTraits::X_monotone_curve_2` are (as in the case of the line-segment traits class, for example). The extended curve types support the additional methods `data()` and `set_data()` for accessing and modifying the data field.

You can create an extended curve (or an extended x-monotone curve) from a basic curve and a curve-data object. When curves are inserted into an arrangement, they may be split, and the decorator handles their data fields automatically as follows:

- When a curve is subdivided into x-monotone subcurves, its data field of type `CurveData` is converted to an object of type `XMonotoneCurveData` using the `Convert` functor. The object is automatically associated with each of the resulting x-monotone subcurves.

  By default the `CurveData` type is identical to the `XMonotoneCurveData` type, and the conversion functor `Convert` is trivially defined. In this case the data field associated with the original curve is just duplicated and stored with the x-monotone subcurves.

---

[21] A model of the Comparable concept must support the equality operator.

- When an x-monotone curve is split into two (typically, when it intersects another curve), the decorator class automatically copies its data field to both resulting subcurves.

- When two x-monotone curves, C_1 and C_2, intersect, the result may include overlapping sections represented as x-monotone curves. In this case the data fields of C_1 and C_2 are merged into a single **XMonotoneCurveData** object using the **Merge** functor, which is supplied as a parameter to the traits class-template. The resulting object is assigned to the data field of the overlapping subcurves.

- Merging two x-monotone curves is allowed only when (i) the two curves are geometrically mergeable—that is, the base-traits class allows them to merge—and (ii) the two curves store the same data field.

Example: Another decorator supported by the *2D Arrangements* package is the **Arr_consolidated_curve_data_traits_2<BaseTraits, Data>** class template. It derives from the **Arr_curve_data_traits_2** class template, and it extends the basic type **BaseTraits::Curve_2** by a single **Data** field, and the basic type **BaseTraits::X_monotone_curve_2** with a *set* of (distinct) data objects. The **Data** type must model the concept **Comparable** to ensure that each set contains only distinct data objects with no duplicates. When a curve with a data field d is subdivided into x-monotone subcurves, each subcurve is associated with a set $S = \{d\}$. In the case of an overlap between two x-monotone curves c_1 and c_2 with associated data sets S_1 and S_2, respectively, the overlapping subcurve is associated with the consolidated set $S_1 \cup S_2$.

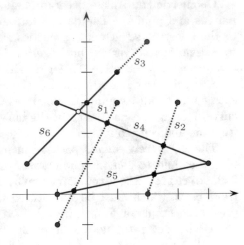

The example below uses **Arr_segment_traits_2** as the base-traits class, attaching an additional *color* field to the segments using the consolidated curve-data traits class. A color may be either *blue* or *red*. Having constructed the arrangement of colored segments, as depicted in the figure on the previous page, we detect the vertices that have incident edges mapped to both blue and red segments. These vertices, drawn as black discs in the figure, correspond to red-blue intersection points. We also locate the edge that corresponds to the overlap between a red line segment and a blue line segment (its endpoints are also drawn as black discs).

```
// File: ex_consolidated_curve_data.cpp

#include <CGAL/basic.h>
#include <CGAL/Arr_consolidated_curve_data_traits_2.h>

#include "arr_exact_construction_segments.h"

enum Segment_color { RED, BLUE };

typedef CGAL::Arr_consolidated_curve_data_traits_2<Traits, Segment_color>
                                                        Data_traits;
typedef Data_traits::Curve_2                            Colored_segment;
typedef CGAL::Arrangement_2<Data_traits>                Colored_arr;

int main()
{
  Colored_arr    arr;
```

```cpp
// Construct an arrangement containing three RED line segments.
insert(arr, Colored_segment(Segment(Point(-1, -1), Point(1, 3)), RED));
insert(arr, Colored_segment(Segment(Point(2, 0), Point(3, 3)), RED));
insert(arr, Colored_segment(Segment(Point(0, 3), Point(2, 5)), RED));

// Insert three BLUE line segments.
insert(arr, Colored_segment(Segment(Point(-1, 3), Point(4, 1)), BLUE));
insert(arr, Colored_segment(Segment(Point(-1, 0), Point(4, 1)), BLUE));
insert(arr, Colored_segment(Segment(Point(-2, 1), Point(1, 4)), BLUE));

//Go over all vertices and print just the ones corresponding to intersection
// points between RED segments and BLUE segments. Skip endpoints of
// overlapping sections.
Colored_arr::Vertex_const_iterator    vit;
for (vit = arr.vertices_begin(); vit != arr.vertices_end(); ++vit) {
  // Go over the current-vertex incident-halfedges and examine their colors.
  bool         has_red = false, has_blue = false;
  Colored_arr::Halfedge_around_vertex_const_circulator eit, first;
  eit = first = vit->incident_halfedges();
  do {
    // Get the color of the current halfedge.
    if (eit->curve().data().size() == 1) {
      Segment_color color = eit->curve().data().front();
      if (color == RED)       has_red = true;
      else if (color == BLUE) has_blue = true;
    }
  } while (++eit != first);

  // Print the vertex only if incident RED and BLUE edges were found.
  if (has_red && has_blue) {
    std::cout << "Red_intersect_blue_at_(" << vit->point() << ")"<<std::endl;
  }
}


// Locate the edges that correspond to a red-blue overlap.
Colored_arr::Edge_iterator    eit;
for (eit = arr.edges_begin(); eit != arr.edges_end(); ++eit) {
//Go over the incident edges of the current vertex and examine their colors.
  bool         has_red = false, has_blue = false;

  Data_traits::Data_container::const_iterator   it;
  for (it=eit->curve().data().begin(); it != eit->curve().data().end();++it)
  {
    if (*it == RED)        has_red = true;
    else if (*it == BLUE) has_blue = true;
  }

  // Print the edge only if it corresponds to a red-blue overlap.
  if (has_red && has_blue)
    std::cout << "Red_overlap_blue_at_[" << eit->curve() << "]"<<std::endl;
}

  return 0;
}
```

Example: The example below uses `Arr_polyline_traits_2` as the base-traits class, attaching an additional *name* field to each polyline using the generic curve-data traits class. It constructs an arrangement of four polylines, named *A*, *B*, *C*, and *D*, as illustrated in the figure to the right. In the case of overlaps, it simply concatenates the names of the overlapping polylines. At the end of the program the curve associated with the edges that correspond to overlapping polylines are replaced with geometrically equivalent curves, but with different data fields.

```
// File: ex_generic_curve_data.cpp

#include <string>

#include <CGAL/basic.h>
#include <CGAL/Arr_curve_data_traits_2.h>

#include "arr_polylines.h"

typedef std::string Name;                    // The name-field type.

struct Merge_names {
  Name operator()(const Name& s1, const Name& s2) const
  { return(s1 + "_" + s2); }
};

typedef CGAL::Arr_curve_data_traits_2<Traits, Name, Merge_names>
                                             Ex_traits;
typedef Ex_traits::Curve_2                   Ex_polyline;
typedef Ex_traits::X_monotone_curve_2        Ex_x_monotone_polyline;
typedef CGAL::Arrangement_2<Ex_traits>       Ex_arrangement;

int main()
{
  // Construct an arrangement of four polylines named A—D.
  Ex_arrangement  arr;

  Point pts1[5] = {Point(0, 0), Point(2, 4), Point(3, 3), Point(4, 4),
                   Point(6, 0)};
  insert(arr, Ex_polyline(Polyline(pts1, pts1 + 5), "A"));

  Point pts2[3] = {Point(1, 5), Point(3, 3), Point(5, 5)};
  insert(arr, Ex_polyline(Polyline(pts2, pts2 + 3), "B"));

  Point pts3[4] = {Point(1, 0), Point(2, 2), Point(4, 2), Point(5, 0)};
  insert(arr, Ex_polyline(Polyline(pts3, pts3 + 4), "C"));

  Point pts4[2] = {Point(0, 2), Point(6, 2)};
  insert(arr, Ex_polyline(Polyline(pts4, pts4 + 2), "D"));

  // Print all edges that correspond to an overlapping polyline.
```

```
Ex_arrangement::Edge_iterator      eit;
std::cout << "The overlapping subcurves:" << std::endl;
for (eit = arr.edges_begin(); eit != arr.edges_end(); ++eit) {
  if (eit->curve().data().length() > 1) {
    std::cout << " [" << eit->curve() << "] "
              << "named: " << eit->curve().data() << std::endl;

    // Modify the curve associated with the edge.
    arr.modify_edge(eit, Ex_x_monotone_polyline(eit->curve(), "overlap"));
  }
}
  return 0;
}
```

5.6 Application: Polygon Orientation

Chapter 8 introduces a package that supports Boolean set operations on point sets bounded by x-monotone curves. These operations expect each input point set to meet a specific set of requirements. In general, point sets with clockwise-oriented boundaries are considered invalid. Chapter 6 introduces an application that not only tests whether point sets are valid, but also attempts to correct invalid points sets. In this chapter, however, we are only interested in the determination of the orientation of polygon boundaries.

The polygon-orientation problem: *Given a sequence of x-monotone segments that compose a closed curve that represents the boundary of a point set, determine whether the orientation of the boundary of a point set is clockwise or counterclockwise.*

The function template `polygon_orientation(begin, end, traits)` accepts a range of x-monotone curve segments that compose a piecewise closed curve that bounds a point set. The target point of every curve segment must be equal to the source point of the next curve segment in the range. The target point of the last curve segment must be equal to the source point of the first curve segment. The type of the input iterators that define the range, namely, `begin` and `end`, must model the concept BidirectionalIterator. The function template also accepts a traits object as a third argument. The type of the traits object must model the concept ArrangementXMonotoneTraits_ 2, and in addition it must provide an operation that lexicographically compares the source and target points of an x-monotone-curve segment. As explained in detail in Chapter 8, this type in fact models the concept ArrangementDirectionalXMonotoneTraits_2, which refines the concept ArrangementXMonotoneTraits_2.

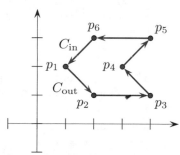

The `polygon_orientation()` function template returns either `CGAL::CLOCKWISE` or `CGAL::COUNTERCLOCKWISE` depending on the orientation of the piecewise closed curve that bounds the point set. The boundary of the polygon depicted in the figure to the right is oriented counterclockwise (thus, the polygon is valid). We search for the lexicographically smallest point p on the boundary of the polygon (p_1 in the figure to the right). The point p, like any other point on the boundary, is incident to two curve segments— the incoming curve segment C_{in} and the outgoing curve segment C_{out}, but in this case we know for a fact that C_{in} and C_{out} lie to the right of p. We compare the y-coordinate of the curve segments infinitesimally close to the right of p. If C_{out} is below C_{in}, the orientation of the boundary is counterclockwise. Otherwise, it is clockwise (and as a convention considered invalid). Notice that subjecting the test above to a point on the boundary that is not lexicographically the smallest may lead to the wrong conclusion, even if the incoming and outgoing curve segments of the point both lie to its right; see p_4 in the figure above.

A call to an instance of the **polygon_orientation()** function template will produce a correct result only if the piecewise closed curve that bounds the point set does not cross itself and does not overlap. However, degenerate cases where several vertices of the input polygon coincide, such as the one depicted in the figure to the right, are correctly handled. The function template is listed below. It is defined in the header file **polygon_orientation.h**.

```
#include <CGAL/basic.h>
#include <CGAL/enum.h>

using CGAL::Comparison_result;

template <typename Traits, typename Input_iterator>
CGAL::Orientation polygon_orientation(Input_iterator begin,
                                      Input_iterator end,
                                      const Traits& traits)
{
  const typename Traits::Compare_xy_2 cmp_xy = traits.compare_xy_2_object();
  const typename Traits::Compare_y_at_x_right_2 cmp_y_at_x_right =
    traits.compare_y_at_x_right_2_object();
  const typename Traits::Compare_endpoints_xy_2 cmp_endpoints =
    traits.compare_endpoints_xy_2_object();
  const typename Traits::Construct_min_vertex_2 ctr_min =
    traits.construct_min_vertex_2_object();

  // Locate the smallest polygon vertex.
  Input_iterator in_min, out_min;
  const typename Traits::Point_2* min_point = NULL;

  // Go over the x-monotone curves that compose the polygon boundary.
  Input_iterator curr = end; --curr;
  Input_iterator next = begin;
  bool next_directed_right = (cmp_endpoints(*curr) == CGAL::SMALLER);
  for (; next != end; curr = next++) {
    bool curr_directed_right = next_directed_right;
    next_directed_right = (cmp_endpoints(*next) == CGAL::SMALLER);

    // It is sufficient to look for a pair of curves with opposite directions.
    if (curr_directed_right || !next_directed_right) continue;

    const typename Traits::Point_2& point = ctr_min(*curr);
    Comparison_result res;
    if ((min_point == NULL) ||
        ((res = cmp_xy(point, *min_point)) == CGAL::SMALLER))
    {
      in_min = curr;
      out_min = next;
      min_point = &point;
      continue;
    }

    // Handle the degenerate case where 2 polygon vertices coincide.
    if (res == CGAL::EQUAL) {
      Comparison_result res_in = cmp_y_at_x_right(*curr, *in_min, point);
```

```
      Comparison_result res_out = cmp_y_at_x_right(*next, *out_min, point);
      CGAL_assertion((res_in != CGAL::EQUAL) && (res_out != CGAL::EQUAL));
      if (((res_in == CGAL::SMALLER) && (res_out == CGAL::SMALLER)) ||
          (((res_in == CGAL::SMALLER) && (res_out == CGAL::LARGER)) &&
           (cmp_y_at_x_right(*curr, *out_min, point) == CGAL::SMALLER)) ||
          (((res_in == CGAL::LARGER) && (res_out == CGAL::SMALLER)) &&
           (cmp_y_at_x_right(*next, *in_min, point) == CGAL::SMALLER)))
      {
        in_min = curr;
        out_min = next;
      }
    }
  }

  // Perform the comparison near the smallest vertex.
  Comparison_result res = cmp_y_at_x_right(*in_min, *out_min, *min_point);
  CGAL_assertion(res != CGAL::EQUAL);
  return (res == CGAL::SMALLER) ? CGAL::CLOCKWISE : CGAL::COUNTERCLOCKWISE;
}
```

The application comes with a **main()** function that extracts an ordered sequence of point set boundary points from an input file. It constructs a list of line segments that comprise the boundary of the point set, and invokes an instance of the **polygon_orientation()** function template above, passing the range of segments as input. The sequence of segments are guaranteed to be consistently oriented, as they are constructed from the ordered sequence of points. Notice that the last line segment connects the first and last points.

```
// File: polygon_orientation.cpp

#include<list>

#include <CGAL/Exact_predicates_exact_constructions_kernel.h>
#include <CGAL/Arr_segment_traits_2.h>

#include "arr_exact_construction_segments.h"
#include "read_objects.h"
#include "polygon_orientation.h"

typedef CGAL::Exact_predicates_exact_constructions_kernel Kernel;
typedef Kernel::FT                                        Number_type;
typedef CGAL::Arr_segment_traits_2<Kernel>                Traits;
typedef Traits::Point_2                                   Point;
typedef Traits::X_monotone_curve_2                        Segment;

int main(int argc, char* argv[])
{
  std::list<Point> points;
  const char* filename = (argc > 1) ? argv[1] : "polygon.dat";
  read_objects<Point>(filename, std::back_inserter(points));
  CGAL_assertion(points.size() >= 3);

  std::list<Segment> segments;
  std::list<Point>::const_iterator it = points.begin();
  const Point& first_point = *it++;
  const Point* prev_point = &first_point;
```

```
  while (it != points.end()) {
    const Point& point = *it++;
    segments.push_back(Segment(*prev_point, point));
    prev_point = &point;
  }
  segments.push_back(Segment(*prev_point, first_point));
  CGAL::Orientation orient =
    polygon_orientation(segments.begin(), segments.end(), Traits());
  std::cout << ((orient == CGAL::COUNTERCLOCKWISE) ?
               "Counterclockwise" : "Clockwise") << std::endl;
  return 0;
}
```

5.7   Bibliographic Notes and Remarks

The separation of the numerical/algebraic part of the computation from the combinatorial part is a fundamental guideline in the design of the *2D Arrangements* package as well as of other parts of CGAL, and it has repeatedly proved useful. It is carried out through the usage of a *traits class*. The original technique introduced by Myers [168] provides a convenient way to associate related types, enumerations, and functions with a type without requiring that they be defined as members of that type. Instead, they are defined as members of a traits class, which substitutes a template parameter of the class template when instantiated. In CGAL the notion of traits classes was expanded [115]. CGAL traits classes are used to associate also predicates and constructions of geometric objects with a type, and they might even carry state.

As in other chapters in this book, we show you how to use traits classes for various curves. The subject of traits-class implementation is only touched upon briefly. Additional discussions appear elsewhere, in particular in the book *Effective Computational Geometry for Curves and Surfaces* [29], Chapter 1, "Arrangements," and Chapter 3, "Algebraic Issues in Computational Geometry."

One of the advantages of the separation of numerics from combinatorics in the *2D Arrangements* package is that you can use the package for your special type of curves, provided that you supply the traits class for these curves. You do not need to know much about algorithms for arrangements to do that. You need, though, to understand the algebra. The two chapters of the book mentioned above [29] can serve as a starting point for that.

Approximation of complex curves simplifies interpolation, facilitates the extraction of numerical features of the curves, and may expedite computation in general. It is applied in areas such as pattern recognition and classification, point-based motion estimation, image understanding, and 3D reconstruction. The level of approximation depends on the application demands. For information on approximation techniques using line segments and circular arcs see, e.g., [121]. Information on approximation techniques using rational functions and polynomial functions can be found in, e.g., [183, Chapter 3].

The large number of geometry-traits models already implemented enables the efficient construction and maintenance of arrangements induced by many different types of curves. The *2D Arrangements* package itself contains several models of the geometry-traits concept; see Table 5.1. A few other models have been developed by other groups of researchers. Models are distinguished not only by the different families of curves they handle, but also by their suitability for constructing and maintaining arrangements with different characteristics. For example, there are two distinct models that handle line segments (see Section 5.2, and [213]) and another one that handles not only (bounded) line segments, but also rays and lines; see Section 5.2.3 and [21, 22]. There are traits models for non-linear curves, such as circular arcs (see Section 5.4.1 and [47, 209, 216]), conic curves (see Section 5.4.2 and [19, 60, 207]), cubic curves [57], and quartic curves that are the projection of the intersection of two quadric surfaces [24], and there are traits classes for arcs of graphs of rational univariate polynomial functions; see Section 5.4.3 and [140, 213]. There is

Table 5.1: Geometry traits models included in the *2D Arrangements* package.
① — ArrangementLandmarkTraits_2
② — ArrangementTraits_2
③ — ArrangementDirectionalXMonotoneTraits_2
④ — ArrangementOpenBoundaryTraits_2.
Recall that both ArrangementLandmarkTraits_2 (①) and ArrangementOpenBoundaryTraits_2 (④)
refine ArrangementBasicTraits_2 (②).

Model Name	Curve Family	Degree	Concepts
`Arr_non_caching_segment_basic_traits_2`	line segments	1	①
`Arr_non_caching_segment_traits_2`	line segments	1	①,②,③
`Arr_segment_traits_2`	line segments	1	①,②,③
`Arr_linear_traits_2`	line segments, rays, and lines	1	①,②,③,④
`Arr_circle_segment_traits_2`	line segments and circular arcs	≤ 2	②,③
`Arr_circular_line_arc_traits_2`	line segments and circular arcs	≤ 2	②
`Arr_conic_traits_2`	circles, ellipses, and conic arcs,	≤ 2	①,②,③
`Arr_rational_function_traits_2`	arcs of rational functions	≤ 2	②,④
`Arr_Bezier_curve_traits_2`	Bézier curves	$\leq n$	②,③
`Arr_algebraic_segment_traits_2`	algebraic curves	$\leq n$	②,③,④
`Arr_polyline_traits_2`	polylines	∞	①,②,③

a traits class that handles Bézier curves; see Section 5.4.4 and [109]. There is even a traits class that handles algebraic curves of arbitrary degree; see Section 5.4.5 and [56]. Michael Kerber, who is one of the authors of this traits class, provided the material for Section 5.4.5, including the two example programs and the two corresponding figures. Finally, there is a model that handles continuous piecewise linear curves, referred to as polylines; see Section 5.3.

As mentioned in Section 5.1.2, the multiplicity of intersection points is directly related to the relative order of the intersecting curves to the right and left of their intersection points, which in turn is exploited to expedite arrangement constructions. For example, the multiplicity of intersection between two curves is odd iff the curves swap their relative order at the intersection point. See [19] and [29, Chapter 2] for a generalization to multiple curves intersecting at a common point.

The traits adaptor class-template (see Section 5.1.6) resembles the *adapter* design pattern (notice the different spelling) as described in [83, Chapter 4]. It converts the interface of a class (a model of a geometry-traits concept in our case) into another interface that clients (the `Arrangement_2` class template, for example) expect.

Generic programming techniques [7] are used to implement the traits adaptor. In particular, we use tag dispatching—a mechanism that enables the controlled dispatching of specific implementations; this mechanism is used by, for example, STL in the implementation of the `std::advance` function template, which accepts an iterator and a number n as input and increments the iterator n times. Here, tag dispatching is used to select the most optimized implementation from a set of specializations; see e.g., [12]. BOOST is another example of a user of the tag dispatching technique.[22]

The traits decorator class-template (see Section 5.5) resembles the *decorator* design pattern as described in [83, Chapter 4]. It attaches additional responsibilities to a model of a geometry-traits concept.

---

[22]See `http://boost-spirit.com/dl_more/fusion_v2/libs/fusion/doc/html/fusion/notes.html`.

5.8   Exercises

5.1 Write a program that constructs arrangements of line segments that are the piecewise line segments of polylines. The graph of the arrangement should be identical to the graph of the arrangement induced by the polylines; see the example program `ex_generic_curve_data`. Compare the performance of the two programs on various inputs. Do you observe a difference?

5.2 Write a program that constructs arrangements of polylines read from an input file. It attaches a string field to each polyline just like the example program coded in `ex_generic_curve_data.cpp` does. This string is referred to as the main name. In addition, it attaches a string field to each segment of each polyline, referred to as the sub-name. Then, issue some vertical ray-shooting queries. For each query point p print the results as follows: If the ray hits an edge e, print the name of the polyline (or polylines in case of an overlap) associated with e and the name of the segment (or segments in case of an overlap) containing p. If the ray hits a vertex v, print the names of the polylines associated with the edge(s) incident to v, and for each polyline the name of the segment whose endpoint is p. Otherwise, report that no feature lies above the query point.

5.3 As mentioned in Section 5.6, a call to an instance of the `polygon_orientation(begin, end, traits)` function template will produce a correct result only if the piecewise closed curve represented by the range `[begin, end)` of x-monotone curves does not cross itself and does not overlap. However, (the interior of) a point set is well defined even if the piecewise closed curve that bounds it is self-overlapping, as long as it does not cross itself; see the point set depicted in the figure to the right.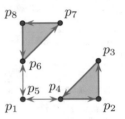

Augment the implementation of the function template `polygon_orientation()` to handle piecewise closed curves that are self-overlapping, but not self-crossing.

5.4 **[Project]** The polylines handled by the current implementation of the polyline traits class-template are restricted to being bounded piecewise linear curves. That is, all the curves in the sequence of piecewise curves of each polyline must be linear, and all the curves, including the first and the last curves in the sequence, must be bounded.

Augment the traits class-template that handles polylines to handle continuous linear curves, where the first and the last linear curve of each polyline can be unbounded.

5.5 **[Project]** Implement a traits class-template that handles poly-curves. A poly-curve is a continuous piecewise general curve that is not necessarily linear.

5.6 **[Project]** Given a family of curves compare the performance of all traits models that are able to handle the given family, for the families and models listed below. Verify that in most cases the traits model dedicated to handling some family is indeed more efficient than the other models that handle the same family. An exception to this rule of thumb is the algebraic traits-model, which is the last one developed. It exploits many advances made to the field of computational geometry that were not available at the time the other traits were developed.

(a) Compare the performance of the various traits models that handle circles, circular arcs, and line segments, namely, `Arr_circle_segment_traits_2`, `Arr_circular_line_arc_traits_2`, `Arr_conic_traits_2`, and `Arr_algebraic_segment_traits_2`.

(b) Compare the performance of the various traits models that handle algebraic curves defined by polynomials of two degrees (of the form $y = ax^2 + bx + c$), namely, `Arr_conic_traits_2`, `Arr_rational_arc_traits_2`, and `Arr_algebraic_segment_traits_2`.

(c) Compare the performance of the two traits models that handle bounded conic curves, namely, `Arr_conic_traits_2` and `Arr_algebraic_segment_traits_2`.

(d) Compare the performance of the two traits models that handle arcs of rational functions, namely, `Arr_rational_arc_traits_2` and `Arr_algebraic_segment_traits_2`.

(e) Compare the performance of the two traits models that handle Bézier curves, namely, `Arr_Bezier_curve_traits_2` and `Arr_algebraic_segment_traits_2`.

5.7 **[Project]** Pattern formations in organisms is a spectacular phenomenon observed in nature. One family of such patterns includes the regular arrangements of outer plant organs such as leaves, blossoms or seeds along branches or in a bud. In the botanical literature these patterns are called phyllotactic patterns. The most common phyllotactic pattern is the spiral pattern, where a single primordium is inserted at each node. A classical example of a spinal pattern is the arrangement of florets (or their mature form, seeds) in the capitula (commonly referred to as heads) of sunflowers. As the name suggests, it is possible to trace spiral curves in the pattern. These curves are called parastichies. The parastichies winding in the same direction around the stem with the same pitch constitute one family of parastichies, and two easily perceived sets of parastichies winding in opposite directions are called a parastichy pair. The numbers of parastichies in the two sets of a parastichy pair are consecutive Fibonacci numbers. The number of parastichies decreases towards the center. For more details see, e.g., [48, 122].

If you look carefully at the sunflower capitulum in the figure to the left, you can, for example, observe a parastichy pair of 55 spirals winding clockwise and 34 spirals winding counterclockwise.

In this exercise you are asked to construct an arrangement that resembles a spiral pattern induced by a parastichy pair. Parastichies have been shown to have the paths of logarithmic spirals. However, the arrangement you are asked to construct is induced by Fibonacci spirals, which well approximate logarithmic spirals on the one hand and are easier to handle on the other. A Fibonacci spiral is a piecewise circular curve comprising circular arcs (quarter circles) connecting the opposite

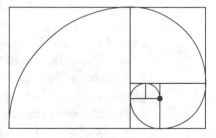

corners of squares in the tiling with squares whose sides are successive Fibonacci numbers in length. The Fibonacci spiral depicted in the figure above uses squares of sizes 1, 1, 2, 3, 5, and 8. The small red circle indicates the spiral source-point.

(a) Develop a function that constructs a single Fibonacci spiral arc. It accepts three

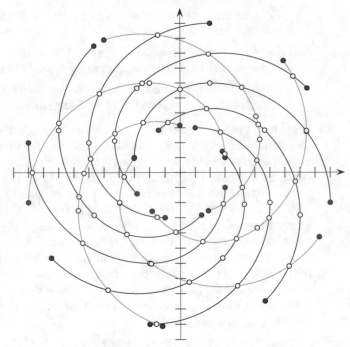

Fig. 5.6: An arrangement induced by a parastichy pair of Fibonacci spirals. The clockwise-winding set (blue) and counterclockwise-winding set (green) consists of eight and five spirals, respectively. Each spiral arc in both sets comprises two circular arcs, the fifth and the sixth circular arcs in the underlying complete Fibonacci spiral.

integers, k, m, and s, that define the resulting Fibonacci spiral arc as follows: Let $c_1, c_2, \ldots$ denote the circular arcs that sequentially compose the (unscaled) Fibonacci spiral. The desired Fibonacci spiral arc comprises $c'_k, c'_{k+1}, \ldots, c'_{k+m}$, where c'_i is the circular arc c_i whose radius is scaled by s.

(b) Develop a program that constructs an arrangement induced by Fibonacci spiral arcs. The program is given seven integers, k_1, m_1, s_1, k_2, m_2, s_2, and n and a rational number ϵ as input. The first integer triplet, k_1, m_1, and s_1, define the clockwise-winding spirals that is the prototype of spirals in the first set of the parastichy pair that induces the arrangement. The second integer triplet, k_2, m_2, and s_2, define the counterclockwise-winding spiral that is the prototype of spirals in the second set. The next input, n, indirectly specifies the number of spirals in the two sets as follows. Let f_i denote the ith Fibonacci number. Then, f_n and f_{n+1} are the numbers of spirals in the two sets, respectively. For example, the prototypes of the clockwise-winding and the counterclockwise-winding sets of spirals that induce the arrangement depicted in Figure 5.6 are both defined by the triplet $\langle 5, 2, 1 \rangle$.

First, the program constructs two prototype spiral arcs, say c_1 and c_2, winding in opposite orientations such that the source points of the underlying complete spirals of which coincide at the origin. Next, it rotates c_1 by the angles $\alpha'_1, \ldots, \alpha'_{f_n}$ about the origin to construct one set of f_n spirals, where α'_i approximates the angle $\alpha_i = i\frac{2\pi}{f_n}$ such that $|\sin\alpha'_i - \sin\alpha_i| \leq \epsilon$ and $|\cos\alpha'_i - \cos\alpha_i| \leq \epsilon$. Then, it rotates c_2 by the angles $\beta'_1, \ldots, \beta'_{f_n}$ about the origin to construct the second set of f_{n+1} spirals, where β'_i approximates the angle $\beta_i = i\frac{2\pi}{f_{n+1}}$ such that $|\sin\beta'_i - \sin\beta_i| \leq \epsilon$ and $|\cos\beta'_i - \cos\beta_i| \leq \epsilon$.

In order to stay in the realm of rational numbers, you may use the free function `rational_rotation_approximation()` provided by CGAL to obtain the desired $\sin\alpha'_i$ and $\cos\alpha'_i$ for $i = 1, \ldots, f_n$, and $\sin\beta'_i$ and $\cos\beta'_i$ for $i = 1, \ldots, f_{n+1}$. The implementation of this function is based on a method presented by Canny, Donald, and Ressler [33].

Chapter 6

Extending the Arrangement

Developing applications that use arrangements to solve problems that are a bit more complicated than the problems presented in previous chapters requires the ability to adapt the arrangement data structure to the application needs. One technique to do this is to extend the arrangement with auxiliary, usually non-geometric, data. In this chapter we describe several ways to extend an arrangement data structure.

6.1   The Notification Mechanism

In some cases it is essential to know exactly what happens inside a specific **Arrangement_2** object. For example, when a new curve is inserted into an arrangement, it may be necessary to keep track of the faces that are split due to this insertion operation. Other important examples are the point-location strategies that require auxiliary data structures (see Section 3.1.1), which must be kept up-to-date when the arrangement changes. The *2D Arrangements* package offers a mechanism that uses *observers* [83, Chapter 5]. The objective behind this mechanism is to define a one-to-many dependency between objects, so that when one object changes state, all its dependents are notified and updated automatically. The observed object does not know anything about the observers. It merely "publishes" information about changes when they occur. In our case observers can be attached to an arrangement object. An attached observer receives notifications about the changes this arrangement undergoes.

An observer object, the type of which is an instance of the **Arr_observer<Arrangement>** class template, stores a pointer to an arrangement object. When the **Arr_observer<Arrangement>** class template is instantiated, the **Arrangement** parameter must be substituted with the type of the arrangement object. The observer receives notifications *just before* a structural change occurs in the arrangement and *immediately after* such a change takes place. **Arr_observer** serves as a base class for other observer classes and defines a set of virtual notification functions, with default empty implementations. The set of functions can be divided into three categories, as follows:

1. Notifiers on changes that affect the entire topological structure of the arrangement. This category consists of two pairs (*before* and *after*) that notify the observer of the following changes:

   - The arrangement is cleared.
   - The arrangement is assigned with the contents of another arrangement.

2. Pairs of notifiers before and after a local change that occurs in the topological structure. Most notifier functions belong to this category. The relevant local changes include:

   - A new vertex is constructed and associated with a point.
   - An edge[1] is constructed and associated with an *x*-monotone curve.

---

[1] The term "edge" refers here to a pair of twin halfedges.

E. Fogel et al., *CGAL Arrangements and Their Applications*, Geometry and Computing 7, DOI 10.1007/978-3-642-17283-0_6, © Springer-Verlag Berlin Heidelberg 2012

- An edge is split into two edges.
- An existing face is split into two faces as a consequence of the insertion of a new edge.
- A hole is created in the interior of a face.
- Two holes are merged to form a single hole as a consequence of the insertion of a new edge.
- A hole is moved from one face to another as a consequence of a face split.
- Two edges are merged into one edge.
- Two faces are merged into one face as a consequence of the removal of an edge that used to separate them.
- One hole is split into two as a consequence of the deletion of an edge that used to connect the two components.
- A vertex is removed.
- An edge is removed.
- A hole is deleted from the interior of a face.

3. Notifiers on a structural change caused by a free function; see Sections 3.4 and 3.5 for a discussion on the free functions. This category consists of a single pair of notifiers, namely, `before_global_change()` and `after_global_change()`. Neither of these functions is invoked by methods of the `Arrangement_2` class template. Instead, they are called by the free functions themselves. It is implied that no point-location queries (or any other queries for that matter) are issued between the call to `before_global_change()` and the call to `after_global_change()`.

See the Reference Manual for a detailed specification of the `Arr_observer` class template and the prototypes of all notification functions.

Each arrangement object stores a list of pointers to `Arr_observer` objects. This list may be empty, in which case the arrangement does not have to notify any external class on the structural changes it undergoes. If, however, there are observers associated with the arrangement object, then whenever one of the structural changes listed in the first two categories above is about to take place, the arrangement object performs a *forward* traversal on this list and invokes the appropriate function of each observer. After the change takes place the observer list is traversed *backward* (from tail to head), and the appropriate notification function is invoked for each observer.

Concrete arrangement-observer classes should inherit from `Arr_observer`. When an observer object is constructed, it is attached to a valid arrangement supplied to the observer constructor, or alternatively the observer can be attached to the arrangement at a later time. When this happens, the observer object inserts itself into the observer list of the associated arrangement and starts receiving notifications whenever this arrangement changes thereafter. Subsequently, the observer object unregisters itself by removing itself from this list just before it is destroyed. Most concrete observer-classes do not need to use the full set of notifications. Thus, the bodies of all notification methods defined in the base class `Arr_observer` are empty. A concrete observer that inherits from `Arr_observer` needs to override only the relevant notification methods. The remaining methods are invoked when corresponding changes occur, but they do nothing.

The trapezoidal map RIC and the landmark point-location strategies both use observers to keep their auxiliary data structures up-to-date. In addition, you can define your own observer classes, inheriting from the base observer-class and overriding the relevant notification functions, as required by your application.

Example: The example below shows how to define and use an observer class. The observer in the example responds to changes in the arrangement faces. It prints a message whenever a face is split into two due to the insertion of an edge and whenever two faces merge into one due to the removal of an edge. The layout of the arrangement is depicted in the figure on the next page;

it comprises six line segments and eight edges (the horizontal segment s_h and the vertical segment s_v induce two edges each). The halfedge e_v is induced by the vertical segment s_v. First, it is associated with the (entire) segment, as obtained by the `insert()` function; see Line 40 in the code excerpt below. After the insertion of s_h (see Line 41) the halfedge is split. After the split e_v (drawn dashed in the figure) is associated with the lower split curve. Eventually, it is removed (along with its twin halfedge). Note the face-split notifications that are invoked as a consequence of the insertion of s_v and s_h and the face-merge notification that is invoked as a consequent of the removal of e_v.

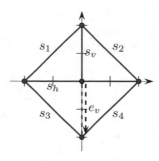

```cpp
1   // File: ex_observer.cpp
2
3   #include <CGAL/basic.h>
4   #include <CGAL/Arr_observer.h>
5
6   #include "arr_exact_construction_segments.h"
7   #include "arr_print.h"
8
9   // An observer that receives notifications of face splits and face mergers.
10  class Faces_observer : public CGAL::Arr_observer<Arrangement> {
11  public:
12    Faces_observer(Arrangement& arr) : CGAL::Arr_observer<Arrangement_2>(arr) {}
13
14    virtual void before_split_face(Face_handle, Halfedge_handle e)
15    {
16      std::cout << "-> The insertion of : [ " << e->curve()
17               << " ]  causes a face to split." << std::endl;
18    }
19
20    virtual void before_merge_face(Face_handle, Face_handle, Halfedge_handle e)
21    {
22      std::cout << "-> The removal of : [ " << e->curve()
23               << " ]  causes two faces to merge." << std::endl;
24    }
25  };
26
27  int main()
28  {
29    // Construct an arrangement containing a single face of a diamond shape.
30    Arrangement      arr;
31    Faces_observer obs(arr);
32    insert_non_intersecting_curve(arr, Segment(Point(-1, 0), Point(0, 1)));
33    insert_non_intersecting_curve(arr, Segment(Point(0, 1), Point(1, 0)));
34    insert_non_intersecting_curve(arr, Segment(Point(1, 0), Point(0, -1)));
35    insert_non_intersecting_curve(arr, Segment(Point(0, -1), Point(-1, 0)));
36
37    // Insert a vertical segment dividing the diamond into two, and a
38    // a horizontal segment further dividing the diamond into four.
39    Segment           s_v(Point(0, -1), Point(0, 1));
40    Halfedge_handle e_v = insert_non_intersecting_curve(arr, s_v);
41    insert(arr, Segment(Point(-1, 0), Point(1, 0)));
42    print_arrangement_size(arr);
43
```

```
44    // Remove a portion of the vertical segment.
45    remove_edge(arr, e_v);                // the observer will make a notification
46    print_arrangement_size(arr);
47
48    return 0;
49 }
```

Observers are especially useful when the Dcel records are extended with additional data-fields, since they help update the data stored in these fields, as the following sections reveal.

6.2   Extending the Dcel

For many applications of the *2D Arrangements* package it is necessary to store additional information (perhaps of non-geometric nature) with the arrangement features. Vertices are associated with `Point_2` objects and edges (halfedge pairs) are associated with `X_monotone_curve_2` objects, both defined by the traits class. Extending the geometric traits-class types by using a traits-class decorator, as explained in Section 5.5, might be a sufficient solution for some applications. However, the Dcel faces are not associated with any geometric object, so traits-class decorators cannot help here. Extending the Dcel face records comes in handy in such cases. As a matter of fact, it is possible to conveniently extend all Dcel records (namely, vertices, halfedges, and faces), which is advantageous in some cases.

All examples presented so far use the default Dcel; namely, they employ the `Arr_default_dcel<Traits>`. This is done implicitly, as an instance of this class template serves as the default for the `Dcel` parameter of the `Arrangement_2<Traits,Dcel>` class template; see Section 2.2. The default Dcel class associates points with vertices and x-monotone curves with halfedges, but nothing more. In this section we show how to use alternative Dcel types to extend the desired Dcel records.

6.2.1   Extending the Dcel Faces

The `Arr_face_extended_dcel<Traits,FaceData>` class template associates an auxiliary data-field of type `FaceData` to each face record in the Dcel. When the `Arrangement_2<Traits,Dcel>` class template is instantiated, substituting the `Dcel` parameter with an instance of this class template results in a nested `Face` type, the interface of which is extended with the access function `data()` and with the modifier `set_data()`. Using these extra functions it is straightforward to access and maintain the auxiliary face-data fields.

Note that the extra data-fields must be maintained by the user application. Users may choose to construct their arrangement, and may only then go over the faces and store data in the appropriate data-fields of the arrangement faces. However, in some cases the face data can only be computed when the face is created (split from another face or merged with another face). In such cases one can use an arrangement observer tailored for this task, which receives updates whenever a face is modified and sets its data fields accordingly.

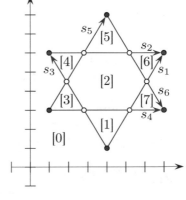

Example: The next example constructs an arrangement that contains seven bounded faces induced by six line segments, $s_1, \ldots, s_6$, as shown in the figure to the right. An observer gets notified each time a new face f is created, and it associates f with a running index, where the index of the unbounded face is 0. As a result, the faces are numbered in their creation order. These numbers are shown in brackets, and their order can easily be verified by examining the insertion order of the segments.

// File: ex_face_extension.cpp

```cpp
#include <CGAL/basic.h>
#include <CGAL/Arr_extended_dcel.h>
#include <CGAL/Arr_observer.h>

#include "arr_exact_construction_segments.h"

typedef CGAL::Arr_face_extended_dcel<Traits, unsigned int> Dcel;
typedef CGAL::Arrangement_2<Traits, Dcel>                  Ex_arrangement;

// An arrangement observer used to receive notifications of face splits and
// to update the indices of the newly created faces.
class Face_index_observer : public CGAL::Arr_observer<Ex_arrangement> {
private:
  unsigned int  n_faces;                  // the current number of faces

public:
  Face_index_observer(Ex_arrangement& arr) :
    CGAL::Arr_observer<Ex_arrangement>(arr), n_faces(0)
  {
    CGAL_precondition(arr.is_empty());
    arr.unbounded_face()->set_data(0);
  }

  virtual void after_split_face(Face_handle old_face, Face_handle new_face,
                                bool)
  {
    new_face->set_data(++n_faces);        // assign index to the new face
  }
};

int main()
{
  // Construct the arrangement containing two intersecting triangles.
  Ex_arrangement          arr;
  Face_index_observer  obs(arr);
  insert_non_intersecting_curve(arr, Segment(Point(4, 1), Point(7, 6)));
  insert_non_intersecting_curve(arr, Segment(Point(1, 6), Point(7, 6)));
  insert_non_intersecting_curve(arr, Segment(Point(4, 1), Point(1, 6)));
  insert(arr, Segment(Point(1, 3), Point(7, 3)));
  insert(arr, Segment(Point(1, 3), Point(4, 8)));
  insert(arr, Segment(Point(4, 8), Point(7, 3)));

  // Go over all arrangement faces and print the index of each face and its
  // outer boundary. The face index is stored in the data field.
  Ex_arrangement::Face_const_iterator            fit;
  Ex_arrangement::Ccb_halfedge_const_circulator  curr;
  std::cout << arr.number_of_faces() << " faces:" << std::endl;
  for (fit = arr.faces_begin(); fit != arr.faces_end(); ++fit) {
    std::cout << "Face no. " << fit->data() << ": ";
    if (fit->is_unbounded()) std::cout << "Unbounded." << std::endl;
    else {
      curr = fit->outer_ccb();
      std::cout << curr->source()->point();
```

```
        do std::cout << " —> " << curr->target()->point();
        while (++curr != fit->outer_ccb());
        std::cout << std::endl;
      }
    }

    return 0;
}
```

 Try: The particular observer used in the example above must be attached to an empty arrangement. Modify the observer class to handle the general case, so that the modified observer can be attached to a nonempty arrangement. In this case, instead of verifying that the arrangement is empty, the observer must reset the indices of all existing faces to 0.

6.2.2   Extending All the DCEL Records

As you continue to use arrangements to solve various problems you will find out that the ability to extend the face records is crucial. Perhaps less common, but also important to satisfy, is the need to extend the vertex and halfedge records as well. The **Arr_extended_dcel<Traits, VertexData, HalfedgeData, FaceData>** class template is used to associate auxiliary data-fields of types **VertexData**, **HalfedgeData**, and **FaceData** with DCEL vertex, halfedge, and face record types, respectively. When the **Arrangement_2<Traits,Dcel>** class template is instantiated, substituting the **Dcel** parameter with an instance of this DCEL class-template results in nested **Vertex**, **Halfedge**, and **Face** types, the interfaces of which are extended with the access function **data()** and with the modifier **set_data()**.

Example: The next example shows how to use a DCEL with extended vertex, halfedge, and face records. In this example each vertex is associated with a color, which is either blue, red, or white, depending on whether the vertex is isolated, represents a segment endpoint, or represents an intersection point. (Notice that the coloring rules suggested here apply only to non-degenerate arrangements, where the sets of isolated points, curve endpoints, and intersection points are mutually exclusive.) In this example segments are treated as directed objects. Each halfedge is associated with a Boolean flag indicating whether its direction is the same as the direction of its associated segment. Each face is also extended to store the size of its outer boundary, that is, the number of halfedges along its outer boundary.

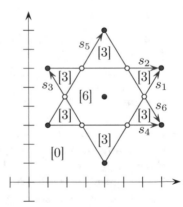

The constructed arrangement is similar to the one constructed in **ex_face_extension.cpp**, given in the previous subsection. In this case, however, we do not use an observer; instead, all auxiliary data-fields are set after the construction phase. Also, note that the data fields are properly maintained when the arrangement is copied to another arrangement object.

```
// File: ex_dcel_extension.cpp

#include <CGAL/basic.h>
#include <CGAL/Arr_extended_dcel.h>

#include "arr_exact_construction_segments.h"

enum Color {BLUE, RED, WHITE};

typedef CGAL::Arr_extended_dcel<Traits, Color, bool, unsigned int>    Dcel;
typedef CGAL::Arrangement_2<Traits, Dcel>                            Ex_arrangement;
```

```cpp
int main()
{
  // Construct the arrangement containing two intersecting triangles.
  Traits traits;
  Ex_arrangement  arr(&traits);
  insert_non_intersecting_curve(arr, Segment(Point(4, 1), Point(7, 6)));
  insert_non_intersecting_curve(arr, Segment(Point(1, 6), Point(7, 6)));
  insert_non_intersecting_curve(arr, Segment(Point(4, 1), Point(1, 6)));
  insert(arr, Segment(Point(1, 3), Point(7, 3)));
  insert(arr, Segment(Point(1, 3), Point(4, 8)));
  insert(arr, Segment(Point(4, 8), Point(7, 3)));
  insert_point(arr, Point(4, 4.5));

  // Go over all arrangement vertices and set their colors according to our
  // coloring convention.
  Ex_arrangement::Vertex_iterator  vit;
  for (vit = arr.vertices_begin(); vit != arr.vertices_end(); ++vit) {
    unsigned int degree = vit->degree();
    vit->set_data((degree == 0) ? BLUE : ((degree <= 2) ? RED : WHITE));
  }

  // Go over all arrangement edges and set their flags.
  // Recall that the value type of the edge iterator is the halfedge type.
  Traits::Equal_2 equal = traits.equal_2_object();
  Ex_arrangement::Edge_iterator  eit;
  for (eit = arr.edges_begin(); eit != arr.edges_end(); ++eit) {
    // Check whether the halfegde has the same direction as its segment.
    bool flag = equal(eit->source()->point(), eit->curve().source());
    eit->set_data(flag);
    eit->twin()->set_data(!flag);
  }

  // Store the size of the outer boundary of every face of the arrangement.
  Ex_arrangement::Face_iterator   fit;
  for (fit = arr.faces_begin(); fit != arr.faces_end(); ++fit) {
    unsigned int boundary_size = 0;
    if (! fit->is_unbounded()) {
      Ex_arrangement::Ccb_halfedge_circulator  curr = fit->outer_ccb();
      boundary_size = std::distance(++curr, fit->outer_ccb())+1;
    }
    fit->set_data(boundary_size);
  }

  // Copy the arrangement and print the vertices along with their colors.
  Ex_arrangement    arr2 = arr;

  std::cout << "The arrangement vertices:" << std::endl;
  for (vit = arr2.vertices_begin(); vit != arr2.vertices_end(); ++vit) {
    std::cout << '(' << vit->point() << ")  ";
    switch (vit->data()) {
      case BLUE  : std::cout << "BLUE."  << std::endl; break;
      case RED   : std::cout << "RED."   << std::endl; break;
      case WHITE : std::cout << "WHITE." << std::endl; break;
```

```
    }
  }

  std::cout << "The_arrangement_outer-boundary_sizes:";
  for (fit = arr2.faces_begin(); fit != arr2.faces_end(); ++fit)
    std::cout << "_" << fit->data();
  std::cout << std::endl;
  return 0;
}
```

———————— *advanced* ————————

The various DCEL classes presented in this section are well suited for most applications based on the *2D Arrangements* package. They are all defined using helper constructs, and in particular the base DCEL class-template `Arr_dcel_base`. However, there are cases where special requirements, not addressed by these DCEL classes, are needed. In such cases you may explicitly extend the base DCEL class-template, as described in the next paragraph, or implement your own DCEL class from scratch and use the resulting DCEL to instantiate the `Arrangement_2` class template. In any case such a class must model the concept ArrangementDcel or its refinement Arrangement-DcelWithRebind. The latter requires a `rebind` struct template, which implements a policy-clone idiom. Here, the DCEL class is the policy class and the `rebind` member template struct is used to pass a different traits type parameter to the policy class template.

In some cases you may want to extend a certain feature type with several fields. You can gather all these fields (and perhaps methods that access and retrieve these fields) in a single construct, and substitute the appropriate parameter of the class template `Arr_extended_dcel<` `Traits, VertexData, HalfedgeData, FaceData>` with this construct. Naturally, you can define three constructs, one for each feature type, and substitute all the three corresponding template parameters appropriately. For example, consider an arrangement that represents a map where features of the same type represent different cartographic entities, e.g., an edge represents a road, a river, or a railway. We would like to associate two strings with each feature, namely, the name and the type of the feature. Following the solution above, accessing or retrieving a specific field will always require an indirection through one of the member functions `set_data()` and `data()`. While this indirection is typically resolved at compile time, and thus has no negative effect on the running time of the generated code, it may have some implication on the space consumption due to compiler padding.[2] Moreover, the code may look cumbersome.

The extended DCEL class that addresses the problem raised above is listed below. Here, each feature type is explicitly extended with two strings, namely, `name` and `type`, eliminating the data constructs.

```
#include <string>

#include <CGAL/basic.h>
#include <CGAL/Arr_dcel_base.h>

// The map-extended dcel vertex.
template <typename Point_2>
class Arr_map_vertex : public CGAL::Arr_vertex_base<Point_2> {
public:
  std::string name, type;
};

// The map-extended dcel halfedge.
template <typename X_monotone_curve_2>
```

---

[2]Compilers add pad bytes into user-defined constructs to comply with alignment restrictions imposed by target microprocessors.

```
class Arr_map_halfedge : public CGAL::Arr_halfedge_base<X_monotone_curve_2> {
public:
  std::string name, type;
};

// The map-extended dcel face.
class Arr_map_face : public CGAL::Arr_face_base {
public:
  std::string name, type;
};

// The map-extended dcel.
template <typename Traits>
class Arr_map_dcel : public
  CGAL::Arr_dcel_base<Arr_map_vertex<typename Traits::Point_2>,
                      Arr_map_halfedge<typename Traits::X_monotone_curve_2>,
                      Arr_map_face>
{};
```

—————— *advanced* ——————

6.2.3 Input/Output for Arrangements with Auxiliary Data

The inserter (<<) and extractor (>>) operators are oblivious to auxiliary data stored with the arrangement cells, regardless of the extension technique. These operators ignore auxiliary data of faces extended using the `Arr_face_extended_dcel` class template and auxiliary data of any DCEL records extended using the `Arr_extended_dcel` class template. Thus, they are suited only for arrangements instantiated using an instance of the default DCEL class-template `Arr_default_dcel`. However, you may find crucial the need to save and restore extended data along the incidence data of the arrangement cells.

The *2D Arrangements* package includes the free functions `write(arr, os, formatter)`, which writes the arrangement `arr` to an output stream `os`, and `read(arr, is, formatter)`, which reads the arrangement `arr` from an input stream `is`. Both operations are performed using a `formatter` object, which defines the I/O format. The package contains three formatter classes, as follows:

- `Arr_text_formatter<Arrangement>` defines a simple textual I/O format for the arrangement topology and geometry, ignoring any auxiliary data that may be associated with the arrangement cells. This is the default formatter used by the arrangement inserter and the arrangement extractor, as defined above.

- `Arr_face_extended_text_formatter<Arrangement>` operates on arrangements the DCEL of which is based on the `Arr_face_extended_dcel<Traits, FaceData>` class template; see Section 6.2.1. It supports reading and writing the auxiliary data-objects stored with the arrangement faces provided that the `FaceData` class supports an inserter and an extractor.

- `Arr_extended_dcel_text_formatter<Arrangement>` operates on arrangements the DCEL of which is based on the `Arr_extended_dcel<Traits, VertexData, HalfedgeData, FaceData>` class template; see Section 6.2.2. It supports reading and writing the auxiliary data-objects stored with the arrangement vertices, edges, and faces, provided that the `VertexData`, `HalfedgeData`, and `FaceData` classes all have inserters and extractors.

Consider the arrangement constructed in the example `ex_dcel_extension.cpp` presented in the previous subsections. In order to perform an I/O operation on this arrangement, you must add the following #`include` statements at the beginning of your program:

`#include <CGAL/IO/Arr_text_formatter.h>`

```
#include <CGAL/IO/Arr_iostream.h>
#include <fstream>
```

Then, instead of inserting the arrangement object **arr** to an output stream using the << operator, use the **write()** function with the appropriate formatter (recall that **Ex_arrangement_2** is the **Arrangement_2** class template instantiated with an extended DCEL class).

```
std::ofstream                                            out_file("arr.dat");
CGAL::Arr_extended_dcel_text_formatter<Ex_arrangement_2>  formatter;
write(arr, out_file, formatter);
out_file.close();
```

Similarly, you can read an arrangement from an input stream using the function **read()** with the appropriate formatter.

```
Ex_arrangement_2                                          arr2;
std::ifstream                                            in_file("arr.dat");
read(arr2, in_file, formatter);
in_file.close();
```

───── *advanced* ─────

You may develop your own formatter classes—models of the ArrangementInputFormatter and ArrangementOutputFormatter concepts, as defined in the Reference Manual. Doing so, you can define other I/O formats, such as an XML-based format or a binary format.

───── *advanced* ─────

6.3 Overlaying Arrangements

You are given two geographic maps represented as arrangements, with some data objects attached to their faces, representing certain geographic information—for instance, a map of the annual precipitation in some country and a map of the vegetation in the same country—and you are asked to locate, for example, places where there is a pine forest *and* the annual precipitation is between 1,000 mm and 1,500 mm. Overlaying the two maps may help you figure out the answer. Computing the overlay of two planar arrangements is also useful for supporting Boolean set operations on polygons or general polygons; see Chapter 8.

Formally, the map overlay of two planar subdivisions S_1 and S_2 is a planar subdivision S, such that there is a face f in S iff there are faces f_1 and f_2 in S_1 and S_2, respectively, such that f is a maximal connected component of $f_1 \cap f_2$.

The overlay of two given arrangements, conveniently referred to as the "blue" and the "red" arrangements, is implemented as a plane-sweep algorithm employing a dedicated visitor; see Section 3.2. The x-monotone curve type is extended with a color attribute (whose value is either blue or red); see Section 5.5. With the help of the extended type unnecessary computations are filtered out while the plane is swept, yielding an efficient process. For example, monochromatic intersections are not computed.

The plane-sweep visitor that concretizes the overlay operation needs to construct a DCEL that properly represents the overlay of two input arrangements. A face in the overlay arrangement corresponds to overlapping regions of the blue and red faces. An edge in the overlay arrangement is due to a blue edge, a red edge, or an overlap of two differently colored edges. An overlay vertex is due to a blue vertex, a red vertex, a coincidence of two differently colored vertices, or an intersection of a blue and a red curve.

The function template **overlay(arr_r, arr_b, arr_o)** accepts two input arrangement objects **arr_r** and **arr_b**, and constructs their overlay object **arr_o**. All three arrangements must use the same geometric primitives. In other words, their types are instances of the **Arrangement_2<Traits,Dcel>** class template, where the **Traits** parameter is substituted with three geometry-traits classes, respectively. The geometry-traits classes of the input arrangements must be convertible to the geometry-traits class of the resulting arrangement.[3] Typically, all three arrangements use the same geometry-traits class.

The **overlay()** function template is suitable for arrangements that do not store any additional data with their DCEL records; namely, arrangements defined using an instance of the default DCEL class-template **Arr_default_dcel**. Typically, the overlay arrangement in this case does not store extra data with its DCEL records as well (or if it does, the additional data-fields cannot be computed by the overlay operation). The overlay arrangement is equivalent to the arrangement induced by all curves of **arr_r** and **arr_b**. Indeed, it is possible to obtain the same result using the standard insertion-operations instead, but, as mentioned above, this is less efficient.

Example: The next program constructs two simple arrangements; each comprises four line segments that form a square, as depicted in the figure to the right. The program computes the overlay of the two arrangements. The resulting arrangement has 16 vertices, 24 edges, and 10 faces (including the unbounded one).

```
// File: ex_overlay.cpp

#include <CGAL/basic.h>
#include <CGAL/Arr_overlay_2.h>

#include "arr_exact_construction_segments.h"
#include "arr_print.h"

int main()
{
  // Construct the first arrangement, containing a square-shaped face.
  Arrangement  arr1;
  insert_non_intersecting_curve(arr1, Segment(Point(2, 2), Point(6, 2)));
  insert_non_intersecting_curve(arr1, Segment(Point(6, 2), Point(6, 6)));
  insert_non_intersecting_curve(arr1, Segment(Point(6, 6), Point(2, 6)));
  insert_non_intersecting_curve(arr1, Segment(Point(2, 6), Point(2, 2)));

  // Construct the second arrangement, containing a rhombus-shaped face.
  Arrangement  arr2;
  insert_non_intersecting_curve(arr2, Segment(Point(4, 1), Point(7, 4)));
  insert_non_intersecting_curve(arr2, Segment(Point(7, 4), Point(4, 7)));
  insert_non_intersecting_curve(arr2, Segment(Point(4, 7), Point(1, 4)));
  insert_non_intersecting_curve(arr2, Segment(Point(1, 4), Point(4, 1)));

  // Compute the overlay of the two arrangements.
  Arrangement  overlay_arr;
  CGAL::overlay(arr1, arr2, overlay_arr);
  print_arrangement_size(overlay_arr);

  return 0;
}
```

Try: Verify that using the overlay operation to obtain the map overlay of two arrangements is more efficient than using the equivalent standard insertion-operations. Modify the program above to replace the call to the instance of the **overlay()** function template with code that contains calls to instances of the **insert()** function templates. Compare the running times using either incremental insertion (see Section 3.4.1) or aggregate insertion (see Section 3.4.2).[4]

---

[3]It is sufficient that all three geometry-traits classes used to instantiate the three types of arrangements derive from a common ancestor that models the geometry-traits concept.

The `overlay()` function template is overloaded with a variant that accepts four arguments, that is, `overlay(arr_r, arr_b, arr_o, ovl_traits)`. The type of the `ovl_traits` additional argument, referred to as the overlay traits, must model the OverlayTraits concept described below. Assume that `arr_r` is of type `Arrangement_2<Traits,Dcel_R>`, `arr_b` is of type `Arrangement_2<Traits,Dcel_B>`, and the resulting `arr_o` is of type `Arrangement_2<Traits,Dcel_O>`. The overlay traits enables the creation of `Dcel_O` records in the overlay arrangement from the features of `Dcel_R` and `Dcel_B` records from the arrangements `arr_r` and `arr_b`, respectively.

We distinguish between (i) an overlay of two arrangements that store additional data-fields only with their faces (e.g., the geographic-map example given at the beginning of this section) and (ii) an overlay of two arrangements that store additional data fields with all their DCEL records (or at least not only with their faces). The arrangement that results from overlaying two face-extended arrangements typically also stores additional data-fields with its faces. The types of such arrangements, for example, could be instances of the `Arrangement_2<Traits,Dcel>` class template, where the `Dcel` parameters are substituted with instances of the `Arr_face_extended_dcel` class template (see Section 6.2.1). The data field that is attached to an overlay face can be computed from the data fields of the two faces (in `arr_r` and `arr_b`) that induce the overlay face. Similarly, the arrangement that results from overlaying two arrangements that store additional data fields with all their DCEL records typically also stores additional data-fields with all its DCEL records. The types of such arrangements, for example, could be instances of the `Arrangement_2<Traits,Dcel>` class template, where the `Dcel` parameters are substituted with instances of the `Arr_extended_dcel` class template (see Section 6.2.2). The data field attached to an overlay feature can be computed from the data fields of the two features (in `arr_r` and `arr_b`) that induce the overlay feature.

As mentioned in the previous paragraph, if any of the DCEL records of your arrangements are extended, you can pass a fourth argument to the `overlay()` call, also referred to as the overlay traits, to control the generation of the extended data in the resulting arrangement. If only the face records are extended, the type of the overlay traits can be an instance of the class template `Arr_face_overlay_traits<ArrangementR,ArrangementB,ArrangementO,OverlayFaceData>`, which models the concept OverlayTraits. An object of this type operates on face-extended arrangements. When instantiated, the `OverlayFaceData` parameter must be substituted with a functor that is capable of combining two face-data fields of types `ArrangementR::Dcel::Face_data` and `ArrangementB::Dcel::Face_data` and computing the output `ArrangementO::Dcel::Face_data` object. The face-overlay traits-class uses this functor to properly construct the overlay faces.

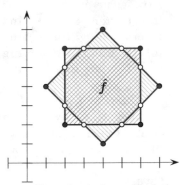

Example: The example below shows how to compute the intersection of two polygons using the `overlay()` function template. It uses a face-extended DCEL type to instantiate the arrangement classes. Each face of the DCEL is extended with a Boolean flag. A polygon is represented as a *marked* arrangement face (whose flag is set). The example uses an instance of the `Arr_face_overlay_traits<ArrR,ArrB,ArrO,OverlayFaceData>` class template as the face-overlay traits class, where the `OverlayFaceData` template parameter is substituted with a functor that simply performs a logical *and* operation on Boolean flags. As a result, a face in the overlay arrangement is marked only when it corresponds to an overlapping region of two marked faces in the input arrangements. Namely, it is part of the intersection of the two polygons. The example computes the intersection between a parallel-axis square and a congruent square rotated 45°. The resulting polygon from the intersection operation is an octagon, which corresponds to the face $\hat{f}$ in the arrangement depicted in the figure above.

// File: ex_face_extension_overlay.cpp

---

[4]The original program constructs small arrangements. Thus, the incurred overhead may overshadow the real time-consumption. You may need to further modify the program to construct larger arrangements; input files of large data can be obtained from `http://acg.cs.tau.ac.il/cgal-arrangement-book`.

```cpp
#include <CGAL/basic.h>
#include <CGAL/Arr_overlay_2.h>
#include <CGAL/Arr_default_overlay_traits.h>

#include "arr_exact_construction_segments.h"

typedef CGAL::Arr_face_extended_dcel<Traits, bool>          Dcel;
typedef CGAL::Arrangement_2<Traits, Dcel>                   Ex_arrangement;
typedef CGAL::Arr_face_overlay_traits<Ex_arrangement, Ex_arrangement,
                                      Ex_arrangement,
                                      std::logical_and<bool> >
                                                            Overlay_traits;

int main()
{
  // Construct the first arrangement, containing a square-shaped face.
  Ex_arrangement  arr1;
  insert_non_intersecting_curve(arr1, Segment(Point(2, 2), Point(6, 2)));
  insert_non_intersecting_curve(arr1, Segment(Point(6, 2), Point(6, 6)));
  insert_non_intersecting_curve(arr1, Segment(Point(6, 6), Point(2, 6)));
  insert_non_intersecting_curve(arr1, Segment(Point(2, 6), Point(2, 2)));
  CGAL_assertion(arr1.number_of_faces() == 2);

  // Mark just the bounded face.
  Ex_arrangement::Face_iterator   fit;
  for (fit = arr1.faces_begin(); fit != arr1.faces_end(); ++fit)
    fit->set_data(fit != arr1.unbounded_face());

  // Construct the second arrangement, containing a rhombus-shaped face.
  Ex_arrangement  arr2;
  insert_non_intersecting_curve(arr2, Segment(Point(4, 1), Point(7, 4)));
  insert_non_intersecting_curve(arr2, Segment(Point(7, 4), Point(4, 7)));
  insert_non_intersecting_curve(arr2, Segment(Point(4, 7), Point(1, 4)));
  insert_non_intersecting_curve(arr2, Segment(Point(1, 4), Point(4, 1)));
  CGAL_assertion(arr2.number_of_faces() == 2);

  for (fit = arr2.faces_begin(); fit != arr2.faces_end(); ++fit)
    fit->set_data(fit != arr2.unbounded_face());   // mark the bounded face.

  // Compute the overlay of the two arrangements, marking only the faces that
  // are intersections of two marked faces in arr1 and arr2, respectively.
  Ex_arrangement  overlay_arr;
  Overlay_traits  overlay_traits;
  CGAL::overlay(arr1, arr2, overlay_arr, overlay_traits);

  // Go over the faces of the resulting arrangement and print the marked ones.
  std::cout << "The intersection is: ";
  for (fit = overlay_arr.faces_begin(); fit != overlay_arr.faces_end(); ++fit)
  {
    if (!fit->data()) continue;
    Ex_arrangement::Ccb_halfedge_circulator curr = fit->outer_ccb();
    std::cout << curr->source()->point();
```

```
      do std::cout << "_—>_" << curr->target()->point();
      while (++curr != fit->outer_ccb());
      std::cout << std::endl;
    }
    return 0;
}
```

Example: The next example demonstrates the face overlay of two arrangements that have unbounded faces as well as bounded ones. The first arrangement (blue) is induced by the two lines $y = x$ and $y = -x$, which subdivide the plane into four unbounded faces, labeled A, B, C, and D. The second arrangement (red) comprises four line segments that form a square-shaped face indexed 1. The unbounded face is indexed 2. When the two arrangements are overlaid, each of the four faces $A, \ldots, D$ is split into an unbounded face (indexed 2) and a bounded face (indexed 1), so the faces of the resulting arrangement are labeled $A_1, A_2, \ldots, D_1, D_2$. **boost::lexical_cast** is used to cast the integral indices into strings to produce the final labels.

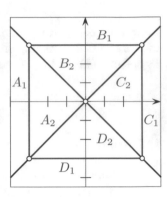

```
// File: ex_overlay_unbounded.cpp

#include <string>
#include <boost/lexical_cast.hpp>

#include <CGAL/basic.h>
#include <CGAL/Arr_extended_dcel.h>
#include <CGAL/Arr_overlay_2.h>
#include <CGAL/Arr_default_overlay_traits.h>

#include "arr_linear.h"

// Define a functor for creating a label from a characer and an integer.
struct Overlay_label {
  std::string operator()(char c, unsigned int i) const
  { return c + boost::lexical_cast<std::string>(i); }
};

typedef CGAL::Arr_face_extended_dcel<Traits, char>        Dcel_dlue;
typedef CGAL::Arrangement_2<Traits, Dcel_dlue>            Arrangement_blue;
typedef CGAL::Arr_face_extended_dcel<Traits, unsigned int> Dcel_red;
typedef CGAL::Arrangement_2<Traits, Dcel_red>             Arrangement_red;
typedef CGAL::Arr_face_extended_dcel<Traits, std::string> Dcel_res;
typedef CGAL::Arrangement_2<Traits, Dcel_res>             Arrangement_res;
typedef CGAL::Arr_face_overlay_traits<Arrangement_blue, Arrangement_red,
                                      Arrangement_res, Overlay_label>
                                                          Overlay_traits;

int main()
{
  // Construct the first arrangement, induced by two line y = x and y = -x.
  Arrangement_blue          arr1;
  insert(arr1, Line(Point(0, 0), Point(1, 1)));
  insert(arr1, Line(Point(0, 0), Point(1, -1)));
```

```
// Label the four (unbounded) faces of the arrangement as 'A' to 'D' by
// traversing the faces incident to the halfedges around the single
// arrangement vertex (0, 0).
char                                                        clabel = 'A';
Arrangement_blue::Halfedge_around_vertex_circulator  first, curr;
curr = first = arr1.vertices_begin()->incident_halfedges();
do curr->face()->set_data(clabel++);
while (++curr != first);

// Construct the second arrangement, containing a single square-shaped face.
Arrangement_red              arr2;
insert(arr2, Segment(Point(-3, -3), Point(3, -3)));
insert(arr2, Segment(Point(3, -3), Point(3, 3)));
insert(arr2, Segment(Point(3, 3), Point(-3, 3)));
insert(arr2, Segment(Point(-3, 3), Point(-3, -3)));

// Give the unbounded face the index 1, and the bounded face the index 2.
Arrangement_red::Face_iterator     fit;
for (fit = arr2.faces_begin(); fit != arr2.faces_end(); ++fit)
   fit->set_data((fit == arr2.unbounded_face()) ? 1 : 2);

// Compute the overlay of the two arrangements.
Arrangement_res   overlay_arr;
Overlay_traits    overlay_traits;
CGAL::overlay(arr1, arr2, overlay_arr, overlay_traits);

// Go over the faces of the overlay arrangement and print their labels.
std::cout << "The_overlay_faces_are:_" << std::endl;
for (Arrangement_res::Face_iterator res_fit = overlay_arr.faces_begin();
     res_fit != overlay_arr.faces_end(); ++res_fit)
{
   std::cout << "__" << res_fit->data().c_str() << "_("
             << (res_fit->is_unbounded() ? "unbounded" : "bounded")
             << ")." << std::endl;
}
return 0;
}
```

If the red and blue arrangements store additional data-fields with all their DCEL records, and the data associated with the overlay DCEL features should be computed from the red and blue DCEL features that induce it, then an appropriate overlay-traits argument must be passed to the **overlay()** call. The overlay-traits type models the **OverlayTraits** concept, which requires the provision of ten functions that handle all possible cases as listed below. Let v_r, e_r, and f_r denote input red features, i.e., a vertex, an edge, and a face, respectively, v_b, e_b, and f_b denote input blue features, and v, e, and f denote output features.

1. A new vertex v is induced by coinciding vertices v_r and v_b.
2. A new vertex v is induced by a vertex v_r that lies on an edge e_b.
3. An analogous case of a vertex v_b that lies on an edge e_r.
4. A new vertex v is induced by a vertex v_r that is contained in a face f_b.
5. An analogous case of a vertex v_b contained in a face f_r.
6. A new vertex v is induced by the intersection of two edges e_r and e_b.
7. A new edge e is induced by the (possibly partial) overlap of two edges e_r and e_b.
8. A new edge e is induced by the an edge e_r that is contained in a face f_b.
9. An analogous case of an edge e_b contained in a face f_r.

10. A new face f is induced by the overlap of two faces f_r and f_b.

The `Overlay_color_traits` class template listed below models the concept OverlayTraits. It assumes that each feature of the input arrangements and of the overlay arrangement is extended with an RGB color stored as an **unsigned int**. It defines ten member functions that correspond to the ten cases listed above. Each of these functions accepts three handles as follows: two handles to the two features of the input arrangements, respectively, that induce a feature of the overlay arrangement and a handle to the induced overlay-arrangement feature. Each of these member functions blends the colors attached to the inducing features and assigns the resulting color to the induced feature. The `Overlay_color_traits` class template is defined in the header file `Overlay_color_traits.h`.

```cpp
template <typename Arrangement> struct Overlay_color_traits {
  typedef unsigned int                            Color;
  typedef typename Arrangement::Vertex_const_handle    V_const_handle;
  typedef typename Arrangement::Halfedge_const_handle  H_const_handle;
  typedef typename Arrangement::Face_const_handle      F_const_handle;
  typedef typename Arrangement::Vertex_handle          V_handle;
  typedef typename Arrangement::Halfedge_handle        H_handle;
  typedef typename Arrangement::Face_handle            F_handle;

  // Compute the average of the red, green, and blue components separately.
  Color blend(Color color1, Color color2) const
  {
    return
      (((( color1 & 0x000000ff) + (color2 & 0x000000ff)) / 2) & 0x000000ff) |
      (((( color1 & 0x0000ff00) + (color2 & 0x0000ff00)) / 2) & 0x0000ff00) |
      (((( color1 & 0x00ff0000) + (color2 & 0x00ff0000)) / 2) & 0x00ff0000);
  }

  void create_face(F_const_handle f1, F_const_handle f2, F_handle f) const
  { f->set_data(blend(f1->data(), f2->data())); }
  void create_vertex(H_const_handle h1, H_const_handle h2, V_handle v) const
  { v->set_data(blend(h1->data(), h2->data())); }
  void create_vertex(V_const_handle v1, V_const_handle v2, V_handle v) const
  { v->set_data(blend(v1->data(), v2->data())); }
  void create_vertex(V_const_handle v1, H_const_handle h2, V_handle v) const
  { v->set_data(blend(v1->data(), h2->data())); }
  void create_vertex(H_const_handle h1, V_const_handle v2, V_handle v) const
  { v->set_data(blend(h1->data(), v2->data())); }
  void create_vertex(F_const_handle f1, V_const_handle v2, V_handle v) const
  { v->set_data(blend(f1->data(), v2->data())); }
  void create_vertex(V_const_handle v1, F_const_handle f2, V_handle v) const
  { v->set_data(blend(v1->data(), f2->data())); }
  void create_edge(H_const_handle h1, H_const_handle h2, H_handle h) const
  {
    h->set_data(blend(h1->data(), h2->data()));
    h->twin()->set_data(blend(h1->data(), h2->data()));
  }
  void create_edge(H_const_handle h1, F_const_handle f2, H_handle h) const
  {
    h->set_data(blend(h1->data(), f2->data()));
    h->twin()->set_data(blend(h1->data(), f2->data()));
  }
  void create_edge(F_const_handle f1, H_const_handle h2, H_handle h) const
```

```
    {
        h->set_data(blend(f1->data(), h2->data()));
        h->twin()->set_data(blend(f1->data(), h2->data()));
    }
};
```

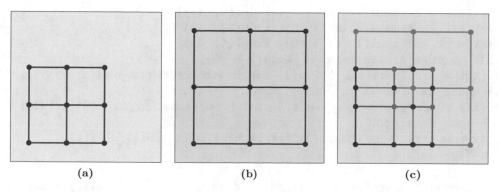

(a)   (b)   (c)

Fig. 6.1: The overlay (c) of two arrangements (a) and (b). Each feature of the arrangements is extended with a color. The color of each feature of the overlay arrangement is the blend of the colors of the two inducing features.

Example: The example program listed below computes the overlay, depicted in Figure 6.1c, of the two arrangements depicted in Figures 6.1a and 6.1b. To see the color blending as produced by the program, view the file **overlay_colored.pdf**. Each feature of the input arrangements and of the overlay arrangement is extended with an RGB color stored as an **unsigned int**. The vertices, halfedges, and faces of the red arrangement are assigned three different shades of red. Similarly, the vertices, halfedges, and faces of the blue arrangement are assigned three different shades of blue. Using an instance of the **Overlay_color_traits** class template as the overlay traits, each feature of the overlay arrangement is assigned a color that is a blend of the colors attached to the inducing features.

// File: ex_overlay_color.cpp

```
#include <CGAL/basic.h>
#include <CGAL/Arr_extended_dcel.h>
#include <CGAL/Arr_overlay_2.h>
#include <CGAL/Arr_default_overlay_traits.h>

#include "arr_exact_construction_segments.h"
#include "Overlay_color_traits.h"

typedef unsigned int                                      Color;
typedef CGAL::Arr_extended_dcel<Traits, Color, Color, Color> Dcel;
typedef CGAL::Arrangement_2<Traits, Dcel>                 Ex_arrangement;

int main()
{
    Ex_arrangement::Vertex_iterator     vit;
    Ex_arrangement::Halfedge_iterator   hit;
    Ex_arrangement::Face_iterator       fit;

    const Color vcol1(0x00000080), hcol1(0x000000ff), fcol1(0x00ccccff);
    const Color vcol2(0x00800000), hcol2(0x00ff0000), fcol2(0x00ffcccc);
```

```
// Construct the first arrangement and assign colors to its features.
Ex_arrangement  arr1;
insert_non_intersecting_curve(arr1, Segment(Point(0, 0), Point(4, 0)));
insert_non_intersecting_curve(arr1, Segment(Point(0, 2), Point(4, 2)));
insert_non_intersecting_curve(arr1, Segment(Point(0, 4), Point(4, 4)));
insert(arr1, Segment(Point(0, 0), Point(0, 4)));
insert(arr1, Segment(Point(2, 0), Point(2, 4)));
insert(arr1, Segment(Point(4, 0), Point(4, 4)));
CGAL_assertion(arr1.number_of_faces() == 5);
for (vit = arr1.vertices_begin(); vit != arr1.vertices_end(); ++vit)
  vit->set_data(vcol1);
for (hit = arr1.halfedges_begin(); hit != arr1.halfedges_end(); ++hit)
  hit->set_data(hcol1);
for (fit = arr1.faces_begin(); fit != arr1.faces_end(); ++fit)
  fit->set_data(fcol1);

// Construct the second arrangement and assign colors to its features.
Ex_arrangement  arr2;
insert_non_intersecting_curve(arr2, Segment(Point(0, 0), Point(6, 0)));
insert_non_intersecting_curve(arr2, Segment(Point(0, 3), Point(6, 3)));
insert_non_intersecting_curve(arr2, Segment(Point(0, 6), Point(6, 6)));
insert(arr2, Segment(Point(0, 0), Point(0, 6)));
insert(arr2, Segment(Point(3, 0), Point(3, 6)));
insert(arr2, Segment(Point(6, 0), Point(6, 6)));
CGAL_assertion(arr2.number_of_faces() == 5);
for (vit = arr2.vertices_begin(); vit != arr2.vertices_end(); ++vit)
  vit->set_data(vcol2);
for (hit = arr2.halfedges_begin(); hit != arr2.halfedges_end(); ++hit)
  hit->set_data(hcol2);
for (fit = arr2.faces_begin(); fit != arr2.faces_end(); ++fit)
  fit->set_data(fcol2);

// Compute the overlay of the two arrangements, while blending the colors
// of their features.
Ex_arrangement  ovl_arr;
Overlay_color_traits<Ex_arrangement> overlay_traits;
CGAL::overlay(arr1, arr2, ovl_arr, overlay_traits);

// Print the overlay-arrangement vertices and their colors.
for (vit = ovl_arr.vertices_begin(); vit != ovl_arr.vertices_end(); ++vit)
  std::cout << vit->point() << ":_0x" << std::hex << std::setfill('0')
            << std::setw(6) << vit->data() << std::endl;
return 0;
}
```

6.4 Storing the Curve History

When you construct an arrangement induced by a set $\mathcal{C}$ of arbitrary planar curves, you end up with a collection $\mathcal{C}''$ of x-monotone subcurves of $\mathcal{C}$ that are pairwise disjoint in their interior; see Section 1.1. These subcurves are associated with the arrangement edges (more precisely, with pairs of DCEL halfedges). The connection between the original input curves and the arrangement edges is lost during the construction process. This loss might be acceptable for some applications.

However, in many practical cases it is important to determine the input curves that give rise to the final subcurves.

The `Arrangement_with_history_2<Traits,Dcel>` class template extends the `Arrangement_` `2` class template with an additional container that represents $\mathcal{C}$ and a cross-mapping between the curves of $\mathcal{C}$ and the arrangement edges they induce. The `Traits` template parameter must be substituted with a model of the ArrangementTraits_2 concept; see Section 3.4.1. It should define the `Curve_2` type and support its subdivision into `X_monotone_curve_2` objects, among the others. The `Dcel` parameter must be substituted with a model of the ArrangementDcelWithRebind concept; see Section 6.2.2. You can use either the default DCEL class or an extended DCEL class (see Section 6.2) based on your needs.

6.4.1   Traversing an Arrangement with History

The `Arrangement_with_history_2` class template extends the `Arrangement_2` class template. Thus, all the iterator and circulator types that are defined by the base class are also available in `Arrangement_with_history_2`. (Refer to Section 2.2.1 for a comprehensive review of this functionality.)

As mentioned above, the `Arrangement_with_history_2` class template maintains a container of input curves, which can be accessed using curve handles. Let `arr` identify some object, the type of which is an instance of this template. The call `arr.number_of_curves()` returns the number of input curves stored in the container, while `arr.curves_begin()` and `arr.curves_end()` return `Arrangement_with_history_2::Curve_iterator` objects that define the valid range of curves that induce the arrangement. The value type of this iterator is `Curve_2`. Moreover, the curve-iterator type is convertible to `Arrangement_with_history_2::Curve_handle`, which is used for accessing the stored curves. For convenience, the corresponding constant-iterator and constant-handle types are also defined.

As mentioned in the previous paragraph, a `Curve_handle` object `ch` serves as a pointer to a curve stored in an arrangement-with-history object `arr`. Using this handle, it is possible to obtain the number of arrangement edges this curve induces by calling `arr.number_of_induced_` `edges(ch)`. The function calls `arr.induced_edges_begin(ch)` and `arr.induced_edges_` `end(ch)` return iterators of type `Arrangement_with_history_2::Induced_edges_iterator` that define the valid range of edges induced by `ch`. The value type of these iterators is `Halfedge_` `handle`. It is thus possible to traverse all arrangement edges induced by an input curve.

The ability to perform the inverse mapping is also important. Given an arrangement edge, you may want to determine which input curve induces it. In case the edge represents an overlap of several curves, you should be able to trace all input curves that overlap over this edge. The `Arrangement_with_history_2` class template is extended with several member functions that enable such an inverse mapping. Given a handle to a halfedge e in an arrangement-with-history object `arr`, the call `arr.number_of_originating_curves(e)` returns the number of curves that induce the edge (which should be 1 in non-degenerate cases, and 2 or more in case of overlaps), while `arr.originating_curves_begin(e)` and `arr.originating_curves_end(e)` return `Arrangement_with_history_2::Originating_curve_iterator` objects that define the range of curves that induce e. The value type of these iterators is `Curve_2`.

Overlaying two arrangement-with-history objects is possible only if their types are instances of the `Arrangement_with_history_2<Traits,Dcel>` class template, where the respective `Traits` parameters are substituted with two traits classes that are convertible to one another. In this case, the resulting arrangement stores a consolidated container of input curves, and automatically preserves the cross-mapping between the arrangement edges and the consolidated curve-set. You may also employ an overlay-traits class to maintain any type of auxiliary data stored with the DCEL cells; see Section 6.3.

6.4.2   Modifying an Arrangement with History

As the `Arrangement_with_history_2` class template extends the `Arrangement_2` class template, it inherits the fundamental modification operations, such as `assign()` and `clear()`, from it. The vertex-manipulation functions are also inherited and supported; see Sections 2.2.2 and 3.4.1 for the details. However, there are some fundamental differences between the interfaces of the two classes, which we highlight next.

The most significant difference between the arrangement-with-history class and the basic arrangement class is the way they handle their input curves. `Arrangement_with_history_2` always stores the `Curve_2` objects that induce it. Thus, it is impossible to insert x-monotone curves into an arrangement with history. The free `insert_non_intersecting_curve()` and the version of `insert()` that accepts an x-monotone curve, as well as their aggregate versions, are therefore not available for arrangement-with-history instances. Only the free overloaded functions `insert()` that accept general curves, namely, the incremental insertion function and the aggregate insertion function, are supported; see Section 3.4.1 for a review of these functions. Notice, however, that while the incremental insertion function `insert(arr, c)` for an `Arrangement_2` object `arr` does not have a return value, the corresponding arrangement-with-history function returns a `Curve_handle` object that points to the inserted curve.

As we are able to keep track of all edges induced by an input curve, we also provide a free function that removes a curve from an arrangement. By calling `remove(arr,ch)`, where `ch` is a valid curve handle, the given curve is deleted from the curve container, and all edges induced solely by this curve (i.e., excluding overlapping edges) are removed from the arrangement. The function returns the number of edges that have been removed.

In some cases, you may need to operate directly on the arrangement edges. We first mention that the specialized insertion-functions (see Section 2.2.2) are not supported, as they accept x-monotone curves. Insertion can only be performed via the free insertion-functions. The other edge-manipulation functions (see Section 2.2.2) are, however, available, but have a different interface that does not use x-monotone curves.

1. Invoking `split_edge(e,p)` splits the edge e at a given point p that lies in its interior.
2. Invoking `merge_edge(e1,e2)` merges the two given edges. There is a precondition that e_1 and e_2 share a common end vertex of degree 2 prior to the merge, and that the x-monotone subcurves associated with these edges are mergeable.
3. It is possible to remove an edge by simply invoking `remove_edge(e)`.

In all cases, the maintenance of cross-pointers for the appropriate input curves is done automatically.

Note that it is possible to attach observers to an arrangement-with-history instance in order to get detailed notifications of the changes the arrangement undergoes; see Section 6.1 for the details.

Example: The example below constructs a simple arrangement of six line segments, as illustrated in the figure to the right, while maintaining the curve history. Note that the input segments s_1 and s_3 overlap over two edges. The example demonstrates the usage of the special traversal functions. It also shows how to issue point-location queries on the resulting arrangement (the query points q_1, q_2, and q_3 are drawn as crosses), using the function `locate_point()` listed on Page 44.

```
// File: ex_curve_history.cpp

#include <CGAL/basic.h>
#include <CGAL/Arrangement_with_history_2.h>
#include <CGAL/Arr_trapezoid_ric_point_location.h>

#include "arr_exact_construction_segments.h"
```

```cpp
#include "point_location_utils.h"

typedef CGAL::Arrangement_with_history_2<Traits>          Arr_with_hist;
typedef Arr_with_hist::Curve_handle                       Curve_handle;
typedef CGAL::Arr_trapezoid_ric_point_location<Arr_with_hist> Point_location;

int main()
{
  // Insert s1, s2, and s3 incrementally.
  Arr_with_hist arr;
  Curve_handle  s1 = insert(arr, Segment(Point(0, 3), Point(4, 3)));
  Curve_handle  s2 = insert(arr, Segment(Point(3, 2), Point(3, 5)));
  Curve_handle  s3 = insert(arr, Segment(Point(2, 3), Point(5, 3)));

  // Insert three additional segments aggregately.
  Segment       segs[] = {Segment(Point(2, 6), Point(7, 1)),
                          Segment(Point(0, 0), Point(2, 6)),
                          Segment(Point(3, 4), Point(6, 4))};
  insert(arr, segs, segs + sizeof(segs)/sizeof(Segment));

  // Print out the curves and the number of edges each one induces.
  Arr_with_hist::Curve_iterator           cit;
  std::cout << "The arrangement contains "
            << arr.number_of_curves() << " curves:" << std::endl;
  for (cit = arr.curves_begin(); cit != arr.curves_end(); ++cit)
    std::cout << "Curve [" << *cit << "] induces "
              << arr.number_of_induced_edges(cit) << " edges." << std::endl;

  // Print the arrangement edges along with the list of curves that
  // induce each edge.
  Arr_with_hist::Edge_iterator            eit;
  Arr_with_hist::Originating_curve_iterator  ocit;

  std::cout << "The arrangement is comprised of "
            << arr.number_of_edges() << " edges:" << std::endl;
  for (eit = arr.edges_begin(); eit != arr.edges_end(); ++eit) {
    std::cout << "[" << eit->curve() << "]. Originating curves: ";
    for (ocit = arr.originating_curves_begin(eit);
         ocit != arr.originating_curves_end(eit); ++ocit)
      std::cout << " [" << *ocit << "]" << std::flush;
    std::cout << std::endl;
  }

  // Perform some point-location queries.
  Point_location    pl(arr);
  locate_point(pl, Point(4, 6));        // q1
  locate_point(pl, Point(6, 2));        // q2
  locate_point(pl, Point(2, 4));        // q3

  return 0;
}
```

Example: The example below demonstrates the usage of the free `remove_curve()` function template. We construct an arrangement of nine circles, while keeping a handle to each inserted circle. We then remove the large circle C_0, which induces 18 edges, as depicted in the figure to the right (note the two vertices induced by splitting the circle into two x-monotone arcs). The example also shows how to use the `split_edge()` and `merge_edge()` member functions when operating on an arrangement-with-history object.

```
// File: ex_edge_manipulation_curve_history.cpp

#include <CGAL/basic.h>
#include <CGAL/Arrangement_with_history_2.h>
#include <CGAL/Arr_walk_along_line_point_location.h>

#include "arr_circular.h"
#include "arr_print.h"

typedef CGAL::Arrangement_with_history_2<Traits>            Arr_with_hist;
typedef Arr_with_hist::Curve_handle                         Curve_handle;
typedef CGAL::Arr_walk_along_line_point_location<Arr_with_hist>Point_location;

int main()
{
  // Construct an arrangement containing nine circles: C[0] of radius 2 and
  // C[1], ..., C[8] of radius 1.
  const Number_type  _7_halves = Number_type(7) / Number_type(2);
  Curve              C[9];
  C[0] = Circle(Kernel::Point_2(_7_halves, _7_halves), 4, CGAL::CLOCKWISE);
  C[1] = Circle(Kernel::Point_2(_7_halves, 6), 1, CGAL::CLOCKWISE);
  C[2] = Circle(Kernel::Point_2(5, 5), 1, CGAL::CLOCKWISE);
  C[3] = Circle(Kernel::Point_2(6, _7_halves), 1, CGAL::CLOCKWISE);
  C[4] = Circle(Kernel::Point_2(5, 2), 1, CGAL::CLOCKWISE);
  C[5] = Circle(Kernel::Point_2(_7_halves, 1), 1, CGAL::CLOCKWISE);
  C[6] = Circle(Kernel::Point_2(2, 2), 1, CGAL::CLOCKWISE);
  C[7] = Circle(Kernel::Point_2(1, _7_halves), 1, CGAL::CLOCKWISE);
  C[8] = Circle(Kernel::Point_2(2, 5), 1, CGAL::CLOCKWISE);

  Arr_with_hist   arr;
  Curve_handle    handles[9];
  for (int k = 0; k < 9; k++) handles[k] = insert(arr, C[k]);

  std::cout << "The initial arrangement size:" << std::endl;
  print_arrangement_size(arr);

  // Remove the large circle C[0].
  std::cout << "Removing C[0] : " << remove_curve(arr, handles[0])
            << " edges have been removed." << std::endl;
  print_arrangement_size(arr);

  // Locate the point q, which should be on an edge e.
  Point_location              pl(arr);
  const Point                 q = Point(_7_halves, 7);
```

```
CGAL::Object                    obj = pl.locate(q);
Arr_with_hist::Halfedge_const_handle  e;
bool                            success = CGAL::assign(e, obj);
CGAL_assertion(success);

// Split the edge into two edges e1 and e2.
Arr_with_hist::Halfedge_handle e1 =
    arr.split_edge(arr.non_const_handle(e), q);
Arr_with_hist::Halfedge_handle e2 = e1->next();
std::cout << "After edge split: " << std::endl;
print_arrangement_size(arr);

arr.merge_edge(e1, e2);                   // merge back the two split edges
std::cout << "After edge merge: " << std::endl;
print_arrangement_size(arr);
return 0;
}
```

The program uses types defined in the header file **arr_circular.h** listed on Page 99.

6.4.3 Input/Output for Arrangements with Curve History

When reading or writing an arrangement-with-history object we would like the curve history to be written into an output stream or read from an input stream alongside with the basic arrangement structure.

The *2D Arrangements* package supplies an inserter and an extractor for any instance of the **Arrangement_with_history_2<Traits,Dcel>** class template. The arrangement is stored using a simple predefined textual format. An object of the **Arrangement_with_history_2<Traits,Dcel>** type can be saved and restored, as long the **Curve_2** type defined by the traits class—as well as the **Point_2** and the **X_monotone_curve_2** types—support the << and >> operators. Thus, you need to add the #**include** statements below to perform I/O operations on an arrangement-with-history object **arr** to and from a stream.

#**include** <CGAL/IO/Arr_with_history_iostream.h>
#**include** <fstream>

Then read or write **arr** simply use the << and >> operators, respectively:

```
std::ofstream    out_file("arr_hist.dat");
out_file << arr;
out_file.close();

std::ifstream    in_file("arr_hist.dat");
in_file >> arr;
in_file.close();
```

┌──────── *advanced* ────────

The *2D Arrangements* package also includes the free function templates **write(arr, os, formatter)** and **read(arr, os, formatter)** that operate on a given arrangement-with-history object **arr**. Both function templates are parameterized by a **formatter** type, which defines the I/O format. The package contains a class template called **Arr_with_hist_text_formatter< ArranagmentFormatter>**, which extends an arrangement formatter class (see Section 6.2.3) and defines a simple textual input/output format.

└──────── *advanced* ────────

6.5   Application: Polygon Repairing and Winding Numbers

Boolean operations on polygons, referred to as Boolean set operations, constitute fundamental tasks in computational geometry. These operations are ubiquitous in computer graphics, computer-aided design and manufacturing (CAD/CAM), electronic design automation, and many more domains. Unfortunately, input data of such operations, namely a set of one or more polygons, used in real-world applications is occasionally corrupted, as it originates from measuring devices that are susceptible to noise and physical disturbances. In some other cases, it contains many degeneracies, which either disable computations based on fixed-precision arithmetic or further slow down computation using exact geometric computation. Chapter 8 introduces a package that supports Boolean set operations on point sets bounded by x-monotone segments in two-dimensional Euclidean space. These operations expect each input point set to meet a specific set of requirements. Naturally, passing point sets that fail to meet these requirements as input to a Boolean set operation must be avoided.

The polygon repairing problem: *Given a sequence of x-monotone segments that compose a closed curve, which represents the boundary of a point set, subdivide the point set into as few as possible simple point sets, each bounded by a counterclockwise-oriented boundary comprising x-monotone segments that are pairwise disjoint in their interior.*

A simple point set is topologically equivalent to a disc, and has a well-defined interior and exterior. Refer to Chapter 8 for the exact definition of this and related terms.

Section 5.6 lists the function template `polygon_orientation()`, which determines the orientation of the boundary of a point set represented by a sequence of x-monotone segments that compose a closed curve. It computes the correct orientation only if the boundary curve does not overlap or cross itself. It can be improved to handle self-overlapping boundaries, but heavier machinery is required to properly handle self-crossing boundaries.

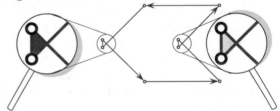

Automatically "fixing" corrupted data, that is, converting invalid point sets, the interiors and exteriors of which are not well defined, to valid ones, is not a simple task. As a matter of fact, it is impossible to come up with a procedure that yields the desired results at all times. Consider, for example, the self-intersecting polygon depicted to the right. It is uncertain which points are contained in the point set and which ones are not, but including the green small triangle on the right and excluding the red small triangle on the left is a good guess. Winding numbers appear to be useful in such cases. The winding number of a point is the number of counterclockwise cycles the oriented boundary makes around the point. It is common to use one of the following three heuristics (although one could come up with more options): (i) Consider a point included in the point set only if its winding number is nonzero. (ii) Consider a point included in the point set only if its winding number is greater than zero. (iii) Consider a point included in the point set only if its winding number is odd. Applying option (ii) results in the guess above, but there is no guarantee that it is the desired result, and so applying any specific option may work well on a number of input cases, but fail on others.

Arrangements can be used to compute the winding numbers of all points in the plane efficiently as follows. We use an arrangement-with-history data structure where each face is extended with an integer that stores the winding number of every point on the face. The use of `Arrangement_with_history_2` enables the retrieval of all curve segments that induce a given edge of the arrangement. We apply a depth-first search (DFS) traversal[5] on all the arrangement faces, starting from the unbounded face, and update the winding-number counters of each face as we visit new faces. Figure 6.2 contains a sketch of the process, where we apply option (iii) above on a star self-intersecting polygon.

We start with the unbounded face. We maintain a counter of the number of counterclockwise cycles the oriented boundary makes. The counter is initialized with zero. Indeed, the winding

---

[5]The Boost Graph Library, BGL, for example, can be employed for this task.

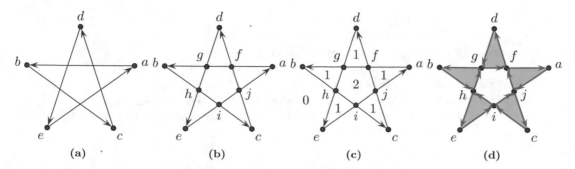

Fig. 6.2: (a) A self-crossing polygon given by $\{a, b, c, d, e\}$. (b) The arrangement data structure constructed from the polygon edges. (c) The arrangement data structure with updated face winding-numbers. (d) The resulting polygons.

number of the unbounded face is zero. When we cross a halfedge e moving from the face f incident to e to the face f' incident to the twin of e, we subtract the number of curve segments inducing e such that f lies to their left, and add the number of remaining curve segments inducing e, that is, the number of curve segments inducing e such that f' lies to their left.

In non-degenerate cases the net absolute contribution of a single edge is 1. However, in degenerate cases, where overlaps occur, the net absolute contribution of a single edge can be any other integer. It is, for example, 2 for every edge out of the four in the figure to the right, as every edge is induced by two curve segments oriented in the same direction. Thus, the winding number of the single bounded face is 2. The functor template 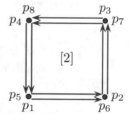 `Winding_number<Arrangement>` listed below, and defined in the header file `Winding_number.h`, computes the winding number of the faces of a given arrangement. When it is instantiated, the template parameter `Arrangement` must be substituted with an instance of the `Arrangement_with_history_2` class template. In addition, the type `Curve_2` nested in the traits class `Arrangement::Traits` must be convertible to the nested type `X_monotone_curve_2`, as the operator of the functor `Compare_endpoints_xy_2` applied to an inducing curve segment c accepts only a curve segment of the latter type, while the type of the inducing curve segments in an arrangement-with-history is `Curve_2`; see Line 22 in the code excerpt below. Moreover, the inducing curve segments must be x-monotone, and thus have a well-defined direction (either left to right or right to left). This allows comparing the direction of an inducing curve segment with the direction of the halfedge induced by it; see Line 35 in the code excerpt below.

```
1   #include <utility>
2
3   #include <CGAL/basic.h>
4   #include <CGAL/enum.h>
5
6   template <typename Arrangement> class Winding_number {
7   private:
8     Arrangement& _arr;
9     typename Arrangement::Traits_2::Compare_endpoints_xy_2 _cmp_endpoints;
10
11     // The Boolean flag indicates whether the face has been discovered already
12     // during the traversal. The integral field stores the winding number.
13     typedef std::pair<bool, int>                              Data;
14
15   public:
16     Winding_number(Arrangement& arr) : _arr(arr)
```

```cpp
17    {
18      // Initialize the winding numbers of all faces.
19      typename Arrangement::Face_iterator fi;
20      for (fi = _arr.faces_begin(); fi != _arr.faces_end(); ++fi)
21        fi->set_data(Data(false, 0));
22      _cmp_endpoints = _arr.traits()->compare_endpoints_xy_2_object();
23      propagate_face(_arr.unbounded_face(), 0);    // compute the winding numbers
24    }
25
26  private:
27    // Count the net change to the winding number when crossing a halfedge.
28    int count(typename Arrangement::Halfedge_handle he)
29    {
30      bool l2r = he->direction() == CGAL::ARR_LEFT_TO_RIGHT;
31      typename Arrangement::Originating_curve_iterator ocit;
32      int num = 0;
33      for (ocit = _arr.originating_curves_begin(he);
34           ocit != _arr.originating_curves_end(he); ++ocit)
35        (l2r == (_cmp_endpoints(*ocit) == CGAL::SMALLER)) ? ++num : --num;
36      return num;
37    }
38
39    // Traverse all faces neighboring the given face and compute the
40    // winding numbers of all faces while traversing the arrangement.
41    void propagate_face(typename Arrangement::Face_handle fh, int num)
42    {
43      if (fh->data().first) return;
44      fh->set_data(Data(true, num));
45
46      // Traverse the inner boundary (holes).
47      typename Arrangement::Hole_iterator hit;
48      for (hit = fh->holes_begin(); hit != fh->holes_end(); ++hit) {
49        typename Arrangement::Ccb_halfedge_circulator cch = *hit;
50        do {
51          typename Arrangement::Face_handle inner_face = cch->twin()->face();
52          if (inner_face == cch->face()) continue;       // discard antenas
53          propagate_face(inner_face, num + count(cch->twin()));
54        } while (++cch != *hit);
55      }
56
57      // Traverse the outer boundary.
58      if (fh->is_unbounded()) return;
59      typename Arrangement::Ccb_halfedge_circulator cco = fh->outer_ccb();
60      do {
61        typename Arrangement::Face_handle outer_face = cco->twin()->face();
62        propagate_face(outer_face, num + count(cco->twin()));
63      } while (++cco != fh->outer_ccb());
64    }
65  };
```

The application presented in this section attempts to "fix" invalid representations of point sets. It is centered around a generic implementation of the following algorithm: The input point set is represented as a sequence of x-monotone segments that compose a closed curve. The program constructs an arrangement that represents the input point set. It calculates the winding numbers associated with each face of the arrangement. Next, it excludes points the winding numbers of which are even (applying the third heuristic above). Finally, it returns the resulting point sets represented by their counterclockwise-oriented boundaries.

The function template `polygon_repairing(begin, end, container, traits)` accepts a range of x-monotone segments that compose a closed curve. The input range `[begin, end)` and the input traits objects must meet the same conditions as the corresponding arguments of the `polygon_orientation(begin, end, traits)` function template listed on Page 122. That is, the target point of every segment must be equal to the source point of the next segment in the range. The target point of the last segment must be equal to the source point of the first segment. The type of the input iterators that define the input range must model the concept BidirectionalIterator.[6] The type of the traits object must model the concept ArrangementDirectionalXMonotoneTraits_ 2. The value type of the input-iterator type must be convertible to the type `Curve_2` nested in the traits class `Traits`, as only this type of curve can be inserted into an arrangement object the type of which is an instance of the `Arrangement_with_history_2` class template. The `container` argument is used to obtain the results. It is a container of point sets the interior and exterior of each of which are well-defined. Each point set is represented by a container of x-monotone curve segments. The function template is listed below, and defined in the header file `polygon_repairing.h`.

```cpp
#include <utility>

#include <CGAL/basic.h>
#include <CGAL/Arrangement_with_history_2.h>
#include <CGAL/Arr_extended_dcel.h>

#include "Winding_number.h"

template <typename Traits, typename Input_iterator, typename Container>
void polygon_repairing(Input_iterator begin, Input_iterator end,
                       Container& res, const Traits& traits)
{
  // Each face is extended with a pair of a Boolean flag and an integral
  // field: The former indicates whether the face has been discovered
  // already during the traversal. The latter stores the winding number.
  typedef std::pair<bool, int>                          Data;
  typedef CGAL::Arr_face_extended_dcel<Traits, Data>    Dcel;
  typedef CGAL::Arrangement_with_history_2<Traits, Dcel>  Arrangement;

  Arrangement arr(&traits);
  insert(arr, begin, end);
  Winding_number<Arrangement> winding_number(arr);

  typename Arrangement::Face_iterator fi;
  for (fi = arr.faces_begin(); fi != arr.faces_end(); ++fi) {
    if ((fi->data().second % 2) == 0) continue;

    CGAL_assertion(!fi->is_unbounded());
    typename Container::value_type polygon;
    typename Arrangement::Ccb_halfedge_circulator cco = fi->outer_ccb();
    do polygon.push_back(cco->curve());
    while (++cco != fi->outer_ccb());
    res.push_back(polygon);
```

---

[6]See `http://www.sgi.com/tech/stl/` for a complete specification of the SGI STL, where the concept BidirectionalIterator is defined, among others.

```
    }
}
```

Like the polygon-orientation example presented in Section 5.6, the application comes with a **main()** function that extracts an ordered sequence of point set boundary vertices from an input file. It constructs a list of line segments that compose the boundary of the polygon as follows. The first segment connects the first point to the second; the second segment connects the second point to the third; and so on; finally, the last segment connects the last and the first points. The function then invokes an instance of the **polygon_repairing()** function template above, passing the range of segments as input. The sequence of segments are guaranteed to be consistently oriented, as they are constructed from the ordered sequence of points.

```cpp
// File: polygon_repairing

#include<list>

#include <CGAL/Exact_predicates_exact_constructions_kernel.h>
#include <CGAL/Arr_segment_traits_2.h>
#include <CGAL/Arrangement_2.h>

#include "read_objects.h"
#include "polygon_repairing.h"

typedef CGAL::Exact_predicates_exact_constructions_kernel Kernel;
typedef CGAL::Arr_segment_traits_2<Kernel>                Traits;
typedef Traits::Point_2                                   Point;
typedef Traits::X_monotone_curve_2                        Segment;

int main(int argc, char* argv[])
{
  std::list<Point> points;
  const char* filename = (argc > 1) ? argv[1] : "polygon.dat";
  read_objects<Point>(filename, std::back_inserter(points));
  CGAL_assertion(points.size() >= 3);

  std::list<Segment> segments;
  std::list<Point>::const_iterator it = points.begin();
  const Point& first_point = *it++;
  const Point* prev_point = &first_point;
  while (it != points.end()) {
    const Point& point = *it++;
    segments.push_back(Segment(*prev_point, point));
    prev_point = &point;
  }
  segments.push_back(Segment(*prev_point, first_point));

  std::list<std::list<Segment> > polygons;
  polygon_repairing(segments.begin(), segments.end(), polygons, Traits());

  std::list<std::list<Segment> >::const_iterator pit;
  for (pit = polygons.begin(); pit != polygons.end(); ++pit) {
    std::copy(pit->begin(), pit->end(),
              std::ostream_iterator<Segment>(std::cout, "\n"));
    std::cout << std::endl;
  }
```

```
        return 0;
}
```

As mentioned above, using winding numbers is a heuristic approach. Consider, for example, the self-intersecting polygon depicted at the top of the figure to the right. The polygon boundary induces two abutting faces with identical (odd and positive) winding numbers. Notice the delicate difference between the polygons in the figure. A small closed portion of the boundary of the top polygon winds once in counterclockwise orientation, i.e., p_4, p_5, p_6, and once again in clockwise orientation, i.e., p_7, p_8, p_9. Naturally, any test involving winding numbers, and in particular the three heuristics described on Page 152, cannot distinguish between the two faces, and thus both faces will either be considered as part of the final point set or be altogether dismissed.

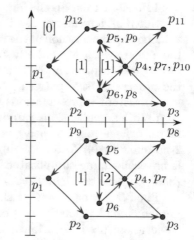

The program listed in this section suffers from a few drawbacks. First, it results in pairs of abutting polygons when fed with polygons such as the one depicted at the top of the figure to the right. Each such pair can be unified into a single polygon by removing their common edge, thereby simplifying the result. Secondly, the directions of the line segments that comprise the output-polygon boundaries are not consistent when fed with polygons such as the one depicted at the bottom of the figure. Finally, the output polygon that results in this case is invalid with respect to the strict definition of valid polygons as stated in Chapter 8.

Try: Enhance the program as follows:
1. Unify abutting polygons by removing their common edge thereby simplifying the result.
2. Reverse the direction of the line segments that comprise the output-polygon boundaries as necessary so that the boundaries are consistently counterclockwise oriented.

In Exercise 8.1 you are asked to further enhance the program and eliminate the last drawback.

6.6   Bibliographic Notes and Remarks

The overlay of two arrangements with m and n vertices, respectively, can be computed using a plane-sweep algorithm in $O(k \log(m + n))$ time, where k is the complexity of the resulting arrangement. The operation is described by de Berg et al. [45, Section 2.3], including details regarding the construction of a Dcel that represents the resulting arrangement.

The need to overlay two or more arrangements is common to many applications. As mentioned in the beginning of this chapter, computing the overlay of two planar arrangements is useful for supporting Boolean set operations on polygons or general polygons as discussed in Chapter 8; see also e.g., [19, 23].

The overlay of two-dimensional arrangements induced by geodesic arcs embedded on the unit sphere (see Section 11.1 and [75, 76]) is a fundamental operation in the computation of Minkowski sums of convex polytopes in $\mathbb{R}^3$ [20], which in turn is used in assembly partitioning of bounded polyhedra in $\mathbb{R}^3$ [72]. Here, the overlay operation is directly applied to two additional variants of such arrangements. While arrangements induced by geodesic arcs embedded on the unit sphere are not covered in this book, it is possible to replace one such arrangement with six (planar) arrangements induced by line segments. These arrangements are embedded on the six planes underlying the six faces of the unit cube (a parallel-axis cube circumscribing the unit sphere) and stitched properly at the edges of the cube [71].

Overlaying two-dimensional arrangements that represent partial envelopes of surfaces in $\mathbb{R}^3$ is the central part of the merge step of the recursive divide-and-conquer algorithm that computes the envelope of the surfaces; see Section 10.3. The overlay operation is also used in the construction of two-dimensional Voronoi diagrams via the construction of envelopes of surfaces in $\mathbb{R}^3$ [191, 192].

Computing the overlay of simply connected planar subdivisions optimally can be done in linear time in the number of features in the overlaid arrangement [67]. It is based on efficient convex subdivision. A convex arrangement, for example, is simply connected. Many applications would benefit from an efficient implementation of an operation that computes the overlay of such restricted arrangements; see Exercise 6.2.

The implementation of the overlay operation provided by the *2D Arrangements* package utilizes the plane sweep algorithmic framework and employs a dedicated visitor, which resembles the *visitor* design-pattern [83, Chapter 5]. Arrangement observers follow a mechanism that resembles the *observer* design-pattern [83, Chapter 5].

Consistent representations of the boundary and interior of three-dimensional point sets are required by applications ranging from interactive visualization to finite element analysis. Hoffmann defines a set of polygons in $\mathbb{R}^3$ to be *consistent* if the union of the polygons is a closed 2-manifold [119]. Automatically constructing a consistent representation from arbitrary data [165] is naturally more complicated in 3-space than in 2-space. Polygon repairing, also referred to as (bad) polygon repairing, is offered to various extents by various software tools.

The *2D Arrangements* package brings together generic programming and design patterns. It is implemented by generic components and other advanced C++ features. One example is the policy-clone idiom used to instantiate a DCEL class with many different possible traits types without ad hoc limitations on the type of the DCEL classes; see Section 6.4. More information on such techniques can be found in [7].

6.7 Exercises

6.1 Develop a program that defines an extended arrangement induced by line segments, such that (i) each halfedge is extended with a real number that represents the length of the curve associated with the halfedge, (ii) each face is extended with a real number that represents the area of the face, and (iii) each vertex is extended with a real number that represents the distance of the point associated with the vertex from the origin. The program constructs the arrangement induced by an input set of line segments. Then, it traverses all the cells of the arrangement and calculates the values of all the fields that extend the cells. Finally, it prints out the arrangement data including the extended values using an output formatter (see Section 6.2.3) that you are required to provide.

6.2 [**Project**] Develop a generic function called `overlay_connected()` that computes the overlay of simply connected arrangements of line segments, as described in [67]. Compare the execution time of the newly developed overlay function and the existing one on arrangements of various sizes.

6.3 [**Project**] Chapter 8 introduces the notion of general polygons, namely point sets the boundaries of which comprise general x-monotone curves. The polygon repairing application presented in Section 6.5 repairs only (linear) polygons, the boundaries of which comprise line segments. With a small change to the main function the application can be extended to support general polygons the boundaries of which are still x-monotone but not necessarily line segments. However, the types `Curve_2` and `X_monotone_curve_2` nested in the traits class must remain convertible to one another. The type `X_monotone_curve_2` must be convertible to the `Curve_2` type, because only curves of type `Curve_2` can be inserted into an arrangement object the type of which is an instance of the `Arrangement_with_history_2` class template. The type `Curve_2` must be convertible to the `X_monotone_curve_2` type, because the operator of the functor `Compare_endpoints_xy_2` accepts only curves of the `X_monotone_curve_2` type.

Augment the functor template `Winding_number` and the `polygon_repairing()` function template to repair general polygons the boundaries of which comprise line segments and circular arcs. Use the traits class `Arr_circle_segment_traits_2` described in Section 5.4.1.

- Apply a straightforward change that explicitly converts curves of type `Curve_2` to curves of type `X_monotone_curve_2` and vice versa where necessary.

- Optimize the upgraded application. Avoid the need to convert between curves of type `Curve_2` and curves of type `X_monotone_curve_2` altogether as follows:

  - Do not use the class template `Arrangement_with_history_2`. Instead, introduce a new class template that represents an arrangement and maintains a mapping from the arrangement halfedges to their inducing x-monotone curves (of type `X_monotone_curve_2`).

  - Alternatively, augment the `Arrangement_with_history_2` class template. Let C be a set of input curves (of type `Curve_2`), let C' be the set of maximal x-monotone subcurves of the curves in C, and let C'' be the set of subcurves of the curves in C' pairwise disjoint in their interiors ; see Section 2.1. Augment the `Arrangement_2` class template with two additional containers that represent C and C', respectively, and (i) a cross-mapping between the curves of C and the curves of C' and (ii) a cross-mapping between the curves of C' and the arrangement halfedges they induce.

Chapter 7

Adapting to Boost Graphs

Boost provides a collection of free peer-reviewed portable C++ source libraries that work well with, and are in the same spirit as, the C++ Standard Template Library (STL). The Boost Graph Library (Bgl) [198], which is one of the libraries in the collection, offers an extensive set of generic graph-algorithms. As our arrangements are embedded as planar graphs, it is only natural to extend the underlying data structure with the interface that the Bgl expects, and gain the ability to perform the operations that the Bgl supports, such as breadth-first search. This section describes how to apply the graph algorithms implemented in the Bgl to `Arrangement_2` instances.

An instance of `Arrangement_2` is adapted as a Boost graph through the provision of a set of free functions that operate on the arrangement features and conform to the relevant Bgl concepts. Besides the straightforward adaptation, which associates a graph vertex with each Dcel vertex and a graph edge with each Dcel halfedge, the package also offers a *dual* adaptor, which associates a graph vertex with each Dcel face such that two vertices are connected iff there is an arrangement edge incident to the two corresponding faces, that is, a halfedge is incident to one face and its twin is incident to the other.

7.1 The Primal Arrangement Representation

Arrangement instances are adapted as Boost graphs by specializing the `boost::graph_traits< Graph>` class template for `Arrangement_2` instances. The graph traits states the graph concepts that the arrangement class models and defines the types and operations required by these concepts.

In this specialization the `Arrangement_2` vertices correspond to the graph vertices, where two vertices are adjacent if there is at least one halfedge connecting them. More precisely, `Arrangement_2::Vertex_handle` is the graph vertex-descriptor, while `Arrangement_2::Halfedge_handle` is the graph edge-descriptor.[1] As halfedges are directed, we consider the graph to be directed as well. Moreover, parallel edges are allowed in our Boost graph, since several interior-disjoint x-monotone curves (e.g., circular arcs) may share two common endpoints, inducing an arrangement with two vertices that are connected with several edges.

As you know, it is possible to efficiently traverse the vertices and halfedges of a given `Arrangement_2` instance. Thus, the arrangement models the graph concepts VertexListGraph and EdgeListGraph that are used for the traversal of all vertex descriptors and all edge descriptors of a given graph, respectively, and are introduced by the Bgl. Using an iterator adaptor of the circulator over the halfedges incident to a vertex, namely `Halfedge_around_vertex_circulator` (see Section 2.2.1), it is possible to go over the incoming and outgoing edges of a vertex in linear time.[2] Therefore, our arrangement also models the Bgl graph concept BidirectionalGraph (this

---

[1] Vertex descriptors and edge descriptors are the terms used by Boost to describe objects that represent vertices and edges, respectively.

[2] The adaptor transforms the circulator into an iterator (so-called Multi-Pass Input Iterator), conforming to the

concept refines IncidenceGraph, which requires only the traversal of outgoing edges).

Notice that the vertex descriptors we use are `Vertex_handle` objects and *not* vertex indices. However, some of the BGL algorithms require (for efficiency reasons) the user to supply the indices $0, 1, \ldots, n-1$ for the vertices, where n is the number of vertices. We therefore introduce the `Arr_vertex_index_map<Arrangement>` class template, which maintains a mapping of vertex handles to indices, as required by the BGL. An object of the appropriate instance of this class template must be attached to a valid arrangement when the map object is created. We use the notification mechanism (see Section 6.1) to dynamically maintain the mapping of vertices to indices, as new vertices might be inserted into the arrangement, and existing vertices might be removed.

Another advantage of having indices for the vertices is the use of *property maps*. Property maps are used by BOOST to generically attach information to features. They are used by BOOST algorithms to receive various parameters that correspond to features, and to output results. For example, when we compute the shortest paths from a given source vertex s to all other vertices, we use three different property maps. The first maps edges to their lengths and is passed as input to the BGL algorithm. The other two property maps are used to hold the output; the first of them maps each vertex to its distance from s, and the second maps each vertex to the descriptor of the vertex that precedes it in the shortest path from s.

Property-map classes supplied by BOOST are based on prevalent data structures. A mapping between vertices and indices allows us to easily use property-map classes supplied by BOOST. For example, the property-map class templates `boost::vector_property_map`, which is based on `std::vector`, and `boost::iterator_property_map`, which can be used to implement a property map based on a native C++ array, require the user to supply such a mapping. The class template `boost::associative_property_map` can be used without such a mapping.

Note, however, that when a vertex is removed from the arrangement, the indices of vertices contained in the `Arr_vertex_index_map` object change. As a consequence the map object is no longer synchronized with the values contained in the property map. Thus, property maps should not be reused in calls to BGL functions if the arrangement is modified in between these calls.

Example: The first example of this chapter demonstrates the application of Dijkstra's shortest path algorithm to compute the Euclidean shortest-path length between a given vertex of an arrangement of linear curves and all other vertices. It uses an instance of the functor template `Edge_length<Arrangement>` to compute the Euclidean length of the linear curve associated with a given halfedge of the arrangement. The functor implements a BOOST property-map that attaches lengths to edges; when the BGL algorithm queries the property map for a length of an edge the property map computes and returns it. The functor template is defined in the header file `Edge_length.h`.

```
#include <boost/property_map.hpp>
#include <CGAL/basic.h>

template <typename Arrangement> class Edge_length {
public:
  // Boost property-type definitions.
  typedef boost::readable_property_map_tag    category;
  typedef double                              value_type;
  typedef value_type                          reference;
  typedef typename Arrangement::Halfedge_handle key_type;

  double operator()(typename Arrangement::Halfedge_handle e) const
  {
    const double diff_x = CGAL::to_double(e->target()->point().x()) -
      CGAL::to_double(e->source()->point().x());
    const double diff_y = CGAL::to_double(e->target()->point().y()) -
```

---

requirements of the BGL graph concepts.

```
     CGAL::to_double(e->source()->point().y());
   return std::sqrt(diff_x*diff_x + diff_y*diff_y);
}


friend double get(const Edge_length& edge_length, key_type key)
{ return edge_length(key); }
};
```

The example below constructs an arrangement of seven line segments, as shown in the figure to the right. Then, it uses the BGL generic implementation of Dijkstra's shortest path algorithm to compute the distances to all vertices from the lexicographically smallest vertex v_0 in the arrangement. Note the usage of the **Arr_vertex_index_map** class template in the call to **boost::dijkstra_shortest_paths()** and in the definition of the distance property-map. We instantiate a property map that attaches a number stored as a double-precision floating-point (**double**) to each vertex. The number represents the distance of the vertex from v_0.

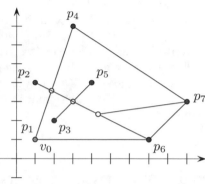

```
// File: ex_bgl_primal_adaptor.cpp

#include <vector>

#include <boost/vector_property_map.hpp>
#include <boost/graph/dijkstra_shortest_paths.hpp>

#include <CGAL/basic.h>
#include <CGAL/graph_traits_Arrangement_2.h>
#include <CGAL/Arr_vertex_index_map.h>

#include "arr_exact_construction_segments.h"
#include "Edge_length.h"

typedef CGAL::Arr_vertex_index_map<Arrangement>   Vertex_index_map;
typedef Edge_length<Arrangement>                  My_edge_length;

int main()
{
  // Construct an arrangement of seven intersecting line segments.
  // We keep a handle for the vertex v0 that corresponds to the point (1,1).
  Point p1(1, 1), p2(1, 4), p3(2, 2), p4(3, 7), p5(4, 4), p6(7, 1), p7(9, 3);
  Arrangement   arr;
  Segment       s(p1, p6);
  Arrangement::Halfedge_handle   e = insert_non_intersecting_curve(arr, s);
  Arrangement::Vertex_handle     v0 = e->source();
  insert(arr, Segment(p1, p4));   insert(arr, Segment(p2, p6));
  insert(arr, Segment(p3, p7));   insert(arr, Segment(p3, p5));
  insert(arr, Segment(p6, p7));   insert(arr, Segment(p4, p7));

  // Create a mapping of the arrangement vertices to indices.
  Vertex_index_map index_map(arr);

  // Create a property map based on std::vector to keep the result distances.
  boost::vector_property_map<double, Vertex_index_map>
```

```
                dist_map(arr.number_of_vertices(), index_map);

    // Perform Dijkstra's algorithm from the vertex v0.
    My_edge_length edge_length;
    boost::dijkstra_shortest_paths(arr, v0, boost::vertex_index_map(index_map).
                             weight_map(edge_length).distance_map(dist_map));

    // Print the distance of each vertex from v0.
    std::cout << "The graph distances of the arrangement vertices from ("
            << v0->point() << ") :" << std::endl;
    Arrangement::Vertex_iterator        vit;
    for (vit = arr.vertices_begin(); vit != arr.vertices_end(); ++vit)
        std::cout << "(" << vit->point() << ") at distance "
                    << dist_map[vit] << std::endl;
    return 0;
}
```

7.2 The Dual Arrangement Representation

An arrangement instance can be represented as a graph other than the one described in the previous section. A dual-graph representation refers to the graph, where each arrangement face corresponds to a graph vertex, and two vertices are adjacent iff the corresponding faces share a common edge on their boundaries. This is done by specializing the `boost::graph_traits` template for `Dual<Arrangement_2>` instances, where `Dual<Arrangement_2>` is a class template specialization that gives a dual interpretation to an arrangement instance.

In the dual representation, `Arrangement_2::Face_handle` is the graph vertex-descriptor, while `Arrangement_2::Halfedge_handle` is the graph edge-descriptor. We treat the graph edges as directed, such that a halfedge is directed from its incident face towards the incident face of its twin halfedge. As two arrangement faces may share more than a single edge on their boundary, we allow parallel edges in our BOOST graph. As in the case of the primal graph, the dual arrangement graph also models the concepts VertexListGraph, EdgeListGraph, and BidirectionalGraph (thus, also IncidenceGraph).

Since we use `Face_handle` objects as the vertex descriptors, we define the `Arr_face_index_map<Arrangement>` class template, which maintains an efficient mapping of face handles to indices.

Example: The next example demonstrates how a property map can be used to update or receive information directly from a feature of the arrangement without the need to search for its index. The example also demonstrates the application of the breadth-first search (BFS) algorithm on a dual arrangement. It uses the functor template `Extended_face_property_map<Arrangement, Type>` to directly access information stored inside the faces. The functor implements a property map that utilizes the `data()` and `set_data()` member functions of the extended face to update or obtain the property. When the property map is instantiated, the `Type` parameter must be substituted with the same type that is used to extend the arrangement face; see Section 6.2.1. The functor template is defined in the header file `Extended_face_property_map.h`.

```
template <typename Arrangement, class Type> class Extended_face_property_map {
public:
    typedef typename Arrangement::Face_handle        Face_handle;

    // Boost property type definitions.
    typedef boost::read_write_property_map_tag       category;
    typedef Type                                     value_type;
    typedef value_type&                              reference;
    typedef Face_handle                              key_type;
```

```
// The get function is required by the property map concept.
friend reference get(const Extended_face_property_map& map, key_type key)
{ return key->data(); }

// The put function is required by the property map concept.
friend void put(const Extended_face_property_map& map,
                key_type key, value_type val)
{ key->set_data(val); }
};
```

The example below constructs the same arrangement constructed by the program coded in **ex_bgl_primal_adaptor.cpp**, given in the previous section section. Then, it performs a BFS traversal on the graph faces, starting from the unbounded face. The DCEL faces are extended each with an unsigned integer indicating the discovered time of the face. The code uses a visitor (i.e., **boost::time_stamper**) that obtains the times and writes them into a property map that updates the faces accordingly. The figure to the right shows the graph dual to the arrangement. It is clear that the unbounded face f_0 is discovered at time 0, the neighboring faces f_1, f_3, and f_4 are discovered at times 1, 2, and 3, and finally f_2 is discovered at time 4.

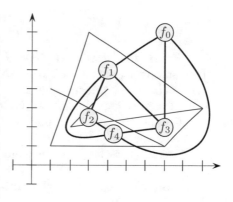

```
// File: ex_bgl_dual_adaptor.cpp
```

```
#include <boost/graph/breadth_first_search.hpp>
#include <boost/graph/visitors.hpp>

#include <CGAL/basic.h>
#include <CGAL/Arr_extended_dcel.h>
#include <CGAL/graph_traits_Dual_Arrangement_2.h>
#include <CGAL/Arr_face_index_map.h>

#include "Extended_face_property_map.h"
#include "arr_exact_construction_segments.h"
#include "arr_print.h"

typedef CGAL::Arr_face_extended_dcel<Traits, unsigned int>    Dcel;
typedef CGAL::Arrangement_2<Traits, Dcel>                     Ex_arrangement;
typedef CGAL::Dual<Ex_arrangement>                            Dual_arrangement;
typedef CGAL::Arr_face_index_map<Ex_arrangement>             Face_index_map;
typedef Extended_face_property_map<Ex_arrangement, unsigned int>
                                                             Face_property_map;

int main()
{
  // Construct an arrangement of seven intersecting line segments.
  Point p1(1, 1), p2(1, 4), p3(2, 2), p4(3, 7), p5(4, 4), p6(7, 1), p7(9, 3);
  Ex_arrangement  arr;
  insert(arr, Segment(p1, p6));
  insert(arr, Segment(p1, p4));   insert(arr, Segment(p2, p6));
  insert(arr, Segment(p3, p7));   insert(arr, Segment(p3, p5));
  insert(arr, Segment(p6, p7));   insert(arr, Segment(p4, p7));
```

```
// Create a mapping of the arrangement faces to indices.
Face_index_map  index_map(arr);

// Perform breadth-first search from the unbounded face, using the event
// visitor to associate each arrangement face with its discover time.
unsigned int     time = 0;
boost::breadth_first_search(Dual_arrangement(arr), arr.unbounded_face(),
                       boost::vertex_index_map(index_map).visitor
                          (boost::make_bfs_visitor
                             (stamp_times(Face_property_map(), time,
                                   boost::on_discover_vertex()))));

// Print the discover time of each arrangement face.
Ex_arrangement::Face_iterator  fit;
for (fit = arr.faces_begin(); fit != arr.faces_end(); ++fit) {
  std::cout << "Discover time " << fit->data() << " for ";
  if (fit != arr.unbounded_face()) {
    std::cout << "face ";
    print_ccb<Ex_arrangement>(fit->outer_ccb());
  }
  else std::cout << "the unbounded face." << std::endl;
}
return 0;
}
```

7.3   Application: Largest Common Point Sets under ϵ-Congruence

The largest common point sets under ϵ-congruence problem: *Given two point sets $A = \{a_1, \ldots, a_m\}$ and $B = \{b_1, \ldots, b_n\}$ in the plane, and a real parameter $\epsilon > 0$, find a translation T and two maximum subsets $\{i_1, \ldots, i_M\} \subseteq \{1, \ldots, m\}$ and $\{j_1, \ldots, j_M\} \subseteq \{1, \ldots, n\}$ such that $\|T(a_{i_k}) - b_{j_k}\| < \epsilon$ for each $1 \le k \le M$.*

The figure to the right depicts an input case where each of the two input point sets consists of four points. We represent the output as a set of pairs of indices, where the first and second elements of each pair identify two points of the two input sets, respectively, and each index appears at most once. These types are listed below, and defined in the header file `Index_pair_set.h`.

```
#include <utility>
#include <set>


typedef std::pair<unsigned int, unsigned int>          Index_pair;
typedef std::set<Index_pair>                           Index_pair_set;
```

We solve this optimization problem in the translation space, which is in our case a plane, where a point (dx, dy) represents a translation of the set A by dx along the x-axis and by dy along the y-axis. Given the input sets A and B, we create a set of mn circles:

$$C_{ij} = \{p \in \mathbb{R}^2 \mid \|p - (b_j - a_i)\| = \epsilon\} .$$

The open disc enclosed by the circle C_{ij} in the translation space represents all translations that bring the point a_i to a distance of up to (but less than) ϵ from the point b_j.[3] If we construct the arrangement of these circles we obtain a subdivision of the translation space into cells, such that each cell is properly covered by a distinct set of discs.

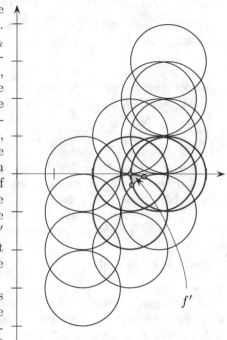

At first glance, it seems that we just need to locate the arrangement face covered by the largest number of discs. However, this set of discs may contain C_{ij} and also C_{ik} for some i, j, and k. Namely, a_i is mapped to an ϵ-environment of both b_j and b_k, which is an invalid solution, as we are looking for a one-to-one mapping. The figure to the right shows the translation space for the input case depicted on the previous page and $eps = 1$. The arrangement is induced by 16 circles; it consists of 118 vertices, 216 edges, and 100 faces. The face denoted by f' in the figure (lightly shaded) is covered by three discs (circled in blue), among others, which are associated with the pairs of indices, respectively, that compose the solution, namely the set $\{\langle 1,1\rangle, \langle 2,2\rangle, \langle 3,3\rangle\}$. Every point located inside the face f' represents a corresponding valid translation. The face f' is covered by another disc (circled in red), which does not contribute to the solution. Notice that there are faces in the arrangement covered by as many as five discs.

For some face f let $\mathcal{I}(f)$ denote the set of pairs of indices that identify the discs that properly cover f. What we are really looking for is a *maximum bipartite matching*—a maximum subset of the index pairs $\{\langle i_1, j_1\rangle, \ldots, \langle i_M, j_M\rangle\} \subseteq \mathcal{I}(f)$, which correspond to the discs that properly cover the face f, in which each index i_k and each index j_k occur exactly once. The function `bipartite_matching()` listed below accepts a set of index pairs and computes the maximum bipartite matching. It exploits the BGL function `boost::edmonds_maximum_cardinality_matching()`, which computes the maximum cardinality matching in general graphs, a generalization of the maximum bipartite matching; see Section 7.4 for more details and references. The function `bipartite_matching()` is defined in the header file `bipartite_matching.h`.

```
#include <vector>
#include <boost/graph/adjacency_list.hpp>
#include <boost/graph/max_cardinality_matching.hpp>

#include "Index_pair_set.h"

unsigned int bipartite_matching(unsigned int m, unsigned int n,
                                const Index_pair_set& index_pair_set,
                                Index_pair_set* match_set = NULL)
{
  typedef boost::adjacency_list<boost::vecS, boost::vecS, boost::undirectedS>
    Graph;

  // Create a graph with (m + n) vertices. The first m vertices and the last n
  // vertices represent the points of the set A and the set B, respectively.
  Graph  G(m + n);

  // Add the graph edges: create an edge between pairs of 'A'
  // and 'B' vertices according to the given index-pair set.
```

---

[3] For the variant we are describing here it is crucial that the distance between matched points is *less than* ϵ. In the exercise we look at the case where this distance is *at most* ϵ.

```
Index_pair_set::const_iterator  it;
for (it = index_pair_set.begin(); it != index_pair_set.end(); ++it)
    boost::add_edge(it->first, it->second + m, G);

// The mate vector holds the result matching.
std::vector<boost::graph_traits<Graph>::vertex_descriptor> mate(m + n);

// Perform the maximum cardinality matching algorithm.
boost::edmonds_maximum_cardinality_matching(G, &mate[0]);
unsigned int match_size = boost::matching_size(G, &mate[0]);

if (match_set != NULL) {
    // Obtain the match.
    boost::graph_traits<Graph>::vertex_iterator vi, vi_end;
    for (tie(vi, vi_end) = vertices(G); vi != vi_end; ++vi)
        if ((mate[*vi] != boost::graph_traits<Graph>::null_vertex()) &&
            (*vi < mate[*vi]))
            match_set->insert(Index_pair(*vi, mate[*vi] - m));
}
return match_size;
}
```

The program below implements the entire algorithm. It reads the two input sets and the value of ϵ, and constructs the circles accordingly. Note that it uses the consolidated curve-data geometry traits to attach a set of pairs of indices $\langle i, j \rangle$ to each circle. It uses a set of pairs rather than a single pair to handle the case of overlapping circles. It then constructs the arrangement of all circles in an aggregate manner.

Once the arrangement is constructed, the program considers the dual graph and traverses the arrangement faces in a breadth-first order. It employs an event visitor that attaches a set of index pairs to each arrangement face. This is done as follows: First, the unbounded face f_0 is trivially associated with an empty index-set $\mathcal{I}(f_0) = \emptyset$. Next, the visitor is notified each time the BFS algorithm crosses an edge e from a face f_1 to a yet undiscovered face f_2. Let $I(e)$ denote the set of index pairs associated with the edge e; then we perform the following operation:

$$\begin{cases} \mathcal{I}(f_2) = \mathcal{I}(f_1) \cup \{I(e)\} & I(e) \cap \mathcal{I}(f_1) = \emptyset \\ \mathcal{I}(f_2) = \mathcal{I}(f_1) \setminus \{I(e)\} & I(e) \cap \mathcal{I}(f_1) = I(e) \end{cases}$$

Namely, if the current face f_1 does not contain $\langle i, j \rangle \in I(e)$, then f_1 is exterior to the corresponding disc C_{ij}. When we cross e we reach f_2, which lies inside the circle, so we insert $I(e)$ into its index set. Otherwise, the current face f_1 lies in the interior of C_{ij}, and we have to remove $I(e)$ from the index set of the neighboring face f_2.

After computing the index-pair set $\mathcal{I}(f)$ associated with each arrangement face f, we go over the faces and compute the largest common point set for each specific face, searching for the face with the largest common point set among all faces.

```
// File: opt_translation.cpp

#include <vector>
#include <iostream>
#include <sstream>
#include <boost/graph/breadth_first_search.hpp>
#include <boost/graph/adjacency_list.hpp>

#include <CGAL/basic.h>
#include <CGAL/Arr_consolidated_curve_data_traits_2.h>
#include <CGAL/Arr_extended_dcel.h>
```

```
#include <CGAL/graph_traits_Dual_Arrangement_2.h>
#include <CGAL/Arr_face_index_map.h>

#include "arr_circular.h"
#include "arr_print.h"
#include "read_objects.h"
#include "Index_pair_set.h"
#include "bipartite_matching.h"

typedef std::vector<Rational_point>                    Point_set;
typedef CGAL::Arr_consolidated_curve_data_traits_2<Traits, Index_pair>
                                                       Data_traits;
typedef Data_traits::Curve_2                           Indexed_circle;
typedef CGAL::Arr_face_extended_dcel<Data_traits, Index_pair_set>  Dcel;
typedef CGAL::Arrangement_2<Data_traits, Dcel>         Ex_arrangement;
typedef CGAL::Dual<Ex_arrangement>                     Dual_arrangement;

// An event visitor class that associates each graph vertex (namely an
// arrangement face) with its set of indices.
struct Index_pair_set_visitor {
public:
  // The visitor is applied (only when a new face is discovered) according
  // to this event tag.
  typedef boost::on_tree_edge                          event_filter;

  // When examining an edge that belongs to the tree, crossing from the
  //current face to an adjacent face, prepare the index-pair set of that face.
  template <typename Edge, typename Graph>
  void operator()(Edge e, const Graph& g)
  {
    // Use the index-pair set of the current face (e's source) as a base set.
    // Check whether the base set contains the index pair associated with the
    // edge. If so, this index should be removed from the set of the adjacent
    // face; otherwise, it should be inserted there.
    Index_pair_set base_set = source(e, g)->data();
    Data_traits::Data_container& edge_set = e->curve().data();
    Data_traits::Data_iterator it;
    for (it = edge_set.begin(); it != edge_set.end(); ++it) {
      Index_pair_set::iterator pos = base_set.find(*it);
      if (pos != base_set.end()) base_set.erase(pos);
      else                       base_set.insert(*it);
      target(e, g)->set_data(base_set);
    }
  }
};

// The main function.
int main(int argc, char* argv[])
{
  if (argc < 4) {
    std::cerr << "Usage: " << argv[0]
              << " <point-set_#1> <point-set_#2> <epsilon>" << std::endl;
    return -1;
  }
```

```
// Read the two input point-sets.
Point_set A, B;
read_objects<Rational_point>(argv[1], std::back_inserter(A));
read_objects<Rational_point>(argv[2], std::back_inserter(B));

// Read the rational error-bound (epsilon) for the match, e.g., 1/1000:
Number_type epsilon;
std::istringstream iss(argv[3], std::istringstream::in); iss >> epsilon;

// For each point a in A and b in B, construct a circle centered at (b - a)
// with radius epsilon. This circle defines a disc in the translation plane
// of all valid translations such that || T(a) - b || < epsilon.
const unsigned int m = A.size(), n = B.size();
std::vector<Indexed_circle>  circs(m*n);
for (unsigned int i = 0; i < m; ++i) {
  for (unsigned int j = 0; j < n; ++j) {
    // Note that we use the constructor for a circle given its radius
    // (instead of the squared radius). The orientation is counterclockwise.
    Rational_point  center(B[j].x() - A[i].x(), B[j].y() - A[i].y());
    Curve           circle(center, epsilon, CGAL::COUNTERCLOCKWISE);
    circs[i*n + j] = Indexed_circle(circle, Index_pair(i, j));
  }
}

// Construct the arrangement of all circles.
Ex_arrangement  arr;
insert(arr, circs.begin(), circs.end());

// Create a mapping of the arrangement faces to indices.
CGAL::Arr_face_index_map<Ex_arrangement>  index_map(arr);

// Perform breadth-first search from the unbounded face. Use the event
// visitor to associate each arrangement face with its index-pair set.
Ex_arrangement::Face_handle             uf = arr.unbounded_face();
boost::breadth_first_search(Dual_arrangement(arr), uf,
                      boost::vertex_index_map(index_map).
                      visitor(boost::make_bfs_visitor
                              (Index_pair_set_visitor())));

// Go over the arrangement faces and find the best one.
Ex_arrangement::Face_const_iterator   fit;
unsigned int                          curr_match;
Ex_arrangement::Face_const_handle     opt_face = uf;
unsigned int                          best_match = 0;
for (fit = arr.faces_begin(); fit != arr.faces_end(); ++fit) {
  // Compute the maximal match of indices pair over the current face.
  curr_match = bipartite_matching(m, n, fit->data());
  if (curr_match > best_match) {
    // Update if we have found a better match than the best so far.
    opt_face = fit;
    best_match = curr_match;
  }
}
```

```
// Recompute and print the best match.
Index_pair_set                              opt_set;
Index_pair_set::const_iterator              ips_it;
bipartite_matching(m, n, opt_face->data(), &opt_set);
std::cout << "Found_" << best_match << "_matching_point_pairs:"<< std::endl;
for (ips_it = opt_set.begin(); ips_it != opt_set.end(); ++ips_it)
  std::cout << "__(" << A[ips_it->first]
            << ")_—>_(" << B[ips_it->second] << ')' << std::endl;
return 0;
}
```

The program uses types defined in the header file `arr_circular.h` listed on Page 99.

 Try: The program coded in `opt_translation.cpp` finds exactly one solution. Alter the code to compute all solutions in case more than one exists. Call the new program `opt_translation_all`. Below, we propose that you generate a family of input sets that have a unique optimal solution each. Use the program `opt_translation_all` to verify that indeed only a single solution exists for each input set.

Generate two sets of points in the plane $A = \{a_1, \ldots, a_n\}$ and $B = \{b_1, \ldots, b_n\}$ of a given size n, and another set of rational numbers $\Upsilon = \{\epsilon_1, \ldots, \epsilon_n\}$, such that the unique solution to the **largest common point sets under ϵ-congruence problem** with input A, B, and $\epsilon = \epsilon_i$ is $\{1, \ldots, i\}$ and $\{1, \ldots, i\}$ for $2 \leq i \leq n$. (For $i = 1$, that is, subsets of cardinality 1, a solution always exists and is never unique.) For example, for $n = 4$, the two quadruple sets of points depicted in the figure on page 167 along with the sequence $\{0.1, 0.5, 1, 2\}$ will do the trick.

Save the two sets of points in two files, say `setA.dat` and `setB.dat`. (Do not forget to place the number of points, n, as the first item in each file.) Then execute the program `opt_translation_all` n times, once for each number $\epsilon_i \in \Upsilon$. If the numbers in Υ are saved in separate lines, in say the file `epsilons.dat`, the desired loop can be performed by executing the following Bash-shell command-line:

```
while read eps; do opt_translation setA.dat setB.dat $eps; done < epsilons.dat
```

7.4 Bibliographic Notes and Remarks

Graphs are mathematical abstractions that are useful for solving many types of problems in computer science. Cormen et al. [41] covers a broad range of algorithms in depth. Consequently, these abstractions must be represented in computer programs. The BGL provides a standardized generic interface for traversing graphs that hides the details of the implementation. The BGL provides some general purpose graph-classes that conform to this interface.

Written by the BGL developers, The BOOST Graph Library: User Guide and Reference Manual [198] gives you all the information you need to take advantage of this library. It presents an in-depth description of the BGL, and provides working examples designed to illustrate the application of the BGL to these real-world problems. The source for the BGL is available as part of the BOOST distribution.

The BGL is one library among many in the collection of free peer-reviewed portable C++ source libraries called BOOST. The examples introduced in the previous chapters, mainly in this chapter, and in the following chapters extensively use STL and BOOST.

In Section 7.3 we provide a complete solution to the problem of computing a largest common point set between two point sets under ϵ-congruence in the plane, where the points are allowed only to translate. This problem comes in many variations (different dimensions, transformations, and metrics), has a long history, and has been the subject of extensive study; see, e.g., [10].

Finding a maximum matching in a bipartite graph is a known problem in graph theory [217, Chapter 3]. It can, for example, be reduced to the well-known *maximum-flow* problem. As mentioned in Section 7.3 the BGL function `boost::edmonds_maximum_cardinality_matching()`

computes the maximum cardinality matching for general graphs. Its implementation closely follows Tarjan's description [201, Chapter 9] of Edmonds' algorithm [55]. It runs in $O(mn\alpha(m,n))$ time, where m and n are the number of edges and vertices in the input graph, and $\alpha()$ is a two-parameter variation of the inverse of Ackerman's function (an extremely slowly growing function). Edmonds' algorithm has subsequently been improved to run in $O(\sqrt{n}m)$ time [157], matching the time for maximum bipartite matching. Another algorithm by Mucha and Sankowski [161] runs in $O(V^{2.376})$ time.

In Exercise 7.5 we ask you to implement a representation of a two-dimensional arrangement as an *incidence graph*. This is a variant of prevailing graph representations of polytopes or arrangements of hyperplanes [53, Chapter 7].

7.5   Exercises

7.1 Given an arrangement induced by a sequence of x-monotone curves that compose a closed curve, which represents the boundary of a point set, the functor template `Winding_number`, introduced in Section 6.5, computes the number of counterclockwise cycles the oriented boundary of every face makes around the face. It applies a depth-first search (DFS) algorithm to compute the winding numbers of all faces. Rewrite the code of `Winding_number` using the BGL implementation of the DFS algorithm and the arrangement dual adaptor.

7.2 In this chapter you were exposed to fundamentally two different methods of implementing property maps. One method requires the extension of the cells of the arrangement with fields that accommodate the property. An example of such a property map is the `Extended_face_property_map` class template listed on Page 164. The other method maintains an external map. Obtaining the property value for a given cell using the first method is done in constant time. However, as the extension is applied to all cells of a given type indiscriminately, the second method is more memory-efficient in cases where the property values of certain cells are irrelevant or become useless and can be discarded.

   (a) Rewrite the program coded in the file `ex_bgl_primal_adaptor.cpp` to use time-efficient property maps. Extend the arrangement cells as necessary. In particular replace the input vertex index-map and the output vertex distance-map with the corresponding new maps. Compare the performance of the original and the modified programs.

   (b) Rewrite the program coded in the file `ex_bgl_dual_adaptor.cpp` to use an external property map. Replace the use of the `Extended_face_property_map` class template with that of a BOOST property-map. Compare the performance of the original and the modified programs.

7.3 Optimize the program coded in the file `opt_translation.cpp` as follows. The original program consists of two phases. During the first phase it stores for each face the list of candidate pairs to participate in the best matching. During the second phase it iterates over all faces, computes the best matching for each face, and maintains the best matching found throughout.

   Improve the program to consist of a single phase that computes the maximum bipartite matching for each face when it is discovered as part of the breadth-first search traversal. There is no need to store extra information in the records of the arrangement faces. Keep only the best match found along the traversal at all times.

7.4 Write an efficient program that solves the variant of the **largest common point sets under ϵ-congruence problem** (see Section 7.3), where one point set is mapped to the *closed ϵ-environment of the other point set. Formally, given two point sets $A = \{a_1, \ldots, a_m\}$ and $B = \{b_1, \ldots, b_n\}$ in the plane with some $\epsilon > 0$, find a translation T and two subsets* *of maximal cardinality, $\{i_1, \ldots, i_M\} \subseteq \{1, \ldots, m\}$ and $\{j_1, \ldots, j_M\} \subseteq \{1, \ldots, n\}$, such that $\|T(a_{i_k}) - b_{j_k}\| \leq \epsilon$ for $1 \leq k \leq M$.*

7.5 **[Project]** Develop a generic adaptor that adapts arrangements as BOOST graphs as follows. Every arrangement cell, that is, a vertex, a halfedge, or a face, is adapted as a vertex of the BOOST graph, and every incidence relation between two cells of the arrangement is adapted as an edge of the BOOST graph. Each arrangement edge associated with a curve C is incident to the two arrangement vertices associated with the endpoints of C. Each arrangement face is incident to the arrangement edges on its boundaries and to the isolated arrangement vertices contained in it if they exist. This is a variant of a fundamental representation of arrangements called the *incidence graph*.

Apply the new adaptor to the program coded in **opt_translation.cpp** to solve the variant of the **largest common point sets under ϵ-congruence problem** (see Section 7.3), where one point set is mapped to the *closed* ϵ-environment of the other point set; see Exercise 7.4.

Chapter 8

Operations on (Curved) Polygons

Once you have mastered the capabilities of the *2D Arrangements* package, you have the tools to apprehend additional CGAL packages that use the *2D Arrangements* package as an infrastructure for performing higher-level operations. We start with the *2D Regularized Boolean Set-Operations* package, which implements Boolean operations on point sets bounded by x-monotone curves in the two-dimensional Euclidean space. In particular, it contains the implementation of *regularized* Boolean set operations, intersection predicates, and point-containment predicates. A regularized Boolean set operation op* can be obtained by first taking the interior of the resulting point set of an *ordinary* Boolean set operation (P op Q) and then by taking the closure of the interior [120]. That is, P op* Q = closure(interior(P op Q)). Regularized Boolean set operations appear in Constructive Solid Geometry (CSG), because regular sets are closed under regularized Boolean set operations, and because regularization eliminates lower-dimensional features, namely, isolated vertices and dangling edges ("antennas"), thus simplifying and restricting the representation to physically meaningful full-dimensional (in our case two-dimensional) objects. Our package provides regularized operations on polygons and general polygons, where the edges of a general polygon may be general x-monotone curves rather than simple line segments. Ordinary Boolean set operations, which distinguish between the interior and the boundary of a polygon, are not implemented within this package.[1] In this chapter we review the *2D Regularized Boolean Set-Operations* package in more depth. We use, unless otherwise stated, the traditional notation to designate regularized operations; e.g., $P \cap Q$ means the *regularized* intersection of P and Q. In Section 8.1 we focus on Boolean set operations on linear polygons, introducing the notions of a polygon with holes and of a polygon set. Section 8.2 properly introduces general polygons. We discuss polygons the edges of which are either line segments or circular arcs, referred to as generalized polygons, before we explain how to construct and use general polygons the edges of which are arbitrary weakly x-monotone curves.

A polygon is said to be *simple* (or *Jordan*) if it is enclosed by a single closed polygonal chain that does not cross itself. In particular, the polygon edges are pairwise disjoint in their interior and the degree of all vertices is 2. Such a polygon is topologically equivalent to a disc, and has a well-defined interior and exterior. A polygon in our context must be simple and its vertices must be ordered counterclockwise around the interior of the polygon.

The counterclockwise cyclic sequence of alternating polygon edges and polygon vertices is referred to as the polygon *boundary*. A polygon the boundary of which is connected through pairwise interior-disjoint edges but contains the same vertex twice or more is called *relatively simple*. For example, the polygon depicted in the figure to the right is relatively simple, but not simple, as the vertices p_2

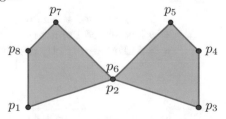

E. Fogel et al., *CGAL Arrangements and Their Applications*, Geometry and Computing 7,   175
DOI 10.1007/978-3-642-17283-0_8, © Springer-Verlag Berlin Heidelberg 2012

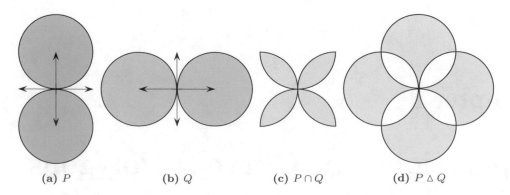

Fig. 8.1: Examples of Boolean set operations on general polygons.

and p_6 coincide.

We extend the notion of a polygon to a point set in $\mathbb{R}^2$ that has the topology of a polygon and boundary edges that map to arbitrary x-monotone curves (not necessarily linear) and refer to it as a *general polygon*. We sometimes use the term *polygon* instead of general polygon for simplicity hereafter.

Let P and Q denote two point sets each composed of general polygons, and let p denote a point. The *2D Regularized Boolean Set-Operations* package supports the following operations:

Intersection: $P \cap Q$.

Union: $P \cup Q$.[2]

Difference: $P \setminus Q$.

Symmetric difference: $P \bigtriangleup Q = (P \setminus Q) \cup (Q \setminus P)$.

Complement: $\overline{P}$.

Intersection predicate: $P \cap Q \neq \emptyset$.

Non-regularized intersection predicate: Do two input sets P and Q overlap, distinguishing between three possible cases:[3]

- P and Q intersect in their interior (that is, their regularized intersection is not empty, or $P \cap Q \neq \emptyset$).
- The boundaries of P and Q intersect but their interiors are disjoint. Namely, they have a finite number of common points or even share boundary curves; in this case $P \cap Q = \emptyset$ still.
- P and Q are strictly disjoint.

Containment predicate: Is the point p contained in a point set P, distinguishing between three cases:[4]

- p is contained in the interior of P.
- p lies on the boundary of P.
- p and P are strictly disjoint.

---

[2]The function that computes the union of two polygons is called `join()`, since the word `union` is reserved in C++.

[3]The function that computes the non-regularized intersection predicate is called `oriented_side()`.

[4]The function that computes the containment predicate is called `oriented_side()`.

8.1   Operations on Linear Polygons

The basic library of CGAL includes the **Polygon_2<Kernel,Container>** class template, which can be used to represent a linear polygon in the plane. The polygon is represented by its vertices stored in a container (a vector by default) of objects of type **Kernel::Point_2**. The polygon edges are line segments (of type **Kernel::Segment_2** objects) connecting adjacent points in the container.

In this section we use the term *polygon* to indicate a **Polygon_2** instance or an object of this type, namely a polygon having straight-line edges. General polygons are only discussed in Section 8.2.

The function template **print_polygon()** listed below, and defined in the header file **pgn_print.h**, demonstrates how to use the basic access functions of the **Polygon_2** class template. It accepts a polygon P, iterates over the range of its points, and inserts each point into the standard output-stream, resulting in a printout of the polygon in a readable format.

```cpp
template <typename Polygon> void print_polygon(const Polygon & pgn)
{
  std::cout << "[ " << pgn.size() << " vertices: (";
  typename Polygon::Vertex_const_iterator  vit;
  for (vit = pgn.vertices_begin(); vit != pgn.vertices_end(); ++vit)
    std::cout << "(" << *vit << ')';
  std::cout << ") ]" << std::endl;
}
```

The basic components of our package are the free-function templates **complement()**, **join()**, **intersection()**, **difference()**, **symmetric_difference()**, **do_intersect()**, and **oriented_side()**. We explain how these functions should be used through several examples presented in the sequel to this section and in the following sections.

Example: Testing whether two polygons intersect results in a Boolean value. The example below tests whether the two triangles T_1 and T_2 depicted in the figure to the right intersect. It uses an instance of the function template **print_polygon()** listed above to print the input triangular polygons.

```cpp
// File: ex_do_intersect.cpp

#include <CGAL/basic.h>
#include <CGAL/Exact_predicates_exact_constructions_kernel.h>
#include <CGAL/Boolean_set_operations_2.h>

#include "pgn_print.h"

typedef CGAL::Exact_predicates_exact_constructions_kernel  Kernel;
typedef Kernel::Point_2                                    Point;
typedef CGAL::Polygon_2<Kernel>                            Polygon;

int main()
{
  // Constuct two triangular polygons and check whether they intersect.
  Polygon T1;
  T1.push_back(Point(1, 1));  T1.push_back(Point(4, 1));
  T1.push_back(Point(1, 3));
  std::cout << "T1 = ";    print_polygon(T1);
```

(a) The union $P \cup Q$ of two polygons, which is a (darkly shaded) point set bounded by a simple polygon and contains a polygonal hole in its interior.

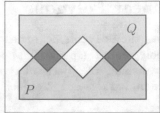

(b) The intersection $P \cap Q$ of two polygons, which is a (darkly shaded) point set comprising of two disjoint polygons.

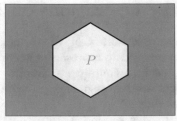

(c) The complement $\overline{P}$ of a simple polygon P, which is a (darkly shaded) point set that has no outer boundary and contains a single polygonal hole.

Fig. 8.2: Operations on simple polygons.

```
Polygon T2;
T2.push_back(Point(2, 2));   T2.push_back(Point(5, 1));
T2.push_back(Point(4, 3));
std::cout << "T2_=_";    print_polygon(T2);

std::cout << (CGAL::do_intersect(T1, T2) ?
            "The_two_polygons_intersect_in_their_interior." :
            "The_two_polygons_do_not_intersect.") << std::endl;
  return 0;
}
```

―――――― *advanced* ――――――

As mentioned in the opening paragraph of this chapter, the *2D Regularized Boolean Set-Operations* package depends on the *2D Arrangements* package. Point sets, which are operands or results of Boolean set operations, are represented by extended arrangements; Chapter 6 explains how the arrangement data structure can be extended with auxiliary data. A Boolean set operation on point sets is performed in steps. First, an arrangement object is constructed for every input point set. Then, the boundary curves of the point sets are inserted into the corresponding arrangements, while data fields that extend the arrangement cells are updated. For example, each face of an arrangement is extended with a Boolean flag that indicates whether the face is included in the interior of the point set. Finally, the arrangements are manipulated, and a new arrangement that represents the resulting point set is constructed. For example, computing the intersection of two point sets reduces to computing the overlay of the two associated arrangements; see Section 6.3.

The function template **do_intersect()** used in the example above has a simple and naive implementation. It applies the intersection operation to compute the intersection between the two input point sets, and tests whether the resulting point set is empty. Even though constructions are not immediately visible to the naked eye when looking at the program code, constructions may occur behind the scene as part of the overlay computation. Thus, a kernel that supports not only exact evaluation of predicates, but also exact construction of geometric objects, is required to ensure robust execution and exact results.

―――――― *advanced* ――――――

8.1.1 Polygons with Holes

A binary operation on two simple polygons may result in a polygon that contains holes in its interior (see Figure 8.2a for in an example where the operation is union), or a set of disjoint polygons; see Figure 8.2b for an example. Moreover, the complement of a simple polygon is an unbounded set that contains a hole; see Figure 8.2c.

Regular sets are closed under regularized Boolean set operations. These operations accept as input, and may produce as output, polygons with holes. A *polygon with holes* represents a point set that may be bounded or unbounded. In the case of a bounded set, its *outer boundary* is represented as a relatively simple (but not necessarily simple) polygon the vertices of which are oriented in counterclockwise order around the interior of the set. In addition, the set may contain *holes*, where each hole is represented as a simple polygon the vertices of which are oriented in clockwise order around the interior of the hole. When you traverse the outer boundary of the polygon, or the boundary of one of its holes, in the given order of the vertices, you should always have the interior of the polygon to your left. Note that an unbounded polygon without holes spans the entire plane. Vertices of holes may coincide with vertices of the boundary or with vertices of other holes; see Figure 8.3.

A point set represented by a polygon with holes is considered to be closed. Therefore, the boundaries of the holes are part of the set (and not part of the holes). The exact definition of the polygon with holes obtained as a result of a Boolean set operation or a sequence of such operations is closely related to the definition of regularized Boolean set operations, being the closure of the interior of the corresponding ordinary operation.

Consider, for example, the regular set depicted in the figure to the right, which is the result of the union of three small triangles T_1, T_2, and T_3. Alternatively, the same set can be obtained by taking the difference between a large triangle T and a small triangle $\tilde{T}$. In general, there are many ways to arrive at a particular point set. However, a canonical representation of a point set is always used to represent the point set regardless of the sequence of operations that were performed in order to arrive at the set. In other words, the set of polygons with holes obtained through the application of any equivalent sequence of operations is *unique*. As a general rule, if two point sets are connected, then they belong to the same polygon with holes. The set depicted in the figure to the right is represented as a single polygon having a triangular outer boundary T with a single triangular hole $\tilde{T}$ in its interior—and not as three triangles that do not have holes at all.

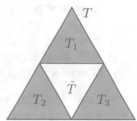

Fig. 8.3: The union of three triangles.

The class template `Polygon_with_holes_2<Kernel,Container>` can be used to represent polygons with holes as described above, where the outer boundary and the hole boundaries are realized as `Polygon_2<Kernel,Container>` objects. Given an object P of an instance of the `Polygon_with_holes_2` class template, you can issue the call `P.is_unbounded()` to check whether P is an unbounded set. If it is bounded, you can obtain the counterclockwise-oriented polygon that represents its outer boundary through the member function `outer_boundary()`. You can also traverse the holes of P using the member functions `holes_begin()` and `holes_end()`. The two functions return iterators of the nested type `Polygon_with_holes_2::Hole_const_iterator` that defines the valid range of P's holes. The value type of this iterator is `Polygon_2`.

Example: The function template `print_polygon_with_holes()` listed below, and defined in the header file `pgn_print.h`, demonstrates how to traverse a polygon with holes. It accepts a `Polygon_with_holes_2` object and uses the free function `print_polygon()` (listed on Page 177 and defined in the same header file) to print all its components in a readable format.

```
template <typename Polygon_with_holes>
void print_polygon_with_holes(const Polygon_with_holes & pwh)
{
  if (! pwh.is_unbounded()) {
    std::cout << "{ Outer boundary = ";
    print_polygon (pwh.outer_boundary());
  }
  else std::cout << "{ Unbounded polygon." << std::endl;

  unsigned int  k = 1;
```

```
  typename Polygon_with_holes::Hole_const_iterator hit;
  std::cout << "  " << pwh.number_of_holes() << " holes:" << std::endl;
  for (hit = pwh.holes_begin(); hit != pwh.holes_end(); ++hit, ++k) {
    std::cout << "    Hole #" << k << " = ";
    print_polygon(*hit);
  }
  std::cout << " }" << std::endl;
}
```

The simple versions of the free set operation functions mentioned above accept one or two **Polygon_2** objects as their input, and result in zero, one, or more **Polygon_with_holes_2** objects as output.

- The complement of a simple polygon P is always an unbounded set with a single polygonal hole. The function **complement(P)** therefore returns a single **Polygon_with_holes_2** object that represents the complement of P.

- The union of two polygons P and Q is always a single connected set, unless of course the two input polygons are completely disjoint. In the latter case $P \cup Q$ trivially consists of the two input polygons. The free function **join(P, Q, R)** therefore returns a Boolean value indicating whether $P \cap Q \neq \emptyset$. If the two polygons are not disjoint, it assigns the polygon with holes object R (which it accepts by reference) to the union of the regularized union operation $P \cup Q$.

- The other three functions, namely, **intersection(P, Q, oi)**, **difference(P, Q, oi)**, and **symmetric_difference(P, Q, oi)**, have a similar interface. As the result of these operations may consist of several disconnected components, they all accept an output iterator **oi**, whose value type is **Polygon_with_holes_2**, and insert the output polygons into its associated container.

The type definitions below are common to all the example programs in this chapter that handle straight-edge polygons. They are kept in a header file named **bops_linear.h**.

#include <list>

#include <CGAL/Exact_predicates_exact_constructions_kernel.h>
#include <CGAL/Boolean_set_operations_2.h>

```
typedef CGAL::Exact_predicates_exact_constructions_kernel  Kernel;
typedef Kernel::Point_2                                     Point;
typedef CGAL::Polygon_2<Kernel>                            Polygon;
typedef CGAL::Polygon_with_holes_2<Kernel>                 Polygon_with_holes;
typedef std::list<Polygon_with_holes>                      Pgn_with_holes_container;
```

Example: The example below demonstrates the usage of the free-function templates **join()** and **intersection()** for computing the union and the intersection of the two simple polygons drawn in Figure 8.2b. Observe that the polygon outer boundaries are oriented in counterclockwise order. The example also uses an instance of the free-function template **print_polygon_with_holes()** listed above.

```
// File: ex_simple_join_intersect.cpp

#include "bops_linear.h"
#include "pgn_print.h"

int main()
{
```

```cpp
// Construct the two input polygons.
Polygon   P;
P.push_back(Point(0, 0));        P.push_back(Point(9, 0));
P.push_back(Point(9, 2));        P.push_back(Point(7, 4));
P.push_back(Point(4.5, 1.5));    P.push_back(Point(2, 4));
P.push_back(Point(0, 2));
std::cout << "P_=_";    print_polygon(P);

Polygon   Q;
Q.push_back(Point(0, 6));        Q.push_back(Point(0, 4));
Q.push_back(Point(2, 2));        Q.push_back(Point(4.5, 4.5));
Q.push_back(Point(7, 2));        Q.push_back(Point(9, 4));
Q.push_back(Point(9, 6));
std::cout << "Q_=_";    print_polygon(Q);

// Compute the union of P and Q.
Polygon_with_holes   unionR;
if (CGAL::join(P, Q, unionR)) {
  std::cout << "The_union:_";
  print_polygon_with_holes(unionR);
}
else
  std::cout << "P_and_Q_are_disjoint_and_their_union_is_trivial."<<std::endl;
std::cout << std::endl;

// Compute the intersection of P and Q.
Pgn_with_holes_container                  intersectionR;
CGAL::intersection(P, Q, std::back_inserter(intersectionR));
Pgn_with_holes_container::const_iterator  it;
std::cout << "The_intersection:" << std::endl;
for (it = intersectionR.begin(); it != intersectionR.end(); ++it) {
  std::cout << "—>_";
  print_polygon_with_holes(*it);
}
return 0;
}
```

8.1.2   Operations on Polygons with Holes

Enabling Boolean operations on point sets that are represented as polygons with holes is not only desired but required, as the domain of simple polygons is not closed under regularized Boolean set operations. Indeed, the *2D Regularized Boolean Set-Operations* package provides overloaded free-functions complement(), intersection(), join(), difference(), symmetric_difference(), and do_intersect() that accept polygons with holes as their input. The prototypes of most functions are the same

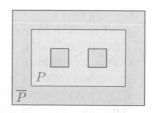

as those of their counterparts that operate on simple polygons. The only exception is the function complement(P, oi), which accepts an input polygon with holes P and an output iterator oi. In contrast to the counterpart function, which accepts a reference to a single polygon with holes for storing the result, this function accepts an output iterator for storing the potentially many polygons with holes that represent the complement of P. For example, the complement of the polygon with holes P in the figure above consists of three polygons—one is unbounded and has a hole that contains the other two.

There are also overloaded free-functions that accept two point sets as input, where one is a simple polygon and the other is a polygon with holes. The complete set of the valid combinations of argument types of all overloaded free-functions that accept two point sets as input is shown in Table 8.1 in Section 8.5.

Example: The example below demonstrates how to compute the symmetric difference between two point sets that contain holes. It also exemplifies how to construct a polygon with holes. Each point set is a rectangle that contains a rectangular hole in its interior. The intersection of the two point sets consists of four squares drawn with a crosshatch pattern in the figure to the right. The symmetric difference between the two point sets is a single polygon that contains five holes—the empty square in the middle and the four squares that constitute the intersection of the two point sets.

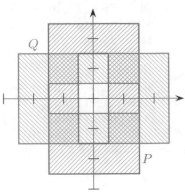

```cpp
// File: ex_symmetric_difference.cpp

#include "bops_linear.h"
#include "pgn_print.h"

int main()
{
  // Construct P - a bounded rectangle that contains a rectangular hole.
  Polygon   outP, holesP;
  outP.push_back(Point(-3, -5));        outP.push_back(Point(3, -5));
  outP.push_back(Point(3, 5));          outP.push_back(Point(-3, 5));
  holesP.push_back(Point(-1, -3));      holesP.push_back(Point(-1, 3));
  holesP.push_back(Point(1, 3));        holesP.push_back(Point(1, -3));
  Polygon_with_holes  P(outP, &holesP, &holesP + 1);
  std::cout << "P_=_"; print_polygon_with_holes(P);

  // Construct Q - a bounded rectangle that contains a rectangular hole.
  Polygon   outQ, holesQ;
  outQ.push_back(Point(-5, -3));        outQ.push_back(Point(5, -3));
  outQ.push_back(Point(5, 3));          outQ.push_back(Point(-5, 3));
  holesQ.push_back(Point(-3, -1));      holesQ.push_back(Point(-3, 1));
  holesQ.push_back(Point(3, 1));        holesQ.push_back(Point(3, -1));
  Polygon_with_holes  Q(outQ, &holesQ, &holesQ + 1);
  std::cout << "Q_=_"; print_polygon_with_holes(Q);

  // Compute the symmetric difference of P and Q.
  Pgn_with_holes_container                     symmR;
  CGAL::symmetric_difference(P, Q, std::back_inserter(symmR));
  std::cout << "The_symmetric_difference:" << std::endl;
  Pgn_with_holes_container::const_iterator   it;
  for (it = symmR.begin(); it != symmR.end(); ++it) {
    std::cout << "_>_";
    print_polygon_with_holes(*it);
  }
  return 0;
}
```

8.1.3   Point Containment and Non-Regularized Operations

The free-function template `oriented_side()` accepts a point and a point set as input, and it returns either `CGAL::ON_POSITIVE_SIDE`, `CGAL::ON_ORIENTED_BOUNDARY`, or `CGAL::ON_NEGATIVE_SIDE` depending on whether the point is contained in the interior of the point set, lies on its boundary, or neither.

Another version of the overloaded free-function template `oriented_side()` accepts two point sets as input. It returns `CGAL::ON_POSITIVE_SIDE` if there exists a point contained in the first point set that is also contained in the interior of the second point set. Otherwise, it returns `CGAL::ON_ORIENTED_BOUNDARY` if there exists a point contained in the first point set that also lies on boundary of the second point set. Otherwise, it returns `CGAL::ON_NEGATIVE_SIDE`. In other words, this version of the function template simply computes the non-regularized intersection predicate.

Example: The example below demonstrates the use of the various versions of the `oriented_side()` function template. The example program defines a square polygon P, as depicted in the figure to the right, and its complement Q. It then locates two points p_1 and p_2 with respect to P and Q.

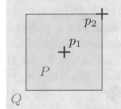

```
// File:  ex_oriented_side.cpp

#include "bops_linear.h"
#include "pgn_print.h"

template <typename Obj1, typename Obj2>
void print_oriented_side(const Obj1& obj1, const Obj2& obj2)
{
  CGAL::Oriented_side os = CGAL::oriented_side(obj1, obj2);
  std::cout << "(" << obj1 << ")"
            << ((os == CGAL::ON_POSITIVE_SIDE) ? " inside " :
               ((os == CGAL::ON_NEGATIVE_SIDE) ? " outside " :" on boundary "))
            << "(" << obj2 << ")" << std::endl;
}

int main()
{
  // Define a square polygon P and its complement Q.
  Polygon P;
  P.push_back(Point(-1, -1));       P.push_back(Point(1, -1));
  P.push_back(Point(1, 1));         P.push_back(Point(-1, 1));

  Polygon_with_holes Q;
  complement(P, Q);

//Define the query points p1 located in P and p2 located on the boundary of P.
  Point p1(0, 0), p2(1, 1);

  print_oriented_side(p1, P);
  print_oriented_side(p2, P);
  print_oriented_side(p1, Q);
  print_oriented_side(p2, Q);
  print_oriented_side(P, Q);
  return 0;
}
```

8.1.4   Connecting Holes

Among the vast number of commercial applications that operate on polygons and polygons with holes there exist some that require an input polygon with holes to be represented as a single sequence of vertices rather than several sequences—one for the outer boundary and one for each hole. The **connect_holes()** function template comes in handy in cases where polygons with holes must be fed to such applications.

The function template **connect_holes()** accepts a polygon with holes P as input and produces a single sequence of vertices that corresponds to the traversal of the outer and inner boundaries of P, where each hole is visited exactly once. The traversal starts from a vertex of the outer boundary. The function locates the topmost vertex of each hole in P, and symbolically connects it to the polygon feature located directly above it, that is, a vertex or an edge of the outer boundary, or of another hole.

Example: The example program below demonstrates the use of the function template **connect_holes()**. It reads an input file that describes a polygon with holes and computes a sequence of points connecting the holes and the outer boundary of the input polygon. The figure to the right depicts the polygon with holes described by the file **polygon_with_holes.dat**, which is used by default. The figure also shows the sequence of 22 points (including repetitions) computed by the example program. The point drawn as a ring is the starting (and the ending) point.

```
// File: ex_connect_holes.cpp

#include <iostream>
#include <list>

#include "bops_linear.h"
#include "read_objects.h"

#include <CGAL/connect_holes.h>

int main(int argc, char* argv[])
{
  // Read a polygon with holes.
  const char* filename = (argc > 1) ? argv[1] : "polygon_with_holes.dat";
  std::ifstream  is;
  if (!open_stream(is, filename)) return -1;
  Polygon_with_holes P;
  is >> P;

  // Connect the outer boundary of the polygon with its holes.
  std::list<Point> pts;
  CGAL::connect_holes(P, std::back_inserter(pts));
  std::copy(pts.begin(), pts.end(),
            std::ostream_iterator<Point>(std::cout, "\n"));
  return 0;
}
```

8.1.5   Operations on Polygon Sets

The result of a regularized operation on two polygons (or polygons with holes) P and Q may be a collection of several disconnected polygons with holes. Thus, it is convenient to represent such a collection in terms of a single object, and enable the use of this object as an operand to Boolean set operations. To this end, we introduce the `Polygon_set_2<Kernel, Container>` class template, a central component in the *2D Regularized Boolean Set-Operations* package. Consider some instance of the template `Polygon_set_2`. An object of this type employs a planar arrangement to represent a point set formed by the collection of several disconnected polygons with holes.

A polygon set object usually represents the result of a sequence of operations that were applied on some input polygons. The representation is unique, regardless of the order of the particular sequence of operations that were performed in order to arrive at it.

A polygon set object is constructible from a single polygon object or from a polygon with holes object. Once constructed, it is possible to insert new polygons (or polygons with holes) into the set using the `insert()` method, as long as the point set represented by the inserted polygons and the point set represented by the existing polygons in the set are disjoint; see Section 8.1.7 for more details about the insertion of disjoint polygons into a polygon set object.

The `Polygon_set_2` class template also provides some access functions and a few queries. You can obtain the unique set of polygons with holes that exclusively cover the point set represented by a polygon set object. The query `S.oriented_side(q)` determines whether the query point q is contained in the interior of the set S, lies on the boundary of the set, or neither (namely, q is exterior to the set S).

The `Polygon_set_2` class template defines the overloaded member-functions `do_intersect()`, `oriented_side()`, `complement()`, `intersection()`, `join()`, `difference()`, and `symmetric_difference()`. The operands of these functions are either simple polygons (`Polygon_2` object), polygons with holes (`Polygon_with_holes_2` objects), or polygon sets (`Polygon_set_2` objects).

Member functions of the `Polygon_set_2` class template that perform Boolean set operations come in two flavors, for example, `S.join(P, Q)` computes the union of P and Q and assigns the result to S, while `S.join(P)` performs the operation $S \longleftarrow S \cup P$. Similarly, `S.complement(P)` sets S to be the complement of P, while `S.complement()` simply produces the complement of the set S.

Finally, the call `S.arrangement()` obtains a reference to the underlying arrangement of S. The faces of the obtained arrangement object are extended with a flag that indicates whether the face is contained in the point set S or not. The flag can be retrieved and modified by the functions `contained()` and `set_contained()`, respectively, which extend the face functionality.

8.1.6 A Sequence of Set Operations

The free functions reviewed in Section 8.1.1 wrap the polygon set class, and are only provided for convenience. A typical such function constructs a pair of `Polygon_set_2` objects, invokes the appropriate method to apply the desired Boolean operation, and transforms the resulting polygon set to the required output format. Thus, when several operations are performed in a sequence, it is much more efficient to use the member functions of the `Polygon_set_2` class template directly, as the time spent on the extraction of the polygons from the internal representation for some operations, and on the reconstruction of the internal representation for the succeeding operations, is avoided.

Example: The function template `print_polygon_set()` listed below and defined in the header file `pgn_print.h` demonstrates how to iterate through the polygons with holes that constitute a polygon set. It accepts a `Polygon_set_2` object and uses the free function `print_polygon_with_holes()` (listed in Section 8.1.1 and defined in the same header file) to print all its polygons with holes in a readable format.

```
template <typename Polygon_set>
void print_polygon_set(const Polygon_set & pgn_set)
{
  typedef typename Polygon_set::Polygon_with_holes_2 Polygon_with_holes;
  typedef std::vector<Polygon_with_holes>            Pgn_with_holes_container;
```

```
    Pgn_with_holes_container res(pgn_set.number_of_polygons_with_holes());
    pgn_set.polygons_with_holes(res.begin());
    std::cout << "The result contains " << res.size() << " components:"
              << std::endl;
    typename Pgn_with_holes_container::const_iterator  it;
    for (it = res.begin(); it != res.end(); ++it) {
      std::cout << "--> ";
      print_polygon_with_holes(*it);
    }
}
```

The next example performs a sequence of three Boolean set operations. First, it computes the union of two concave quadrilaterals P and Q; see Figure 8.2a for a similar operation. Then, it computes the complement of the result of the union operation. Finally, it intersects the result of the complement operation with a rectangle R, confining the final result to the area of the rectangle. The resulting set S comprises two components: a polygon with a hole, and a simple polygon contained in the interior of this hole; see the figure to the right for an illustration; the result S is darkly shaded.

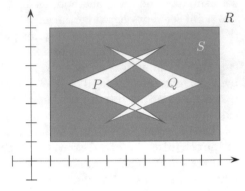

```
// File: ex_sequence.cpp

#include "bops_set_linear.h"
#include "pgn_print.h"

int main()
{
  // Construct the two initial polygons and the clipping rectangle.
  Polygon      P, Q, R;

  P.push_back(Point(2, 4));          P.push_back(Point(7, 2));
  P.push_back(Point(4, 4));          P.push_back(Point(7, 6));

  Q.push_back(Point(4, 2));          Q.push_back(Point(9, 4));
  Q.push_back(Point(4, 6));          Q.push_back(Point(7, 4));

  R.push_back(Point(1, 1));          R.push_back(Point(10, 1));
  R.push_back(Point(10, 7));         R.push_back(Point(1, 7));

  // Perform a sequence of set operations and print the result.
  Polygon_set  S;
  S.insert(P);
  S.join(Q);                    // Compute the union of P and Q.
  S.complement();               // Compute the complement.
  S.intersection(R);            // Intersect with the clipping rectangle.
  print_polygon_set(S);
  return 0;
}
```

This example uses the types defined in the header file **bops_linear.h**; see Section 8.1.1. In addition, it uses the type **Polygon_set_2** listed below, and defined in the header file **bops_set_linear.h**.

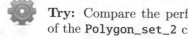

#include <CGAL/Polygon_set_2.h>

typedef CGAL::Polygon_set_2<Kernel>   Polygon_set;

Try: Compare the performance of the code block in the example above that uses an instance of the `Polygon_set_2` class template to perform Boolean set operations with the performance of equivalent code that consists of free-function calls. You may need to alter the code of the example so that it can read large input sets to make the two options distinct.

8.1.7 Inserting Non-Intersecting Polygons

If you wish to compute the union of a polygon (or a polygon with holes) P with a general-polygon set R and store the result in R, you can construct a polygon set S and apply the union operation as follows:

```
Polygon_2 S(P);
R.join(S);
```

As a matter of fact, you can apply the union operation directly:

```
R.join(P);
```

However, if you know that the polygon does not intersect any of the polygons with holes represented by R, you can use the more efficient method **insert()** as follows:

```
R.insert(P);
```

As **insert()** assumes that $P \cap R = \emptyset$, it does not try to compute intersections between the boundaries of P and of R. This fact significantly speeds up the insertion process in comparison with the insertion of a non-disjoint polygon that intersects R.

The **insert()** function is overloaded with four versions that accept the following arguments, respectively:

1. A simple polygon, that is, a `Polygon_2` object.
2. A polygon with holes, that is, a `Polygon_with_holes_2` object.
3. A range of simple polygons.
4. A range of simple polygons and a range of polygons with holes.

When a range of polygons or polygons with holes are inserted into a polygon set R, all the polygons in the range and all the polygons represented by R must be pairwise disjoint. However, there is a subtle difference between the precondition that applies to input polygons and the precondition that applies to input polygons with holes. An input polygon must be strictly disjoint from all other input polygons, all other input polygons with holes, and all the polygons with holes represented by R. On the other hand, an input polygon with holes must be disjoint from all other polygons except perhaps at the vertices. Recall that a vertex of a polygon with holes may appear more than once in the cyclic chains of outer and inner boundary vertices, as the outer boundary of a polygon with holes is represented as a relatively simple (but not necessarily simple) polygon.

Example: The example below is a benchmark that compares the performance of the insertion and union operations. It constructs a polygon set from $\frac{n(n+1)}{2}$ equilateral triangles stored as objects of type `Polygon_2`, where $n = 4$ by default, as depicted in the figure to the right. The triangles are pairwise disjoint except at the vertices. The resulting polygon set consists of one polygon-with-holes~~polygon with holes~~ object that has $\frac{(n-1)n}{2}$ holes. The polygon set is constructed twice using two different methods as follows: (i) For each triangle we construct an equivalent `Polygon_with_holes` object, and insert it into the polygon set. (ii) For each triangle we compute the union

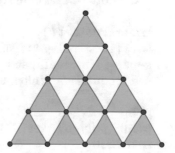

of the triangle with the point set, retaining the result in the point set. The time consumption of the two methods is measured and reported.

```cpp
// File: ex_insertion.cpp

#include <list>
#include <boost/lexical_cast.hpp>
#include <boost/timer.hpp>

#include "bops_set_linear.h"
#include "pgn_print.h"

//Generate a "pyramid" made of n(n+1)/2 (approximately) equilateral triangles.
// The base of each triangle is 2 and the height is sqrt(3) ~ 1.732.
template <typename Output_iterator>
Output_iterator generate(unsigned int n, Output_iterator oi)
{
  for (unsigned int j = 0; j != n; ++j) {
    for (unsigned int i = 0; i != n-j; ++i) {
      Polygon polygon;
      Kernel::FT yb(CGAL::Gmpq(1732 * j, 1000));
      Kernel::FT yt(CGAL::Gmpq(1732 * (j + 1), 1000));
      polygon.push_back(Point(Kernel::FT(j + 2 * i), yb));
      polygon.push_back(Point(Kernel::FT(j + 2 * i + 2), yb));
      polygon.push_back(Point(Kernel::FT(j + 2 * i + 1), yt));
      *oi++ = polygon;
    }
  }
  return oi;
}

int main(int argc, char* argv[])
{
  unsigned int n = (argc >= 2) ? boost::lexical_cast<unsigned int>(argv[1]):4;
  std::list<Polygon> polygons;
  generate(n, std::back_inserter(polygons));

  std::list<Polygon>::const_iterator it;
  Polygon_set R1, R2;

  boost::timer timer;
  for (it = polygons.begin(); it != polygons.end(); ++it)
    R1.insert(Polygon_with_holes(*it));
  double secs = timer.elapsed();
  std::cout << "Insertion took " << secs << " seconds." << std::endl;

  timer.restart();
  for (it = polygons.begin(); it != polygons.end(); ++it) R2.join(*it);
  secs = timer.elapsed();
  std::cout << "Union took " << secs << " seconds." << std::endl;

  print_polygon_set(R1);
  print_polygon_set(R2);
  return 0;
}
```

8.1.8 Performing Multiway Operations

There are several ways to compute the union of a set of polygons $P_1, \ldots P_m$. You can do it incrementally as follows: At each step compute the union of $S_{k-1} = \bigcup_{i=1}^{k-1} P_i$ with P_k and obtain S_k. Namely, if the polygon set is given as a range [begin, end), the following loop computes their union in S:

```
Input_iterator  iter = begin;
Polygon_set_2  S(*iter++);
while (iter != end) S.join(*iter++);
```

A second option is to use a divide-and-conquer approach; namely, bisect the set of polygons into two sets, compute the union of each set recursively, obtain the partial results in S_1 and S_2, and finally, compute the union $S_1 \cup S_2$. The union operation can also be performed more efficiently for sparse polygons, which have a relatively small number of intersections, using a third option that simultaneously computes the union of all polygons. This is done by constructing a planar arrangement of all input polygons, utilizing the plane-sweep algorithm, and then extracting the result from the arrangement. Similarly, it is also possible to aggregately compute the intersection $\bigcap_{i=1}^{m} P_i$ of a set of input polygons.

Our package provides the overloaded free-function templates `join()` and `intersection()` that compute the union and the intersection of many input polygons, respectively. The package also provides the overloaded free-function template `do_intersect()` that determines whether the intersection of many input polygons is not empty. For each of these operations there are three versions; one accepts a range of polygons, a second one accepts a range of polygons with holes, and a third one accepts two ranges—one of polygons and another of polygons with holes. There is no restriction on the polygons in the ranges; naturally, they may intersect each other.

The implementation of each of these function templates exploits both the divide-and-conquer approach and the plane-sweep algorithm. It considers the polygons in the input ranges as members of a single set of input polygons. At each invocation of the recursive divide-and-conquer procedure, the number of input polygons is examined. If it exceeds a certain predetermined constant,[5] it proceeds as usual. Otherwise, it applies the plane-sweep algorithm on the input polygons. Notice that the merge step applies the plane-sweep algorithm nonetheless.

The class template `Polygon_set_2` also provides equivalent member functions that aggregately operate on a range of input polygons or polygons with holes. When such a member function is called, the polygons represented by the polygon set object serve as operands as well. Thus, you can easily compute the union of your polygon range as follows:

```
Polygon_set_2  S;
S.join(begin, end);
```

An elaborate example that utilizes multiway operations on general polygons is given at the end of this chapter; see Section 8.3.

Try: The program coded in `ex_insertion.cpp` listed in the previous subsection compares the performance of the insertion operation and the incremental union operation. Augment the program to compare also the aggregate union operation.

8.2 Operations on Curved Polygons

So far you have been exposed only to straight-edge polygons and polygon sets, namely closed point sets bounded by piecewise linear curves. The *2D Regularized Boolean Set-Operations* package allows a more general geometric mapping of the polygon edges. The operations provided by the package operate on point sets bounded by x-monotone segments of general curves (e.g., conic

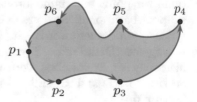

---

[5]This constant is set to 5 by default.

arcs and segments of graphs of polynomial functions). We refer to any singly-connected point set, the boundary of which is a simple curve comprising a finite number of x-monotone subcurves, as a general polygon. The x-monotone segments of a valid general polygon are *directed*; the target vertex of every edge is the source of the next edge. The interior of the polygons lies to the left of each of the directed curves. For example, the point set depicted in the figure on the previous page is a valid general polygon bounded by six curved x-monotone segments.

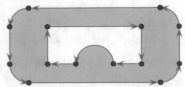

A general polygon with holes is defined in an analogous way to a linear polygon with holes. A general polygon with holes represents a bounded or an unbounded point set that may contain holes. The point set may not be singly connected. Recall, that a polygon with holes is valid only if (i) its outer boundary bounds a relatively simple polygon oriented counterclockwise, (ii) each one of its holes is a simple polygon oriented clockwise, (iii) all holes are pairwise disjoint, except perhaps at the vertices, and (iv) all holes are contained in the polygon bounded by the outer boundary, and their boundaries and the outer boundary are disjoint, except perhaps at the vertices. The point set depicted in the figure above, for example, is a valid general polygon with holes that contains a single hole. It is bounded by line segments and circular arcs.

Consider an instance of the class template `General_polygon_set_2<Traits,Dcel>`. An object of this type represents point sets that comprise a finite number of general polygons with holes that are pairwise disjoint, and provides various Boolean set operations on such sets.

The template parameter `Dcel` must be substituted with a model of the concept GeneralPolygonSetDcel. It is instantiated by default with the type `Gps_default_dcel<Traits>`. You can override this default with a different `Dcel` class, typically an extension of the `Gps_default_dcel` class template. Overriding the default is necessary only if you intend to obtain the underlying internal arrangement and process it further.

The `Traits` parameter of the `General_polygon_set_2<Traits,Dcel>` class template must be substituted with a model of the GeneralPolygonSetTraits_2 concept. A model of this concept has the means to (i) construct general polygons and general polygons with holes, (ii) traverse their boundaries, and (iii) handle the geometric objects (points and x-monotone curves) that compose their boundaries. A model of the concept GeneralPolygonSetTraits_2 is tailored to handle a specific family of curves.

Two additional concepts play a role in this context, namely, GpsTraitsGeneralPolygon_2 and GpsTraitsGeneralPolygonWithHoles_2. Models of these concepts represent general polygons and general polygons with holes, respectively. They must be default-constructible, copy-constructible, and assignable, but no other formal requirements are imposed on them beyond the variant above. The `General_polygon_set_2<Traits>::Polygon_2` and `General_polygon_set_2<Traits>::Polygon_with_holes_2` types nested in `General_polygon_set_2` must model the concepts GpsTraitsGeneralPolygon_2 and GpsTraitsGeneralPolygonWithHoles_2, respectively. Operations on these types are provided by the traits class described in detail in the next section.

8.2.1 The Traits-Class Concepts

As edges of general polygons are represented by x-monotone curves, a traits class that substitutes the `Traits` parameter of the `General_polygon_set_2<Traits,Dcel>` must nest definitions for point and x-monotone curve types, and support basic predicates involving objects of these types. In other words, it should model the ArrangementBasicTraits_2 concept. Recall the hierarchy of traits-class concepts mentioned in Section 3.4.1 and described in more detail in Section 5.1.

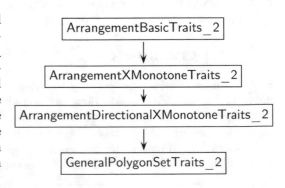

In most applications the curves that represent

edges of general polygons may intersect each other when Boolean set operations are performed on the general polygons. Thus, the traits class that substitutes the `Traits` parameter must also model the ArrangementXMonotoneTraits_2 concept.

The ArrangementXMonotoneTraits_2 concept does not require the x-monotone curve type to represent a directed curve. Observe, that none of the traits requirements refer to the *source* or *target* points of x-monotone curves (in contrast to the pairs of directed halfedges that are associated with x-monotone curves). The only required predicate that relates to ordering is the predicate that lexicographically compares points in general, and endpoints of x-monotone curves in particular. The concept ArrangementDirectionalXMonotoneTraits_2 refines the ArrangementXMonotoneTraits_2 concept by treating its x-monotone curves as directed objects. It thus requires the two additional operations on x-monotone curves defined below.

`Compare_endpoints_xy_2:` Lexicographically compare the source and target points of an x-monotone curve.

`Construct_opposite_2:` Given an x-monotone curve, construct its opposite curve; namely, swap its source and target points.

In addition, the operations `Intersection_2`, `Split_2`, `Are_mergeable_2`, and `Merge_2`, which are required by the concept ArrangementXMonotoneTraits_2, (see Section 5.1.2) are refined as follows:

`Intersection_2:` Compute all intersection points and overlapping sections of two given x-monotone curves, C_1 and C_2. If possible, compute also the multiplicity of each intersection point. In the case C_1 and C_2 overlap, the overlapping subcurves are given the direction of C_1 and C_2 if their directions are identical. Otherwise, the overlapping subcurves are given an arbitrary direction.

`Split_2:` Split a given curve C at a given point p that lies in the interior of C, into two interior-disjoint subcurves, C_1 and C_2, oriented in the same direction as C. That is, the source and target points of C become the source point of one curve and the target point of the other, respectively.

`Are_mergeable_2:` Determine whether two x-monotone curves C_1 and C_2 can be merged into a single x-monotone curve representable by the traits class. C_1 and C_2 are mergeable if their underlying curves are identical and they do not bend to form a non-x-monotone curve. As a precondition the source point of one curve and the target point of the other must coincide.

`Merge_2:` Merge two mergeable curves C_1 and C_2 into a single curve C. Without loss of generality, assume that the target point of C_1 and the source point of C_2 coincide; then, the source point of C_1 and the target point of C_2 become the source and target points of C, respectively.

The traits classes `Arr_segment_traits_2`, `Arr_non_caching_segment_traits_2`, `Arr_circle_segment_traits_2`, `Arr_conic_traits_2`, `Arr_rational_arc_traits_2`, and `Arr_Bezier_curve_traits_2`, which are included in the *2D Arrangements* package and distributed with CGAL, are all models of the refined concept ArrangementDirectionalXMonotoneTraits_2.[6]

The concept GeneralPolygonSetTraits_2 refines the ArrangementDirectionalXMonotoneTraits_2 concept by adding a set of requirements that, when fulfilled, enables the construction of general polygons and general polygons with holes and the traversal of such polygons. A model of this concept needs to define the nested types `Polygon_2` and `Polygon_with_holes_2` that model the GpsTraitsGeneralPolygon_2 and GpsTraitsGeneralPolygonWithHoles_2 concepts, respectively. In addition, it has to support the following operations:

---

[6]The `Arr_polyline_traits_2` class is *not* a model of the ArrangementDirectionalXMonotoneTraits_2 concept, as the x-monotone curve it defines is always directed from left to right. Thus, an opposite curve cannot be constructed. However, there is a natural alternative in the form of the traits classes for line segments.

Construct_polygon_2: Construct a general polygon from a range of polygon boundary directional x-monotone curves, with the precondition that the target point of one curve coincides with the source point of the next curve in the range, and the target point of the last curve coincides with the source point of the first curve, forming a cyclic order.

Construct_polygon_with_holes_2: Construct a general polygon with holes given its outer boundary and the range of its holes.

Construct_curves_2: Obtain the range of outer-boundary curves of a general polygon.

Construct_outer_boundary: Obtain the outer boundary of a general polygon with holes represented as a **Polygon_2** object.

Construct_holes: Obtain the range of holes of a general polygon with holes represented as **Polygon_2** objects.

Is_unbounded: Determine whether a general polygon with holes in unbounded.

As in the case of computations using models of the ArrangementXMonotoneTraits_2 concept, operations are robust only when exact arithmetic is used. When inexact arithmetic is used, degenerate and nearly-degenerate configurations may result in abnormal termination of the program or yield incorrect results.

8.2.2   Operations on Polygons with Circular Arcs

Several predefined models of the GeneralPolygonSetTraits_2 concept are distributed with CGAL. The **Gps_segment_traits_2** class template is one of them. It is used to construct (linear) polygons and (linear) polygons with holes, traverse their boundaries, and handle points and line segments that comprise their boundaries. The class template **Polygon_set_2** introduced in Section 8.1.5 is a specialization of the **General_polygon_set_2** class template. It uses the **Gps_segment_traits_2** traits model to perform Boolean set operations on (linear) polygons and polygons with holes.

Another model of the concept GeneralPolygonSetTraits_2, called **Gps_circle_segment_traits_2**, is used to construct generalized polygons and generalized polygons with holes, traverse their boundaries, and handle points, line segments, and circular arcs that comprise their boundaries. Consider an instance of the **General_polygon_set_2<Traits,Dcel>** class template obtained by substituting the **Traits** parameter with an instance of the **Gps_circle_segment_traits_2** class template. An object of this type represents a collection of generalized polygons with holes, the boundaries of which comprise line segments and circular arcs. The circle-segment traits class-template (see Section 5.4.1) provides the necessary constructions of line segments and circular arcs and predicates on such objects. It is parameterized by a geometric kernel, and is efficient, as it uses only rational arithmetic. Substituting the kernel parameter of the traits class-template when it is instantiated with a filtered kernel increases the efficiency of the traits class. The type definitions listed below are defined in the header file **bops_circular.h**.

```
#include <CGAL/Exact_predicates_exact_constructions_kernel.h>
#include <CGAL/Gps_circle_segment_traits_2.h>
#include <CGAL/General_polygon_set_2.h>

typedef CGAL::Exact_predicates_exact_constructions_kernel Kernel;
typedef CGAL::Gps_circle_segment_traits_2<Kernel>          Traits;

typedef CGAL::General_polygon_set_2<Traits>                Polygon_set;
typedef Traits::Polygon_2                                  Polygon;
typedef Traits::Polygon_with_holes_2                       Polygon_with_holes;
```

Example: The example listed below uses the `General_polygon_set_2<Traits>` class template, where its `Traits` parameter is substituted with an instance of the `Gps_circle_segment_traits_2` class template, to compute the union of four circles $C_1, \ldots, C_4$ and four rectangles $R_1, \ldots, R_4$, as depicted in the figure to the right. Each circle is represented as a general polygon bounded by two x-monotone circular arcs, which comprise the upper half and the lower half of the circle. Each rectangle is bounded by four line segments. The union is computed incrementally. Note that the union of the four circles is computed with the `insert()` method, as the circles are disjoint, while the union of the rectangles is computed using the `join()` method. The result, a single polygon with a single hole, is inserted into the standard output-stream with the inserter (<<) operator.

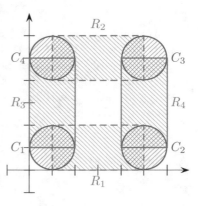

The *2D Regularized Boolean Set-Operations* package provides *inserter* (<<) and *extractor* (>>) operators. The former inserts general-polygon objects, general polygon-with-holes objects, and general point set objects into output streams. The latter extracts objects of the same types from input streams. The format is a simple plain-text encoding of the topology and, geometry of the objects.

```cpp
// File: ex_circle_segment.cpp

#include <list>
#include <iostream>

#include <CGAL/IO/Gps_iostream.h>

#include "bops_circular.h"

typedef Traits::Rational_point_2                 Point;
typedef Traits::Rational_circle_2                Circle;

// Construct a polygon from a circle.
Polygon ctr_polygon(const Circle& circle)
{
  // Subdivide the circle into two x-monotone arcs.
  Traits traits;
  Traits::Curve_2 curve(circle);  // circle orientation is counterclockwise
  std::list<CGAL::Object>  objects;
  traits.make_x_monotone_2_object()(curve, std::back_inserter(objects));
  CGAL_assertion(objects.size() == 2);

  // Construct a polygon that comprises the two x-monotone arcs.
  Polygon                        polygon;
  std::list<CGAL::Object>::iterator  it = objects.begin();
  Traits::X_monotone_curve_2         arc;
  CGAL::assign(arc, *it++);  polygon.push_back(arc);
  CGAL::assign(arc, *it);    polygon.push_back(arc);
  return polygon;
}

// Construct a quadrilateral polygon.
Polygon ctr_polygon(const Point& p1, const Point& p2,
```

```
                        const Point& p3,  const Point& p4)
{
  Polygon              polygon;
  polygon.push_back(Traits::X_monotone_curve_2(p1, p2));
  polygon.push_back(Traits::X_monotone_curve_2(p2, p3));
  polygon.push_back(Traits::X_monotone_curve_2(p3, p4));
  polygon.push_back(Traits::X_monotone_curve_2(p4, p1));
  return polygon;
}


// The main program:
int main()
{
  // Insert four non-intersecting circles.
  Polygon_set  S;
  S.insert(ctr_polygon(Circle(Point(1, 1), 1)));        // C1
  S.insert(ctr_polygon(Circle(Point(5, 1), 1)));        // C2
  S.insert(ctr_polygon(Circle(Point(5, 5), 1)));        // C3
  S.insert(ctr_polygon(Circle(Point(1, 5), 1)));        // C4

  // Compute the union with four rectangles.
  S.join(ctr_polygon(Point(1, 0), Point(5, 0), Point(5, 2), Point(1, 2)));
  S.join(ctr_polygon(Point(1, 4), Point(5, 4), Point(5, 6), Point(1, 6)));
  S.join(ctr_polygon(Point(0, 1), Point(2, 1), Point(2, 5), Point(0, 5)));
  S.join(ctr_polygon(Point(4, 1), Point(6, 1), Point(6, 5), Point(4, 5)));

  std::cout << S;
  return 0;
}
```

8.2.3 General-Polygon Set Traits-Adaptor

An instance of the class template `Gps_traits_2<Traits,GeneralPolygon>` is yet another model of the concept GeneralPolygonSetTraits_2. Thus, an instance of this template can substitute the `Traits` parameter of the class template `General_polygon_set_2<Traits>`. However, the `Gps_traits_2` class template does not support a specific families of curves. Instead, it adapts traits classes that do support specific family of curves to the GeneralPolygonSetTraits_2 concept. The `Gps_circle_segment_traits_2` class template introduced in the prevision section is in fact defined in terms of this adaptor and the `Arr_circle_segment_traits_2` class template, which supports the specific family of circular arcs and line segments; see Section 5.4.1. All this is done with the help of yet another concept, namely GeneralPolygon_2, as explained next.

The class template `Gps_traits_2` together with the concept GeneralPolygon_2 and its model `General_polygon_2<Traits>` facilitates the production of models of the GeneralPolygonSetTraits_2 concept.

A model of the concept GeneralPolygon_2 is used to represent a simple point set in the plane bounded by x-monotone curves. It must fulfill the following requirements:

- It must be constructible from a range of pairwise interior-disjoint x-monotone curves $c_1, \ldots, c_n$. The target point of one curve coincides with the source point of the next curve in the range, and the target point of the last curve c_n coincides with the source point of the first curve c_1, forming a cyclic order.

- It is possible to iterate through the range of x-monotone curves that form the cyclic order.

In particular, it must define the type **X_monotone_curve_2**, which is used to represent an x-monotone curve of the point set boundary. It must provide a constructor from a range of such curves, and a pair of methods, namely, **curves_begin()** and **curves_end()**, that can be used to iterate over the point set boundary x-monotone curves.

The class template **General_polygon_2<Traits>** models the concept GeneralPolygon_2. Its sole template parameter must be substituted with a model of the concept ArrangementDirectionalX-MonotoneTraits_2, from which it obtains the **Point_2** and **X_monotone_curve_2** types. It uses the geometric operations on objects of these types provided by such a model to maintain a container of directed curves of type **X_monotone_curve_2**, which represents the boundary of a general polygon.

The definition of the adaptor **Gps_traits_2<Traits,GeneralPolygon>** is rather simple. It is derived from the traits class that substitutes the **Traits** parameter, which must model the concept ArrangementDirectionalXMonotoneTraits_2, inheriting its necessary types and methods. It further exploits the methods provided by the type that substitutes the parameter **GeneralPolygon**, which must model the concept GeneralPolygon_2. By default, the **GeneralPolygon** parameter is substituted with **General_polygon_2<Traits>**.

The code excerpt listed below defines a general-polygon set type that can be used to perform Boolean set operations on point sets bounded by the x-monotone curve type defined by the arrangement-traits class **Arr_traits_2**, which is a representative model of the concept Arrangement-DirectionalXMonotoneTraits_2.

```
#include <CGAL/General_polygon_2.h>
#include <CGAL/Gps_traits_2.h>

typedef CGAL::General_polygon_2<Arr_traits_2>               General_polygon_2;
typedef CGAL::Gps_traits_2<Arr_traits_2,General_polygon_2> Traits_2;
typedef CGAL::General_polygon_set_2<Traits_2>
  General_polygon_set_2;
```

The concept GeneralPolygonWithHoles_2 and its model **General_polygon_with_holes_2** are described next. A model of the GeneralPolygonWithHoles_2 concept is used to represent a bounded or an unbounded point set that may contain holes. The point set may be not simply connected. A model of this concept must satisfy the following requirements:

- It must be constructible from a general polygon that represents the outer boundary and a range of general polygons that represent the holes.

- It must be possible to access the general polygon that represents the outer boundary (in the case of a bounded set) using the method **outer_boundary()**.

- It must be possible to iterate through the general polygons that represent the holes. In particular, every model must define the pair of access methods **holes_begin()** and **holes_end()**, which return two iterators, respectively, that define the range.

When the class template **General_polygon_with_holes_2<GeneralPolygon_2>** is instantiated, its template parameter **GeneralPolygon_2** must be substituted with a model of the concept GeneralPolygon_2. The class template **Gps_traits_2<ArrTraits_2,GeneralPolygon_2>**, being a model of the concept GeneralPolygonSetTraits_2, defines the required nested type **Polygon_with_holes_2** to be **General_polygon_with_holes_2<GeneralPolygon_2>**. This implies that with respect to the definitions on the previous page both types **Gps_traits_2::Polygon_with_holes_2** and **General_polygon_set_2::Polygon_with_holes_2** refer to **General_polygon_with_holes_2<General_polygon_2>**.

Example: The program below demonstrates how to perform Boolean set operations on general polygons that are bounded by chains of Bézier curves. Note how it uses the **Gps_traits_2<Traits,GeneralPolygon>** adaptor, where the **Traits** template parameter is substituted with the **Arr_Bezier_curve_traits_2** class template, as defined in the header file **arr_Bezier.h**; see Section 5.4.4.

Recall that every Bézier curve is defined by a sequence of control points (see Section 5.4.4) to form chains, the last control point of every curve must be identical to the first control point of its successor. The function **read_Bezier_polygon()** included in the example reads the curves from an input file until they form a closed chain, which is assumed to be the outer boundary of the polygon. If more curves are present, it constructs polygons that correspond to holes in the area bounded by the outer boundary. The function ensures that the outer boundary is counterclockwise-oriented and every hole is clockwise-oriented. Note that this function is also responsible for subdividing the input Bézier curves into x-monotone subcurves, as required by the **Gps_traits_2** adaptor.

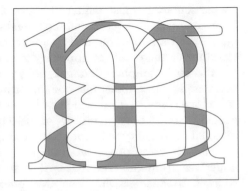

The default input files used by the example, namely, **char_g.dat** and **char_m.dat**, represent the characters **g** and **m** in the Times New Roman font. Their intersection comprises nine simple polygons as depicted in the figure above.

```
// File: ex_traits_adaptor.cpp

#include <iostream>
#include <list>
#include <boost/timer.hpp>

#include "arr_Bezier.h"
#include "read_objects.h"

#include <CGAL/Gps_traits_2.h>
#include <CGAL/Boolean_set_operations_2.h>

typedef CGAL::Gps_traits_2<Traits>              Gps_traits;
typedef Gps_traits::General_polygon_2           Polygon;
typedef Gps_traits::General_polygon_with_holes_2  Polygon_with_holes;
typedef std::list<Polygon_with_holes>           Polygon_set;

// Read a general polygon with holes formed by Bezier curves from the file.
bool read_Bezier_polygon(const char* filename, Polygon_with_holes& P)
{
  // Read the Bezier curves.
  std::list<Bezier_curve>  curves;
  read_objects<Bezier_curve>(filename, std::back_inserter(curves));

  // Read the curves and construct the general polygon these curves form.
  std::list<Traits::X_monotone_curve_2> xcvs;
  bool                    first = true;
  Rat_point               p0;
  std::list<Polygon>      pgns;
  Rat_kernel kernel;
  const Rat_kernel::Equal_2 equal = kernel.equal_2_object();
  Traits traits;
  Traits::Make_x_monotone_2 mk_x_monotone = traits.make_x_monotone_2_object();
  std::list<Bezier_curve>::const_iterator it;
  for (it = curves.begin(); it != curves.end(); ++it) {
    std::list<CGAL::Object> x_objs;
```

```
      mk_x_monotone(*it, std::back_inserter(x_objs));

      std::list<CGAL::Object>::const_iterator  xoit;
      for (xoit = x_objs.begin(); xoit != x_objs.end(); ++xoit) {
        Traits::X_monotone_curve_2 xcv;
        if (CGAL::assign(xcv, *xoit)) xcvs.push_back(xcv);
      }

      // Check whether the current curve closes a polygon; namely, whether its
      // target point (the last control point) equals the source of the first
      // curve in the current chain.
      if (first) {
        // This is the first curve in the chain — store its source point.
        p0 = it->control_point(0);
        first = false;
        continue;
      }
      if (equal(p0, it->control_point(it->number_of_control_points() − 1))) {
        // Push a new polygon into the polygon list. Make sure that the polygon
        // is counterclockwise−oriented if it represents the outer boundary
        // and clockwise−oriented if it represents a hole.
        Polygon          pgn(xcvs.begin(), xcvs.end());
        CGAL::Orientation   orient = pgn.orientation();

        if ((pgns.empty() && orient == CGAL::CLOCKWISE) ||
            (! pgns.empty() && orient == CGAL::COUNTERCLOCKWISE))
          pgn.reverse_orientation();

        pgns.push_back(pgn);
        xcvs.clear();
        first = true;
      }
    }

    if (! xcvs.empty()) return false;

    // Construct the polygon with holes.
    std::list<Polygon>::iterator pit = pgns.begin();
    // The first polygon is the outer boundary and the rest are the holes.
    P = Polygon_with_holes(pgns.front(), ++pit, pgns.end());
    return true;
  }


  // The main program.
  int main(int argc, char* argv[])
  {
    // Get the name of the input files from the command line, or use the default
    // char_g.dat and char_m.dat files if no command−line parameters are given.
    const char* filename1 = (argc > 1) ? argv[1] : "char_g.dat";
    const char* filename2 = (argc > 2) ? argv[2] : "char_m.dat";

    // Read the general polygons from the input files.
    Polygon_with_holes  P1, P2;
```

```
boost::timer timer;
if (! read_Bezier_polygon(filename1, P1)) {
  std::cerr << "Failed to read " << filename1 << " ..." << std::endl;
  return -1;
}

if (! read_Bezier_polygon(filename2, P2)) {
  std::cerr << "Failed to read " << filename2 << " ..." << std::endl;
  return -1;
}
double secs = timer.elapsed();
std::cout << "Constructed the input polygons in " << secs << " seconds."
          << std::endl;

// Compute the intersection of the two polygons.
Polygon_set R;
timer.restart();
CGAL::intersection(P1, P2, std::back_inserter(R));
secs = timer.elapsed();

std::cout << "The intersection polygons are of sizes: {";
for (Polygon_set::const_iterator rit = R.begin(); rit != R.end(); ++rit)
  std::cout << ' ' << rit->outer_boundary().size();
std::cout << " }" << std::endl;
std::cout << "The intersection computation took " << secs << " seconds."
          << std::endl;
return 0;
}
```

8.3   Application: Multiway Operations on General Polygons

The multiway-operation problem: *Given a set of general polygons $P_1, \ldots P_k$, the edges of which are either straight-line segments or circular arcs, compute the union of all polygons $\bigcup_{i=1}^{k} P_i$ (or their intersection $\bigcap_{i=1}^{k} P_i$) in an efficient manner.*

The application presented in this section solves the problem above. It uses an instance of the function template `read_polygons<Traits_2, Output_iterator>()`, which reads the description of circles and general polygons bounded by straight-line segments and circular arcs from an input file and constructs them. It accepts the name of a file that contains the plain-text description of the input general polygons, and an output iterator for storing the newly constructed general polygons. The function template also accepts a traits object as a third argument. The type of the traits object must model the concept ArrangementDirectionalXMonotoneTraits_2.

All values (point coordinates, radii, and bulges) are given in floating-point representations. This input is common in the fields of computer-aided design and computer-aided manufacturing. For instance, the AutoCAD DXF, the *Drawing Exchange Format*, which enables the interchange of drawings between AutoCAD and other programs,[7] uses this representation to store general polygons. We convert the floating-point numbers to rational numbers using an instance of the `numerical_cast()` function template introduced in Section 4.2.1. The full circles specified in the input file are clearly rational. We next show that the edges of the general polygons are also supported by rational curves, and they therefore comply with the input requirements of the `Gps_circle_segment_traits_2` class.

---

[7]See http://www.autodesk.com/techpubs/autocad/acadr14/dxf for a complete specification of this format.

A full circle in the input is specified by its center coordinates and its radius. The representation of a general polygon is a bit more involved. It is defined by a list of points, denoted by $p_0, \ldots p_{n-1}$, that represent the polygon vertices. An additional parameter called *bulge* defines the curve connecting adjacent vertices. Its default value is 0, which indicates a straight segment. The bulge is the tangent of $\frac{1}{4}$ of the included angle for an arc, made negative if the arc goes clockwise from p_i to p_{i+1} and positive otherwise; a bulge of 1 indicates a semicircle.

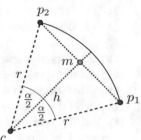

Although the definition of the bulge seems a bit strange at first glance, it allows for computing the supporting circle of the circular arc without using trigonometric functions. Assume that we are given the points $p_1 = (x_1, y_1)$ and $p_2 = (x_2, y_2)$, with a bulge λ between them, where $\lambda = \tan \frac{\alpha}{4}$ and α is the angle that defines the circular arc. If $\lambda = 0$, there is no circular arc defined in between p_1 and p_2, and we are done. Let us therefore assume that $\lambda > 0$, so the arc is counterclockwise-oriented from p_1 to p_2. We wish to compute the center $c = (x_0, y_0)$ of the supporting circle of the corresponding circular arc and its squared radius r^2. Let $m = (\frac{x_1+x_2}{2}, \frac{y_1+y_2}{2})$ be the midpoint of p_1 and p_2. Note that $\angle p_1 m c$ is a right angle; hence, we have $\sin \frac{\alpha}{2} = \frac{d}{2r}$, where $d = \sqrt{(x_2 - x_1)^2 + (y_2 - y_1)^2}$ is the distance between p_1 and p_2. As for any angle φ, $\sin \varphi = \frac{2t}{1+t^2}$, where $t = \tan \frac{\varphi}{2}$, we obtain

$$
r^2 = \frac{d^2}{4 \sin^2 \frac{\alpha}{2}} = \frac{\left((x_2 - x_1)^2 + (y_2 - y_1)^2\right)(1 + \lambda^2)^2}{16 \lambda^2} . \tag{8.1}
$$

Let h be the distance between c and m. It is clear that $\tan \frac{\alpha}{2} = \frac{d}{2h}$. At the same time, we have $\tan \frac{\alpha}{2} = \frac{2\lambda}{1-\lambda^2}$, so $h = \frac{d(1-\lambda^2)}{4\lambda}$. If we denote the angle that the vector $\vec{p_1 p_2}$ forms with the positive direction of the x-axis by θ, we have $\sin \theta = \frac{1}{d}(y_2 - y_1)$ and $\cos \theta = \frac{1}{d}(x_2 - x_1)$. Note that if we add a vector of length h that forms an angle of $(\theta + \frac{\pi}{2})$ with the x-axis to the midpoint m, we reach the circle center c. Since $\sin(\theta + \frac{\pi}{2}) = \cos \theta$ and $\cos(\theta + \frac{\pi}{2}) = -\sin \theta$, we obtain

$$
\begin{aligned}
c &= m + \left(h \cos(\theta + \frac{\pi}{2}), h \sin(\theta + \frac{\pi}{2})\right) = \\
&= \left(\frac{x_1 + x_2}{2} + \frac{(1 - \lambda^2)(y_1 - y_2)}{4\lambda}, \frac{y_1 + y_2}{2} + \frac{(1 - \lambda^2)(x_2 - x_1)}{4\lambda}\right) .
\end{aligned} \tag{8.2}
$$

Note that Equations (8.1) and (8.2) also hold for the case where $\lambda < 0$ and the arc is clockwise-oriented. We conclude that the supporting circle of the arc has rational squared radius, and its center has rational coordinates.

The function template `read_polygons()` is listed below and defined in the header file **read_polygons.h**.

```
#include <fstream>
#include <vector>
#include <list>

#include <CGAL/basic.h>

#include "numerical_cast.h"
#include "read_objects.h"

// Convert a double-precision value to a rational number.
template <typename Number_type> Number_type to_rational(const double& val)
{
    // Assume there are at most 6 decimal digits after the decimal point.
    return numerical_cast<Number_type, 1000000>(val);
}
```

```cpp
// Read a set of input polygons from the given input stream.
template <typename Traits, typename Output_iterator>
Output_iterator read_polygons(const char* filename, Output_iterator oi,
                              Traits& traits)
{
  std::ifstream is;
  if (! open_stream(is, filename)) return oi;

  typedef typename Traits::NT                 NT;
  typedef typename Traits::Rational_point_2   Rational_point;
  typedef typename Traits::Rational_circle_2  Circle;
  typedef typename Traits::Point_2            Point;
  typedef typename Traits::X_monotone_curve_2 X_monotone_curve;
  typedef typename Traits::Curve_2            Curve;

  // Read the polygons one by one.
  typename Traits::Make_x_monotone_2 mk_x_monotone =
    traits.make_x_monotone_2_object();
  X_monotone_curve xcv;

  char ctype;
  while (is >> ctype) {
    // Read the polygon according to its type: 'p' (polygon) or 'c' (circle).
    typename Traits::Polygon_2 pgn;
    if (ctype == 'c' || ctype == 'C') {
      // Read a circle, given by its center (x, y) and its radius r.
      double dx, dy, dr;
      is >> dx >> dy >> dr;
      const NT x0 = to_rational<NT>(dx);
      const NT y0 = to_rational<NT>(dy);
      const NT rad = to_rational<NT>(dr);

      // Break the circle into two x-monotone arcs.
      std::list<CGAL::Object> objects;
      mk_x_monotone(Curve(Rational_point(x0, y0), rad),
                    std::back_inserter(objects));
      CGAL_assertion(objects.size() == 2);
      std::list<CGAL::Object>::iterator it = objects.begin();
      CGAL::assign(xcv, *it++);    pgn.push_back(xcv);
      CGAL::assign(xcv, *it);      pgn.push_back(xcv);
    }
    else if (ctype == 'p' || ctype == 'P') {
      // Read the number of vertices in the generalized polygon, then read
      // all points and bulge values.
      unsigned int n;
      is >> n;

      std::vector<std::pair<std::pair<double, double>, double> > verts(n);
      for (unsigned int i = 0; i < n; ++i) {
        double dx, dy, dbulge;
        is >> dx >> dy >> dbulge;
        verts[i] = std::make_pair(std::make_pair(dx, dy), dbulge);
      }
```

```
     // Now compute the polygon edges.
  for (unsigned int i = 0; i < n; ++i) {
    const NT  x1 = to_rational<NT>(verts[i].first.first);
    const NT  y1 = to_rational<NT>(verts[i].first.second);
    const NT  x2 = to_rational<NT>(verts[(i + 1) % n].first.first);
    const NT  y2 = to_rational<NT>(verts[(i + 1) % n].first.second);
    const NT  bulge = to_rational<NT>(verts[i].second);
    const CGAL::Sign    sign_bulge = CGAL::sign(bulge);

    if (sign_bulge == CGAL::ZERO) {
      // Zero bulge value: the current edge is a line segment.
      if (CGAL::compare(x1, x2) != CGAL::EQUAL ||
          CGAL::compare(y1, y2) != CGAL::EQUAL)
        pgn.push_back(X_monotone_curve(Rational_point(x1,y1),
                                       Rational_point(x2,y2)));
    }
    else {
      // A non-zero bulge: ps and pt are connected by a circular arc.
      // Compute the center and the squared radius of its supporting circle.
      CGAL_assertion((CGAL::compare(x1, x2) != CGAL::EQUAL) ||
                     (CGAL::compare(y1, y2) != CGAL::EQUAL));

      const NT common = (1 - CGAL::square(bulge)) / (4*bulge);
      const NT x0 = (x1 + x2)/2 + common*(y1 - y2);
      const NT y0 = (y1 + y2)/2 + common*(x2 - x1);
      const NT sqr_bulge = CGAL::square(bulge);
      const NT sqr_dist = CGAL::square(x2 - x1) + CGAL::square(y2 - y1);
      const NT sqr_radius = sqr_dist * (1/sqr_bulge + 2 + sqr_bulge) / 16;

      // Construct the arc: A positive (resp. negative) bulge implies a
      // counterclockwise- (resp. clockwise-) oriented arc.
      Circle supp_circ = (sign_bulge == CGAL::POSITIVE) ?
        Circle(Rational_point(x0, y0), sqr_radius,
               CGAL::COUNTERCLOCKWISE) :
        Circle(Rational_point(x0, y0), sqr_radius, CGAL::CLOCKWISE);

      Curve circ_arc(supp_circ, Point(x1, y1), Point(x2, y2));

      // Break the arc into x-monotone subarcs (there can be at most
      // three subarcs) and add them to the polygon.
      std::list<CGAL::Object> objects;
      mk_x_monotone(circ_arc, std::back_inserter(objects));
      CGAL_assertion(objects.size() <= 3);
      std::list<CGAL::Object>::iterator  it = objects.begin();
      for (it = objects.begin(); it != objects.end(); ++it)
        if (CGAL::assign (xcv, *it)) pgn.push_back (xcv);
    }
  }
}
else CGAL_error();              // illegal type

*oi++ = pgn;                    // write the polygon to the output iterator.
}
```

```
    close_stream(is);
    return oi;
}
```

The application comes with a **main()** function that first constructs the input polygons. It calls an instance of the **read_polygons()** function template, as described above. Then, it reverses the orientation of polygons oriented clockwise. An instance of the template function **polygon_ orientation()** described in Section 5.6 is used to detect such polygons. Finally, it uses the multiway set operation function, which is based on the divide-and-conquer approach, to compute their union. This variant is recommended when the number of intersections of the input polygons is of the same order of magnitude as the complexity of the result. If this is not the case, computing the result incrementally may prove more efficient.

The file **vlsi.dat** is used as default. It describes a partial layout of a printed circuit board (PCB) courtesy of ManiaBarco Incorporated. The file contains the description of 2,926 general polygons, 645 out of which are full circles. Some of the input general polygons are oriented clockwise and must be repaired before being processed further; see Figure 8.4a. The union consists of 615 general polygons with holes; see Figure 8.4b. Some of the output polygons indeed have holes. For example, the digit 0, which appears at the top portion of the lens, is made up of eight different input polygons in Figure 8.4a, but only one output polygon with a single hole in Figure 8.4b.

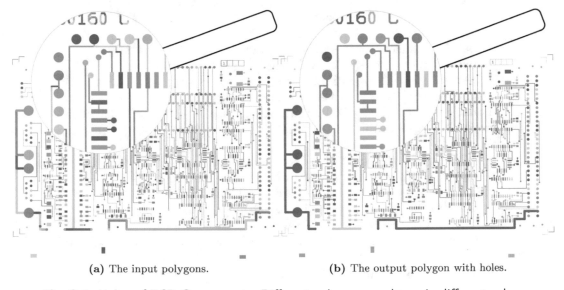

(a) The input polygons.   (b) The output polygon with holes.

Fig. 8.4: Union of PCB Components. Different polygons are drawn in different colors.

```
// File: operation_on_ccb.cpp

#include <list>
#include <boost/timer.hpp>

#include <CGAL/basic.h>
#include <CGAL/IO/Gps_iostream.h>

#include "bops_circular.h"
#include "read_polygons.h"
#include "polygon_orientation.h"

int main(int argc, char* argv[])
```

```
{
    // Read the circular polygons from the input file.
    const char* filename = (argc >= 2) ? argv[1] : "vlsi.dat";
    std::list<Polygon>         pgns;
    Traits traits;
    std::cout << "Reading_<" << filename << ">_..._" << std::flush;
    boost::timer    timer;
    read_polygons(filename, std::back_inserter(pgns), traits);
    double secs = timer.elapsed();
    std::cout << "Read_" << pgns.size() << "_polygons_(" << secs << "_seconds)."
              << std::endl;

    // Repair polygon orientation.
    std::list<Polygon>::iterator it;
    for (it = pgns.begin(); it != pgns.end(); ++it) {
      if (CGAL::CLOCKWISE ==
          polygon_orientation(it->curves_begin(), it->curves_end(), traits))
        it->reverse_orientation();
    }

    // Compute the union of the polygons.
    Polygon_set S(traits);
    std::cout << "Computing_the_union_..._" << std::flush;
    timer.restart();
    S.join(pgns.begin(), pgns.end());
    secs = timer.elapsed();
    std::cout << "Done!_(" << secs << "_seconds)." << std::endl;

    return 0;
}
```

8.4   Application: Obtaining Silhouettes of Polyhedra

We conclude this chapter with an application that obtains the silhouettes of closed polyhedra in $\mathbb{R}^3$.

The polyhedra-silhouette problem: *Given a closed polyhedron P and a direction d, compute the silhouette of P as observed along d.*

Given a polyhedron P and a direction d, the program obtains the outline of the shadow of P cast onto a plane perpendicular to d, where the scene is illuminated by a light source at infinity directed along d. Section 2.3 lists an example program that obtains the silhouettes of convex polytopes, where the light source is directed vertically downwards. However, the main challenge presented by the problem introduced in this section is handling not only convex, but also concave (closed) polyhedra. A silhouette is represented as an arrangement—the same representation used in Section 2.3.

Figure 8.5a shows a hand model obtained from the AIM@SHAPE Shape Repository (id 785).[8] Figure 8.5b shows the shadow of the hand cast onto the yz-plane as computed by the program below. Thanks to the rich feature-set offered by

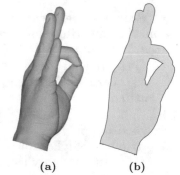

(a)   (b)

Fig. 8.5: Laurent's hand.

---

[8]see http://shapes.aim-at-shape.net/viewgroup.php?id=788#.

CGAL in general and the *2D Regularized Boolean Set-Operations* package in particular, the code is made very simple. First, the input-direction coordinates given as floating-point numbers as part of the command line are converted into rational numbers, and the target plane is calculated accordingly. Next, the program parses a file in the Object File Format (OFF)[9] that describes the input closed polyhedron, constructs the polytope, and stores the result in a temporary object, the type of which is an instance of the CGAL **Polyhedron_3** class template. Next, the program traverses the polyhedron facets and culls the facets not facing the target plane; see Line 55 in the code excerpt below. For each facet facing the target plane, it projects the facet-boundary vertices onto the target plane, and constructs the corresponding two-dimensional polygons. Finally, it constructs a polygon set that represents the union of all the two-dimensional polygons.

```
1   // File: polyhedron_projection.cpp
2
3   #include <boost/lexical_cast.hpp>
4
5   #include "bops_set_linear.h"
6   #include "Normal_equation.h"
7
8   #include <CGAL/Arrangement_2.h>
9   #include <CGAL/Polyhedron_3.h>
10  #include <CGAL/Polyhedron_traits_with_normals_3.h>
11  #include <CGAL/IO/Polyhedron_iostream.h>
12  #include <CGAL/IO/Gps_iostream.h>
13
14  typedef CGAL::Polyhedron_traits_with_normals_3<Kernel> Polyhedron_traits;
15  typedef CGAL::Polyhedron_3<Polyhedron_traits>        Polyhedron;
16  typedef Kernel::Point_3                              Point_3;
17  typedef Kernel::Plane_3                              Plane_3;
18  typedef Kernel::Direction_3                          Direction_3;
19
20  int main(int argc, char* argv[])
21  {
22    // Read the direction from the command line.
23    CGAL_assertion(argc > 3);
24    Kernel::FT x = boost::lexical_cast<double>(argv[1]);
25    Kernel::FT y = boost::lexical_cast<double>(argv[2]);
26    Kernel::FT z = boost::lexical_cast<double>(argv[3]);
27    Direction_3 direction(x, y, z);
28
29    // Read the polyhedron from the specified input file.
30    const char* filename = (argc > 4) ? argv[4] : "hand.off";
31    std::ifstream in_file(filename);
32    if (!in_file.is_open()) {
33      std::cerr << "Failed to open " << filename << "!" << std::endl;
34      return -1;
35    }
36    Polyhedron polyhedron;
37    in_file >> polyhedron;
38    std::transform(polyhedron.facets_begin(), polyhedron.facets_end(),
39                   polyhedron.planes_begin(), Normal_equation());
40
41    // Go over the polyhedron facets and project them onto the plane.
42    std::list<Polygon> polygons;
```

---

[9]See, for example, http://www.fileinfo.com/extension/off/.

```
43    Kernel kernel;
44    Kernel::Compare_z_3 cmp_z = kernel.compare_z_3_object();
45    Kernel::Construct_projected_xy_point_2 proj =
46      kernel.construct_projected_xy_point_2_object();
47    Kernel::Construct_translated_point_3 translate =
48      kernel.construct_translated_point_3_object();
49    Point_3 origin = kernel.construct_point_3_object()(CGAL::ORIGIN);
50    Plane_3 plane =  kernel.construct_plane_3_object()(origin, direction);
51    Point_3 r = translate(origin, direction.vector());
52    Polyhedron::Facet_const_iterator fit;
53    for(fit = polyhedron.facets_begin(); fit != polyhedron.facets_end(); ++fit){
54      // Discard facets facing the given direction.
55      if(CGAL::angle(translate(origin,fit->plane()), origin, r) == CGAL::OBTUSE)
56
57        continue;
58
59      // Go over the facet vertices and project them.
60      Polygon polygon;
61      Polyhedron::Halfedge_around_facet_const_circulator hit=fit->facet_begin();
62      do {
63        const Point_3& point = hit->vertex()->point();
64        polygon.push_back(proj(plane, point));
65      } while (++hit != fit->facet_begin());
66      polygons.push_back(polygon);
67    }
68    polyhedron.clear();
69    Polygon_set S;
70    S.join(polygons.begin(), polygons.end());
71
72    std::cout << S;
73    return 0;
74 }
```

8.5 Bibliographic Notes and Remarks

Boolean set operations on geometric objects are fundamental and frequently used in all areas of geometric computing. They are the basis of Constructive Solid Geometry (CSG), they are central to algorithmic motion planning and the computation of Minkowski sums (see Chapter 9), and they are relevant to computer graphics and molecular modeling, to mention just a few areas.

Table 8.1: The valid combinations of argument types of the overloaded free-functions intersection(), join(), difference(), symmetric_difference(), and do_intersect().

First Argument Type	Second Argument Type
Polygon_2	Polygon_2
Polygon_2	Polygon_with_holes_2
Polygon_with_holes_2	Polygon_2
Polygon_with_holes_2	Polygon_with_holes_2
General_polygon_2	General_polygon_2
General_polygon_2	General_polygon_with_holes_2
General_polygon_with_holes_2	General_polygon_2
General_polygon_with_holes_2	General_polygon_with_holes_2

As explained in the chapter, if you have an arrangement-overlay operation at your disposal, then you can compute any Boolean set operation on objects of any type of shape, provided you have the traits class for the curves that comprise the boundaries of the shapes. Since CGAL supports arrangements of algebraic curves of arbitrary degree (see Chapter 5), you can compute Boolean set operations for a wide range of planar shapes. Indeed the *2D Regularized Boolean Set-Operations* package of CGAL [78] offers a rich interface to assist you with this task. The interface includes, for example, a large set of overloaded free functions; see Table 8.1. In Exercise 8.7 you are asked to extend the set of overloaded free functions with variants that accept additional types. Despite the flexibility gained by the rich interface, several issues still need to be discussed and addressed.

Suppose you wish to exclude certain lower-dimensional features of your planar shape; say you are only interested in the interior of one shape, the closure of another shape, and you take their union. In such a case, regularized Boolean set operations as described in the chapter will not suffice. You need to use the arrangement-overlay operation directly to produce refined representations of this type. The *2D Boolean Operations on Nef Polygons* package of CGAL [189] allows you to do exactly that for (linear) polygonal regions.

Then comes the major issue of the efficiency of the Boolean set operations. The case of *union* has been intensively studied in computational geometry, as it raises interesting combinatorial and algorithmic questions [2]. Consider, for example, a set of n triangles. The maximum combinatorial complexity of their union is $\Theta(n^2)$ and in that sense the sweep-based overlay solution is near-optimal, as it runs in $O(n^2 \log n)$ time. One can compute this union in worst-case optimal time $O(n^2)$ by incrementally constructing the arrangement of lines supporting the triangle edges; see Chapter 4. Practically, one is better off using the sweep-based algorithm as the line-arrangement algorithm will compute many spurious intersections. An alternative solution is to use an incremental algorithm for line segments (see, e.g., [196, Section 6.6]), which runs in time $O(n^2\alpha(n))$ in the worst case, where $\alpha()$ is the inverse of Ackermann's function (an extremely slowly growing function).

At the other extreme, some extra property of these triangles, such as their being *fat* [150, 174] or their being a collection of pseudo-discs [45, Chapter 13], [127], is helpful: We apply divide-and-conquer on the set of triangles, and recursively compute the union of subsets with a plane-sweep merge step. Then we get an algorithm that runs in near linear time (near linear in n).

The discussion above applies almost verbatim to n simple polygons the complexity of each pf which is bounded by some constant. The extreme-case algorithms (near-quadratic or near-linear) apply also, with slight changes in the running times, to curved objects, each of constant maximum descriptive complexity.

The big question is what happens in between these two extremes. Can we compute the union of, say, triangles in an output-sensitive manner? It turns out that even deciding whether a set of triangles covers another triangle is a so-called 3-sum hard problem [82], which means we do not know of an algorithm running in time $o(n^2)$ to solve it. We proposed a heuristic speedup of computing the union of triangles in [62]. Under the assumption that only a subset of size $o(n/\log^2 n)$ of the input triangles contributes to the union boundary, a sub-quadratic algorithm is proposed in [63]. In general, finding efficient algorithms in practice for computing the union of planar regions (even for triangles) remains a major challenge.

Planar set intersection also appears frequently in geometric computing. If our input is a set of n convex polygons the complexity of each is bounded by some constant, then their intersection has at most $O(n)$ edges and it can be computed in worst-case optimal time $O(n \log n)$. Here, the challenge in the context of arrangements is often to maintain the intersection dynamically as objects are added or removed while traversing the arrangement; see, for example, [27, 148, 194].

Symmetric difference, the Boolean set operation counterpart of the exclusive disjunction (XOR), is interesting in the context of shape matching; see, e.g., [9].

8.6   Exercises

8.1 Consider the program coded in the files **ex_polygon_repairing.cpp**, **polygon_repairing. h**, and **Winding_number.h** listed in Section 6.5. When fed with the polygon depicted at the bottom of the figure on Page 157 it results in a single polygon that is topologically not equivalent to a disc and thus invalid. (p_4 and p_7 in the figure coincide at a vertex, the degree of which is 4.) Enhance the program so that it always results in valid polygons or valid polygons with holes.

8.2 Using only **complement()** and **intersection()**, provide an alternative implementation for the following free-function templates:

template <**typename** Kernel, **typename** Container>
inline bool join(**const** Polygon_2<Kernel, Container>& pgn1,
                  const Polygon_2<Kernel, Container>& pgn2);

template <**typename** Kernel, **typename** Container>
inline bool difference(**const** Polygon_2<Kernel, Container>& pgn1,
                  const Polygon_2<Kernel, Container>& pgn2);

template <**typename** Kernel, **typename** Container>
inline bool symmetric_difference(**const** Polygon_2<Kernel, Container>& pgn1,
                  const Polygon_2<Kernel, Container>& pgn2);

template <**typename** Kernel, **typename** Container>
inline bool oriented_side(**const** Polygon_2<Kernel, Container>& pgn1,
                  const Polygon_2<Kernel, Container>& pgn2);

Compare the performance of the implementation of the function templates provided by the package with the performance of your implementation.

8.3 The implementation of the function templates that compute multiway Boolean set operations exploits both the divide-and-conquer approach and the plane-sweep algorithm; see Section 8.1.8. For example, consider the following member of the **General_polygon_set_2** class template:

template <**typename** Input_iterator>
void join(Input_iterator begin, Input_iterator end, **unsigned int** k = 5)

It computes the union of the polygon in the range **[begin, end)** and initializes the hosting **General_polygon_set_2** object with the results. At each invocation of the recursive divide-and-conquer procedure, the number of input polygons is examined. If it is larger than k it proceeds as usual. Otherwise, it applies the plane-sweep algorithm on the input polygons.

Find the optimal values of k for the applications listed in Sections 8.3 and 8.4 when they are executed on the respective default inputs. These applications are coded in the files **operations_on_pcb.cpp** and **polyhedron_projection.cpp**, respectively.

8.4 Write a program that reads a set of polygons with holes from an input file, computes the union of the input polygons, and then efficiently computes the winding numbers of all the faces of the underlying arrangement. Extend the arrangement faces properly, and avoid copying and converting the data (which is unnecessary).

8.5 Develop a function template called **connect_holes_hor()** that can serve as an alternative to the function template **connect_holes()** (see Section 8.1.4), and has the same return and argument types, that is,

template <**typename** Kernel, **typename** Container, **typename** Output_iterator>
Output_iterator
connect_holes_hor(**const** Polygon_with_holes_2<Kernel, Container>& pwh,
                    Output_iterator oi);

The new function, like the existing one, should connect the holes of an input polygon-with-holes object. As a result it should compute a sequence of points that spans the outer and inner boundaries of the input polygon with holes. However, instead of computing the topmost vertex of each hole and symbolically connecting it to the feature located above it, it computes the leftmost vertex of each hole and symbolically connects it to the feature located directly to its left. The sequence should start (and end) with the leftmost vertex of the outer boundary.

8.6 Develop a function template called `disconnect_holes()` that reverts the operation of the function template `connect_holes()`; see Section 8.1.4. It accepts a sequence of points representing a traversal of the outer and inner boundaries of a polygon with holes, as computed by a call to an instance of the function template `connect_holes()`. The function template `disconnect_holes()` should restore the original polygon with holes.

template <**typename** Input_iterator, **typename** Kernel, **typename** Container>
void disconnect_holes(Input_iterator begin, Input_iterator end,
                    CGAL::Polygon_with_holes_2<Kernel, Container>& pwh);

8.7 Develop additional variants of the overloaded free-function templates `intersection()`, `join()`, `difference()`, `symmetric_difference()`, and `do_intersect()` that accept arguments of the following types: `Triangle_2`, `Iso_rectangle_2`, and `Circle_2`. The prototype below is just one example out of many possible combinations.

template <**typename** Kernel, **typename** Container, **typename** Output_iterator>
Output_iterator intersection(Triangle_2<Kernel>& tri,
                    Polygon_2<Kernel, Container>& p,
                    Output_iterator oi);

Chapter 9

Minkowksi Sums and Offset Polygons

Given two sets $A, B \in \mathbb{R}^2$, their *Minkowski sum*, denoted by $A \oplus B$, is their point-wise sum, namely the set

$$A \oplus B = \{a + b \mid a \in A, b \in B\} \ .$$

Minkowski sums are used in many applications, such as motion planning and computer-aided design and manufacturing. The *2D Minkowski Sums* package of CGAL contains functions that compute the planar Minkowski sum of two simple polygons (see Chapter 8 for the precise definition of a simple polygon) and the planar Minkowski sum of a simple polygon and a disc—an operation also referred to as *offsetting* or *dilating* a polygon.[1] This package, like the *2D Regularized Boolean Set-Operations* package, is implemented on top of the arrangement infrastructure. The two packages are integrated well to allow mixed operations. For example, it is possible to apply Boolean set operations on objects that are the result of Minkowski sum computations.[2]

9.1 Computing the Minkowski Sum of Two Polygons

Computing the Minkowski sum of two convex polygons P and Q with m and n vertices, respectively, is rather easy. Observe that $P \oplus Q$ is a convex polygon bounded by copies of the $m + n$ edges ordered according to the angle they form with the x-axis. As the two input polygons are convex, their edges are already sorted by the angle they form with the x-axis; see the figure to the right. The Minkowski sum can therefore be computed using an operation similar to the merge step of the merge-sort algorithm[3] in $O(m + n)$ time, starting from the two bottommost vertices in P and in Q and merging the ordered list of edges.

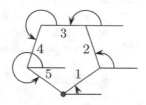

If the polygons are not convex, you can utilize either the *decomposition* or the *convolution* approaches described below. Regarding the implementation of the two approaches, applications of Minkowski sum operations are restricted to polygons that are simple and in particular do not contain holes. (Resulting sums may contain holes though.) However, this restriction can be relaxed when the decomposition approach is applied.

Decomposition: We decompose P and Q into convex sub-polygons. Namely, we obtain two sets of convex polygons $P_1, \ldots, P_k$ and $Q_1, \ldots, Q_\ell$, such that $\bigcup_{i=1}^k P_i = P$ and $\bigcup_{j=1}^\ell Q_j = Q$.

---

[1]The family of valid types of summands is slightly broader for certain operations, e.g., a degenerate polygon consisting of two points only is a valid operand for the approximate-offsetting operation.

[2]The operands of the Minkowski sum operations supported by the *2D Minkowski Sums* package must be (linear) polygons, as opposed to the operands of the Boolean set operations supported by the *2D Regularized Boolean Set-Operations* package. The latter belong to the broader family of general polygons.

[3]See, for example, http://en.wikipedia.org/wiki/Merge_sort.

E. Fogel et al., *CGAL Arrangements and Their Applications*, Geometry and Computing 7, DOI 10.1007/978-3-642-17283-0_9, © Springer-Verlag Berlin Heidelberg 2012

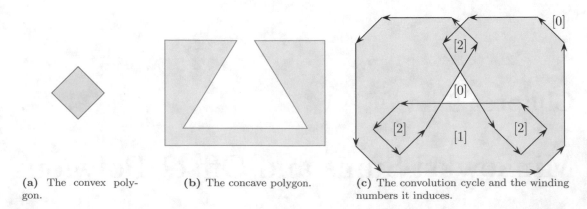

(a) The convex polygon.   **(b)** The concave polygon.   **(c)** The convolution cycle and the winding numbers it induces.

Fig. 9.1: The convolution of a convex polygon and a non-convex polygon. The convolution consists of a single self-intersecting cycle, drawn as a sequence of directed line segments. Each face of the arrangement induced by the segments forming the cycle contains its winding number. The Minkowski sum of the two polygons is shaded.

We then calculate the pairwise sums $S_{ij} = P_i \oplus Q_j$ using the simple procedure described above, and finally compute the union $P \oplus Q = \bigcup_{ij} S_{ij}$; see Section 8.1.8.

This approach relies on a successful decomposition of the input polygons into convex pieces, and its performance depends on the quality and performance of the decomposition. The supplied decomposition methods do not handle point sets that are not simple or contain holes.

Remark: This restriction can be relaxed by the introduction of new decomposition methods that are applicable to a broader family of point sets. However, in the remainder of this chapter, we only refer to simple polygons, with one exception—the family of valid operands of the convolution-based offset operation includes line segments.

Convolution: Let $P = (p_0, \ldots, p_{m-1})$ and $Q = (q_0, \ldots, q_{n-1})$ denote the vertices of the input polygons. We assume that both P and Q have positive orientations (i.e., their boundaries wind in a counterclockwise order around their interiors), and compute the convolution of the two polygon boundaries. The *convolution* of these two polygons, denoted $P * Q$, is a collection of line segments of the form $[p_i + q_j, p_{i+1} + q_j]$,[4] where the vector $\overrightarrow{p_i p_{i+1}}$ lies between $\overrightarrow{q_{j-1} q_j}$ and $\overrightarrow{q_j q_{j+1}}$, and,[5] symmetrically, of segments of the form $[p_i + q_j, p_i + q_{j+1}]$, where the vector $\overrightarrow{q_j q_{j+1}}$ lies between $\overrightarrow{p_{i-1} p_i}$ and $\overrightarrow{p_i p_{i+1}}$.

The segments of the convolution form a number of closed (not necessarily simple) polygonal curves called *convolution cycles*. (In simpler cases, for example, when at least one of the summands is convex, there is only one convolution cycle and its construction is straightforward. How to create more than one cycle when necessary is not described here; see references in Section 9.4.) The Minkowski sum $P \oplus Q$ is the set of points having a nonzero winding number with respect to the cycles of $P * Q$. Informally speaking, the winding number of a point $p \in \mathbb{R}^2$ with respect to some planar curve γ is the count of how many times γ winds in a counterclockwise orientation around p; see Section 6.5 for another use of winding numbers; see Figure 9.1 for an illustration.

The number of segments in the convolution of two polygons is usually smaller than the number of segments that constitute the boundaries of the sub-sums S_{ij} of the decomposition approach.

---

[4] Throughout this chapter, we increment or decrement an index of a vertex modulo the number of vertices of the polygon.

[5] We say that a vector $\vec{v}$ lies between two vectors $\vec{u}$ and $\vec{w}$, if $\vec{v}$ is reached strictly before $\vec{w}$ is reached, when all three vectors are moved to the origin, and $\vec{u}$ is rotated counterclockwise. Note that this also covers the case where $\vec{u}$ has the same direction as $\vec{v}$.

As both approaches construct the arrangement of these segments and extract the sum from this arrangement, computing Minkowski sums using the convolution approach usually generates a smaller intermediate arrangement; hence, it is faster and consumes less space.

9.1.1   Computing Minkowski Sum Using Convolutions

The function template `minkowski_sum_2(P, Q)` accepts two simple polygons P and Q and computes their Minkowski sum $S = P \oplus Q$ using the convolution method. The types of the operands are instances of the `Polygon_2` class template; see Section 8.1. As the input polygons may not be convex, their Minkowski sum may not be simply connected and may contain polygonal holes; see, for example, Figure 9.1. The type of the returned object S is therefore an instance of the `Polygon_with_holes_2` class template. The outer boundary of S is a polygon that can be accessed using `S.outer_boundary()`, and its polygonal holes are given by the range [`S.holes_begin()`, `S.holes_end()`) (where S contains `S.number_of_holes()` holes in its interior); see Section 8.1.1.

Example: The example program below constructs the Minkowski sum of a square and a triangle, as depicted in the figure to the right. The result in this case is a convex hexagon. This program, like other example programs in this chapter, includes

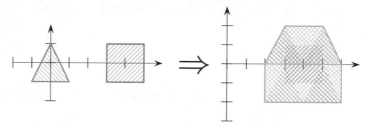

the header file **bops_linear.h**, which defines the polygon types; see Page 180 for the code listing.

```
// File: ex_sum_triangles.cpp

#include <CGAL/basic.h>
#include <CGAL/minkowski_sum_2.h>

#include "bops_linear.h"

int main()
{
  // Construct the triangle.
  Polygon   P;
  P.push_back(Point(-1, -1));  P.push_back(Point(1, -1));
  P.push_back(Point(0, 1));
  std::cout << "P_=_" << P << std::endl;

  // Construct the square.
  Polygon   Q;
  Q.push_back(Point(3, -1));   Q.push_back(Point(5, -1));
  Q.push_back(Point(5, 1));    Q.push_back(Point(3, 1));
  std::cout << "Q_=_" << Q << std::endl;

  // Compute the Minkowski sum.
  Polygon_with_holes  sum = CGAL::minkowski_sum_2(P, Q);
  CGAL_assertion(sum.number_of_holes() == 0);
  std::cout << "P_(+)_Q_=_" << sum.outer_boundary() << std::endl;
  return 0;
}
```

Example: The program below computes the Minkowski sum of two polygons that are read from an input file. In this case the sum may contain holes. The Minkowski sum, for example, of the

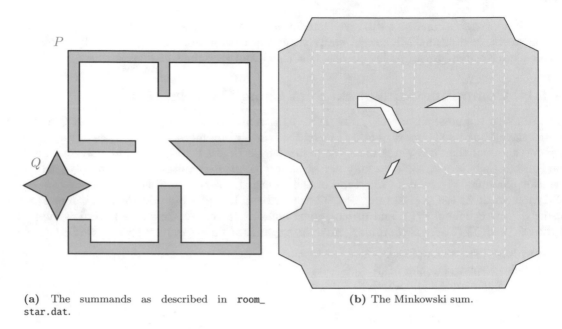

(a) The summands as described in `room_star.dat`.

(b) The Minkowski sum.

Fig. 9.2: The Minkowski sum of two non-convex polygons P and Q.

polygons described in the default input file **room_star.dat** is not simple and contains four holes, as illustrated in Figure 9.2.

```
// File: ex_sum_with_holes.cpp

#include <fstream>

#include <CGAL/basic.h>
#include <CGAL/minkowski_sum_2.h>

#include "bops_linear.h"
#include "pgn_print.h"

int main(int argc, char* argv[])
{
  // Open the input file and read the two polygons from it.
  const char* filename = (argc > 1) ? argv[1] : "rooms_star.dat";
  std::ifstream    in_file(filename);
  if (! in_file.is_open()) {
    std::cerr << "Failed to open the input file." << std::endl;
    return -1;
  }
  Polygon   P, Q;
  in_file >> P >> Q;
  in_file.close();

  // Compute and print the Minkowski sum.
  Polygon_with_holes  sum = CGAL::minkowski_sum_2(P, Q);
  std::cout << "P (+) Q = ";
  print_polygon_with_holes(sum);
  return 0;
}
```

While in general the convolution approach to computing Minkowski sums runs faster, we observed that when the proportion of reflex vertices in both summands is large, the decomposition approach runs faster. Hence, we also describe how to employ the decomposition-based Minkowski sum procedure.

9.1.2 Decomposition Strategies

In order to compute the Minkowski sum of two polygons P and Q using the decomposition method, issue the call `minkowski_sum_2(P, Q, decomp)`, where `decomp` is an object of a type that models the concept PolygonConvexDecomposition, which in turn refines a Functor concept variant. Namely, it provides a function operator (`operator()`) that accepts a planar polygon and returns a range of convex polygons that represents its convex decomposition.

The *2D Minkowski Sums* package includes four models of the PolygonConvexDecomposition concept as follows:

- The `Optimal_convex_decomposition<Kernel>` class template uses the dynamic programming algorithm by Greene for computing an optimal decomposition of a polygon into a minimal number of convex sub-polygons. While this algorithm results in a small number of convex polygons, it consumes rather many resources, as it runs in $O(n^4)$ time and $O(n^3)$ space in the worst case, where n is the number of vertices in the input polygon.

- The `Hertel_Mehlhorn_convex_decomposition<Kernel>` class template implements the approximation algorithm suggested by Hertel and Mehlhorn, which triangulates the input polygon and then discards unnecessary triangulation edges. After triangulation (carried out by the constrained-triangulation procedure of CGAL) the algorithm runs in $O(n)$ time and space, and guarantees that the number of sub-polygons it generates is not more than four times the optimum.

- The `Greene_convex_decomposition<Kernel>` class template is an implementation of Greene's approximation algorithm, which computes a convex decomposition of the polygon based on its partitioning into y-monotone polygons. This algorithm runs in $O(n \log n)$ time and $O(n)$ space, and has the same guarantee on the quality of approximation as Hertel and Mehlhorn's algorithm.

- The `Small_side_angle_bisector_convex_decomposition<Kernel>` class template is based on the angle-bisector decomposition method suggested by Chazelle and Dobkin, which runs in $O(n^2)$ time. In addition, it applies a heuristic by Flato that reduces the number of output polygons in many common cases. The convex decompositions that it produces usually yield efficient running times for Minkowski sum computations. It starts by examining each pair of reflex vertices in the input polygon, such that the entire interior of the diagonal connecting these vertices is contained in the polygon. Out of all available pairs, the vertices p_i and p_j are selected, such that the number of reflex vertices encountered when traversing the boundary of the polygon from p_i to p_j in clockwise order is minimal. The polygon is split by the diagonal $p_i p_j$. This process is repeated recursively on both resulting sub-polygons. In case it is not possible to eliminate two reflex vertices at once any more, each reflex vertex is eliminated by a diagonal that is closest to the angle bisector emanating from this vertex and having rational-coordinate endpoints on both sides.

Example: The example below demonstrates the computation of the Minkowski sum of the same input polygons used in `ex_sum_with_holes.cpp` (depicted in Figure 9.2), using the small-side angle-bisector decomposition strategy.

// File: ex_sum_by_decomposition.C

#include <fstream>

```
#include <CGAL/Exact_predicates_exact_constructions_kernel.h>
#include <CGAL/minkowski_sum_2.h>
#include <CGAL/Small_side_angle_bisector_decomposition_2.h>

#include "pgn_print.h"

typedef CGAL::Exact_predicates_exact_constructions_kernel Kernel;
typedef Kernel::Point_2                                   Point;
typedef CGAL::Polygon_2<Kernel>                           Polygon;
typedef CGAL::Polygon_with_holes_2<Kernel>                Polygon_with_holes;

int main(int argc, char* argv[])
{
  // Open the input file and read two polygons from it.
  const char* filename = (argc > 1) ? argv[1] : "rooms_star.dat";
  std::ifstream    in_file(filename);
  if (! in_file.is_open()) {
    std::cerr << "Failed to open the input file." << std::endl;
    return -1;
  }
  Polygon    P, Q;
  in_file >> P >> Q;
  in_file.close();

  // Compute the Minkowski sum using the decomposition approach.
  CGAL::Small_side_angle_bisector_decomposition_2<Kernel> ssab_decomp;
  Polygon_with_holes  sum = CGAL::minkowski_sum_2(P, Q, ssab_decomp);
  std::cout << "P (+) Q = "; print_polygon_with_holes(sum);
  return 0;
}
```

 Try: The convolution approach is typically more efficient than the decomposition approach, as the underlying arrangement computed by the former tends to be smaller than the arrangement computed by the latter using any decomposition strategy. Try to generate two input polygons such that computing their Minkowski sum using some decomposition strategy consumes less time than computing it using the convolution approach. Try to generate a pair of input polygons P and Q for each decomposition strategy $\mathcal{S}$ such that computing $P \oplus Q$ using $\mathcal{S}$ is the most efficient among all methods.

9.2   Offsetting a Polygon

The operation of computing the Minkowski sum $P \oplus B_r$ of a polygon P with a disc B_r of radius r centered at the origin is widely known as *offsetting* the polygon P by a radius r.

Let P be a simple polygon, and let $p_0, \ldots, p_{n-1}$ be the vertices of P oriented counterclockwise around the interior of P. If P is a convex polygon the offset is easily computed by shifting each polygon edge by r away from the polygon, namely to the right side of the edge. As a result we obtain a collection of n disconnected *offset edges*. Each pair of adjacent offset edges induced by $p_{i-1}p_i$ and p_ip_{i+1} are connected by a circular arc of radius r, whose supporting circle is centered at p_i. The angle that defines such a circular arc equals $180° - \angle p_{i-1}p_ip_{i+1}$; see Figure 9.3a for an illustration. The running time of this simple process is naturally linear with respect to the size of the polygon.

If P is not convex, its offset can be obtained by decomposing it into convex sub-polygons $P_1, \ldots P_k$ such that $\bigcup_{i=1}^{k} P_i = P$, computing the offset of each sub-polygon, and finally calculating

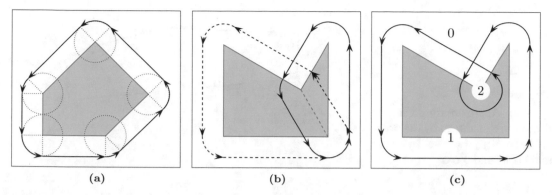

Fig. 9.3: (a) The offset of a convex polygon. (b) The offset of a non-convex polygon as computed by decomposing it into two convex sub-polygons. (c) The offset of a non-convex polygon as computed using the convolution approach. The convolution cycle induces an arrangement with three faces, whose winding numbers are indicated.

the union of these offset sub-polygons; see Figure 9.3b. However, as with the case of the Minkowski sum of a pair of polygons, it is also more efficient to compute the *convolution cycle* of the polygon and the disc B_r,[6] which can be constructed by adapting the process described in the previous paragraph for convex polygons: The only difference is that a circular arc induced by a reflex vertex p_i is defined by an angle $180° + \angle p_{i-1} p_i p_{i+1}$; see Figure 9.3c for an illustration. Recall that the last step consists of computing the winding numbers of the faces of the arrangement induced by the convolution cycle and discarding the faces with zero winding numbers.

9.2.1 Approximating the Offset with a Guaranteed Error Bound

Let P by a counterclockwise-oriented simple polygon all vertices of which $p_0, \ldots, p_{n-1}$ have *rational* coordinates, i.e., for each vertex $p_i = (x_i, y_i)$ we have $x_i, y_i \in \mathbb{Q}$. Consider the Minkowski sum of P with a disc of radius r, where r is also a rational number. The boundary of this sum comprises line segments and circular arcs, where:

- Each circular arc is supported by a circle of radius r centered at a polygon vertex p_i. The equation of this circle, $(x - x_i)^2 + (y - y_i)^2 = r^2$, has only rational coefficients.

- Each line segment is supported by a line parallel to a polygon edge $p_i p_{i+1}$ at distance r from this edge. Let $A, B, C \in \mathbb{Q}$ denote the coefficients of the equation $Ax + By + C = 0$ of the supporting line of $p_i p_{i+1}$. The locus of all points that lie at distance r from the line $Ax + By + C = 0$ is given by:

$$\frac{(Ax + By + C)^2}{A^2 + B^2} = r^2 \ .$$

Thus, the linear offset edges are segments of an algebraic curve of degree 2 (a conic curve) with rational coefficients. This curve is actually a pair of the parallel lines $Ax + By + (C \pm r \cdot \sqrt{A/B + 1}) = 0$. The offset edge is supported by the line $Ax + By + C' = 0$, where $C' = C + r \cdot \sqrt{A/B + 1}$ is in general *not* a rational number. Therefore, the line segments that compose the offset boundaries cannot be represented as segments of lines with rational coefficients. In Section 9.2.2 we use the line-pair representation to construct the offset polygon in an exact manner using the traits class for conic arcs.

The class template `Gps_circle_segment_traits_2<Kernel>`, included in the *2D Regularized Boolean Set-Operations* package, is used for representing generalized polygons the edges of which

---

[6] As the disc is convex, it is guaranteed that the convolution curve comprises a single cycle.

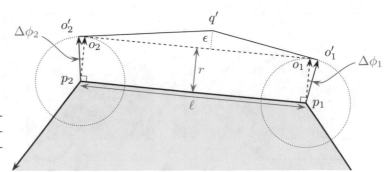

Fig. 9.4: Approximating the offset edge $o_1 o_2$ induced by the polygon edge $p_1 p_2$ by two line segments $o_1' q'$ and $q' o_2'$.

are circular arcs or line segments, and for applying Boolean set operations (e.g., intersection and union) on such generalized polygons. When it is instantiated, the template parameter `Kernel` must be substituted with a geometric kernel that employs exact rational arithmetic. The curves that compose the polygon edges are then arcs of circles with rational coefficients or segments of lines with rational coefficients. However, the line segments that result from the (exact) offsetting operation may be irrational. As we still wish to use the polygons defined by the `Gps_circle_segment_traits_2` class template to compute Boolean set operations more efficiently, we apply a simple approximation scheme such that each irrational line segment is approximated by two rational segments.

Consider the example depicted in Figure 9.4, where the exact offset edge $o_1 o_2$ is obtained by shifting the polygon edge $p_1 p_2$ by a vector of length r that forms an angle ϕ with the x-axis. We select two points o_1' and o_2' with rational coordinates that lie on the two circles of radius r centered at p_1 and p_2, respectively. These points are selected such that $\phi_1' < \phi < \phi_2'$, where $\phi_j', j = 1, 2$ is the angle that the vector $\overrightarrow{p_j o_j'}$ forms with the x-axis. Then, we construct two tangents to the two circles at o_1' and o_2', respectively. The tangent lines have rational coefficients. Finally, we compute the intersection point of the two tangents, denoted by q'. The two line segments $o_1' q'$ and $q' o_2'$ approximate the original offset edge $o_1 o_2$.

The `approximated_offset_2(P, r, epsilon)` function template accepts a polygon P that is either simple or degenerate (consisting of two points only), an offset radius r, and (a floating-point number) $\epsilon > 0$. It constructs an approximation of the offset of P by the radius r using the procedure described above. Furthermore, it is guaranteed that the approximation error, namely the distance of the point q' from $o_1 o_2$, is bounded by ϵ. Using this function, it is possible to use the `Gps_circle_segment_traits_2` class template, which considerably speeds up the (approximate) construction of the offset polygon and the application of Boolean set operations on such polygons; see Section 5.4.1 for the exploitation of the efficient square root extension number type. The function returns an object of the nested type `Gps_circle_segment_traits_2<Kernel>` `::Polygon_with_holes_2` representing the approximated offset polygon. Recall that if P is not convex, its offset may not be simple and may contain holes, the boundaries of which are also formed by line segments and circular arcs.

Example: The example below demonstrates the construction of an approximated offset of a non-convex polygon, as depicted in Figure 9.5. The program uses types defined in the header file `bops_circular.h` listed on Page 192.

// File: ex_approx_offset.cpp

#include <fstream>
#include <boost/timer.hpp>

#include <CGAL/basic.h>
#include <CGAL/approximated_offset_2.h>

#include "bops_circular.h"

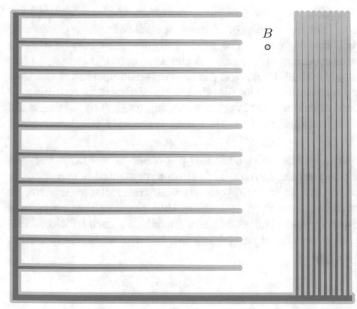

Fig. 9.5: The offset (lightly shaded) of a polygon P (darkly shaded) and a disc B. The input file **spiked.dat**, used by default by the example programs **ex_approx_offset.cpp** and **ex_exact_offset.cpp**, describes the polygon P.

```cpp
typedef CGAL::Polygon_2<Kernel>                              Linear_polygon;

int main(int argc, char* argv[])
{
  // Open the input file and read a polygon.
  const char* filename = (argc > 1) ? argv[1] : "spiked.dat";
  std::ifstream    in_file(filename);
  if (! in_file.is_open()) {
    std::cerr << "Failed to open the input file." << std::endl;
    return -1;
  }
  Linear_polygon  P;
  in_file >> P;
  in_file.close();
  std::cout << "Read an input polygon with " << P.size() << " vertices."
            << std::endl;

  // Approximate the offset polygon with radius 5 and error bound 0.00001.
  boost::timer timer;
  Polygon_with_holes offset = CGAL::approximated_offset_2(P, 5, 0.00001);
  double secs = timer.elapsed();

  std::cout << "The offset polygon has " << offset.outer_boundary().size()
            << " vertices, " << offset.number_of_holes() << " holes."
            << std::endl;
  std::cout << "Offset computation took " << secs << " seconds." << std::endl;
  return 0;
}
```

9.2.2   Computing the Exact Offset

As mentioned in the previous section, it is possible to represent offset polygons in an exact manner if the edges of the polygons are represented as arcs of conic curves with rational coefficients. The `offset_polygon_2(P, r, traits)` function template computes the offset of a given polygon P by a rational radius r in an exact manner. The input polygon P must be either simple or degenerate consisting of two vertices (representing a line segment). The `traits` argument should be capable of handling conic arcs in an exact manner—using an instance of the `Arr_conic_traits_2` class template with the number types provided by the CORE library is the preferred option (at the time of this writing); see Section 5.4.2 for more details. The function template returns an object of the nested type `Gps_traits_2<ArrConicTraits>::Polygons_with_holes_2` (see Section 8.2.3 for more details on the traits-class adaptor `Gps_traits_2`), which represents the exact offset polygon.

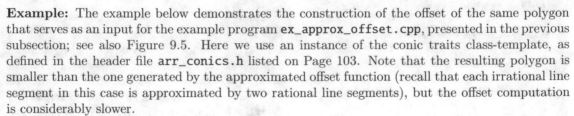

Example: The example below demonstrates the construction of the offset of the same polygon that serves as an input for the example program `ex_approx_offset.cpp`, presented in the previous subsection; see also Figure 9.5. Here we use an instance of the conic traits class-template, as defined in the header file **arr_conics.h** listed on Page 103. Note that the resulting polygon is smaller than the one generated by the approximated offset function (recall that each irrational line segment in this case is approximated by two rational line segments), but the offset computation is considerably slower.

```cpp
// File: ex_exact_offset.cpp

#include <fstream>
#include <boost/timer.hpp>

#include <CGAL/basic.h>
#include <CGAL/Gps_traits_2.h>
#include <CGAL/offset_polygon_2.h>

#include "arr_conics.h"

typedef CGAL::Polygon_2<Rat_kernel>        Polygon;
typedef CGAL::Gps_traits_2<Traits>         Gps_traits;
typedef Gps_traits::Polygon_with_holes_2   Offset_polygon_with_holes;

int main(int argc, char* argv[])
{
  // Open the input file, and read the input polygon.
  const char* filename = (argc > 1) ? argv[1] : "spiked.dat";
  std::ifstream    in_file(filename);
  if (! in_file.is_open()) {
    std::cerr << "Failed to open the input file." << std::endl;
    return -1;
  }
  Polygon  P;
  in_file >> P;
  in_file.close();
  std::cout << "Read an input polygon with " << P.size() << " vertices."
            << std::endl;

  // Compute the offset polygon.
  Traits          traits;
  boost::timer timer;
  Offset_polygon_with_holes offset = CGAL::offset_polygon_2(P, 5, traits);
  double secs = timer.elapsed();

  std::cout << "The offset polygon has " << offset.outer_boundary().size()
```

```
        << "_vertices,_" << offset.number_of_holes() << "_holes."
        << std::endl;
std::cout << "Offset_computation_took_" << secs << "_seconds." << std::endl;
return 0;
}
```

┌────── *advanced* ──────

Both function templates `approximated_offset_2()` and `offset_polygon_2()` also have over-loaded versions that accept a decomposition strategy and use the polygon-decomposition approach to compute (or approximate) the offset. These functions are typically considerably slower than their counterparts that employ the convolution approach. However, similar to the functions that compute the general Minkowski sum, they are able to compute the offset of polygons with holes, given a decomposition strategy that handles polygons with holes. The decomposition methods that we describe in this chapter are not implemented to handle polygons with holes.

└────── *advanced* ──────

9.3 Motion-Planning Applications

A set P translated by a vector t is denoted by P^t. *Collision detection* is a procedure that determines whether two sets P and Q overlap. The *separation distance* $\pi(P, Q)$ and the *penetration depth* $\delta(P, Q)$, defined as

$$\pi(P, Q) = \min\{\|t\| \mid P^t \cap Q \neq \emptyset, t \in \mathbb{R}^d\} \text{ and}$$
$$\delta(P, Q) = \inf\{\|t\| \mid P^t \cap Q = \emptyset, t \in \mathbb{R}^d\},$$

are the minimum distances by which P has to be translated so that P and Q intersect or become interior-disjoint, respectively. The problems of finding the distances above can also be posed given a normalized vector r that represents a direction, in which case the minimum distance sought is in direction r. The *directional penetration depth*, for example, is defined as

$$\delta_r(P, Q) = \inf\{\alpha \mid P^{\alpha\vec{r}} \cap Q = \emptyset, \alpha \in \mathbb{R}\}.$$

Minkowski sums can be used to detect collision between and compute relative placement of two sets. These basic operations play an important role in solving motion-planning problems. The study of motion-planning problems was originally motivated by the design of autonomous robots that have to find their way in a known environment. Thus, we refer to the moving entity as a *robot*. In a typical motion-planning problem we are given a robot moving in an environment cluttered with static obstacles, and have to plan a collision-free path from a given start position to a given goal position, or determine that such a path does not exist. In this section we limit ourselves to robots moving in a *planar* environment amidst *polygonal* obstacles.

When considering motion-planning problems, it is often more convenient to talk about the *configuration space* rather than to consider the physical *workspace* (which is two-dimensional in our case). In the configuration space, the robot is shrunk to a single point (hence, every point in this space represents a single *configuration* of the robot), whereas the obstacles are inflated in the following manner. We choose an arbitrary but fixed point Q to be the *reference point* of Q. Assume O is an obstacle in the original workspace, then $C(O)$ is the corresponding configuration-space obstacle, such that a configuration q is contained in $C(O)$ iff the robot interferes with O when its reference point is positioned at q in the workspace.

Let P and Q be two point sets in the plane having a nonempty intersection, and let $p \in P \cap Q$. Let $-Q$ denote the reflection of Q about the origin. Note that since $p \in Q \iff -p \in -Q$, then $P \oplus (-Q)$ contains the origin as $(0, 0) = p + (-p)$. The converse is also true: If $(0, 0) \in P \oplus (-Q)$, then $P \cap Q \neq \emptyset$. Thus, the sum $P \oplus (-Q)$ is the configuration-space obstacle induced by P with respect to the robot Q, with the origin as the reference point.

Detecting collision between P and Q and computing their relative placement can be conveniently done in the configuration space, where their Minkowski sum $M = P \oplus (-Q)$ resides. These problems can be solved in many ways, and not all require the explicit representation of the Minkowski sum M. However, having it available is attractive, especially when the robot is restricted to translations only, as the combinatorial structure of the Minkowski sum M is invariant to translations of P or Q. Motion-planning applications are based on the following well-known observations:

$$P^u \cap Q^w \neq \emptyset \Leftrightarrow w - u \in M = P \oplus (-Q) ,$$
$$\pi(P^u, Q^w) = \min\{\|t\| \mid (w - u + t) \in M, t \in \mathbb{R}^2\} ,$$
$$\delta_r(P^u, Q^w) = \inf\{\alpha \mid (w - u + \alpha\vec{r}) \notin M, \alpha \in \mathbb{R}\} .$$

In the remainder of this section we consider two specific motion-planning variants, namely, the case of a polygonal robot translating amidst polygonal obstacles, and the case of coordinating the motion of two disc robots amidst obstacles. In the former case we are able to construct a complete representation of the configuration space and solve the problem in an exact manner. In the latter case we combine a sampling approach together with partial exact representation to capture the structure of the configuration space, which is too complicated for us to fully represent in an exact manner.

9.3.1   Application: A Translating Polygonal Robot

The translating polygon problem: *Given a simple-polygon robot Q, which can translate (but not rotate) in a room cluttered with pairwise interior-disjoint polygonal obstacles, devise a data structure that can efficiently answer queries of the following form: Given a start position s of some reference point in Q and a goal position g of the same reference point, plan a collision-free path of the robot from s to g.*

Let $P_1, \ldots, P_m$ be the polygonal obstacles. As the robot is usually moving in some bounded area, say a closed room, we consider the walls of this room as polygonal obstacles as well. Let $q_0 \in Q$ be our reference point on the robot. Let us place the robot Q such that q_0 coincides with the origin, and reflect the robot about the origin. We denote the reflected polygon by $-Q$. Recall that Q, when located such that its reference point q_0 coincides with $q \in \mathbb{R}^2$, interferes with some polygon P iff $q \in P \oplus (-Q)$. Thus, the configuration-space obstacles in our case are given by $C(P_i) = P_i \oplus (-Q)$, and the *forbidden-configuration space*, denoted by $\mathcal{C}_{\mathrm{forb}}$ and formally defined below, is the union of these obstacles.

$$\mathcal{C}_{\mathrm{forb}} = \bigcup_{i=1}^m P_i \oplus (-Q) .$$

Note that in this case a configuration of the robot is uniquely determined by two parameters, namely, the x- and y-coordinates of the reference point q_0. Indeed, the configuration space is two-dimensional, and $\mathcal{C}_{\mathrm{forb}}$ specifies all the forbidden locations for the reference point.

Answering the decision problem—namely, given a start location s and a goal location g, determine whether there exists a collision-free path connecting s and g—is easy. We compute $\mathcal{C}_{\mathrm{forb}}$, and consider its complement, $\mathcal{C}_{\mathrm{free}} = \mathbb{R}^2 \setminus \mathcal{C}_{\mathrm{forb}}$, called the *free-configuration space*. A path exists between s and g only if these two points lie in the same connected component of $\mathcal{C}_{\mathrm{free}}$. If we represent the configuration space as an arrangement containing *free* and *forbidden* faces, this amounts to issuing two point-location queries with s and g, respectively, and checking whether the two points lie in the same *free* face. Observe that in this particular variant of motion-planning problem, we do not allow the robot to touch the obstacles. Thus, the forbidden-configuration space is closed and the free-configuration space is open.

Planning a collision-free path (properly contained in the free face) in case such a path exists is a bit more involved, since the face containing s and g may be concave. We resort to a simple procedure that produces a path strictly contained in the free face, if it exists, as follows. We decompose the arrangement faces into (convex) pseudo trapezoids by applying vertical decomposition; see Section 3.6. Given the decomposed arrangement, it is easy to plan a path between two end configurations s and g. Observe that if either s or g coincide with a vertex or with an edge that is not added by the vertical decomposition process, we can immediately determine that a collision-free path does not exist, since all vertices are contained in the forbidden-configuration space. In the remainder of this section, a *face* is a pseudo trapezoid in the vertical decomposition of the arrangement.

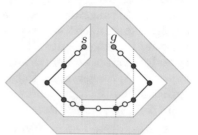

If s (respectively g) is contained in a face, let f_s (respectively f_g) denote the containing face. s (respectively g) may lie on a vertical edge added by the decomposition. In this case, let f_s (respectively f_g) denote one of the two free faces incident to the vertical edge. If $f_s = f_g$, we simply stretch a straight-line segment from s to g. As the faces are convex, the line is contained in the face. If, however, $f_s \neq f_g$, we need to find a sequence of free adjacent faces that connect f_s to f_g. As two adjacent free faces are separated by a vertical edge added by the decomposition (other edges always separate free faces from forbidden faces), we traverse the free faces of the decomposed arrangement, starting from f_s, in a breadth-first order. For each face we examine, we consider the vertical edges along its boundaries that connect it with other free faces. If we reach f_g, we terminate the search and conclude that a path exists. Furthermore, we can reconstruct the sequence of faces we visited, interleaved with the vertical edges we crossed. We transform this sequence into a polyline that comprises alternating interior points of visited faces and midpoints of crossed vertical edges. Finally, we connect s and g to the respective polyline ends. As our faces are all convex, the resulting polyline forms a collision-free path between s and g. Observe that the introduction of face-interior points is not always necessary. An example is depicted in the figure above. Redundant points are drawn as small rings.

Other variants of the motion-planning problem impose requirements on the path, which are stated as an objective function expressing some quantity, such as the length of the path, some measure of the distance between the path and the boundaries of the free face, or a combination of both.

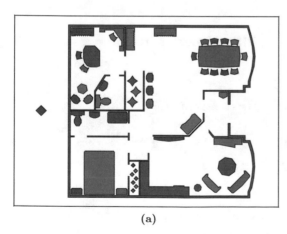

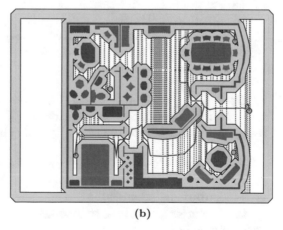

(a)   (b)

Fig. 9.6: An illustration of motion planning. (a) The workspace, which comprises a diamond-shaped robot (darkly shaded), translating in a house amidst polygonal obstacles. The robot and the obstacles are described in the files `diamond.dat` and `house.dat`, respectively. (b) The configuration space, which consists of the forbidden-configuration space (lightly shaded on top of the original obstacles, which are also shown), and the free-configuration space, which is decomposed into trapezoidal faces using vertical segments (dotted). The queries, given in the file `house_queries.dat`, are drawn as circles, and the resulting paths, if they exist, are drawn as polylines.

The program `mp_tr_polygon.cpp` implements the motion-planning algorithm described above.

It uses the Minkowski sum operation and the multiway union operation of the *2D Regularized Boolean Set-Operations* package (see Section 8.1.8) to compute the forbidden-configuration space. As it requires the arrangement representation of the resulting polygon set, it uses the following type definitions:

```
typedef Polygon_set::Arrangement_2                          Arrangement;
typedef Arrangement::X_monotone_curve_2                     Segment;
typedef Arrangement::Vertex_handle                         Vertex_handle;
typedef Arrangement::Vertex_const_handle                   Vertex_const_handle;
typedef Arrangement::Halfedge_handle                       Halfedge_handle;
typedef Arrangement::Halfedge_const_handle                 Halfedge_const_handle;
typedef Arrangement::Face_handle                           Face_handle;
typedef Arrangement::Face_const_handle                     Face_const_handle;
typedef CGAL::Arr_landmarks_point_location<Arrangement>    Landmarks_pl;
typedef std::pair<Point, Point>                            Point_pair;
```

The definitions of the `Polygon_2` and `Polygon_set_2` types are given in the header file `bops_linear.h` listed on Page 180. Note that we use the `Arrangement_2` type defined by the `Polygon_set_2` class template. This is an arrangement of line segments, but when instantiated its `Dcel` template parameter is substituted with an extended Dcel such that each arrangement face is marked if it is contained in the polygon set. Formally, let f be a face. `f->contained()` evaluates to **true** if f is contained in the polygon set (a forbidden face in our case), and to **false** otherwise. We properly maintain the containment flag for faces that arise as a result of the vertical decomposition application.

Our motion-planning application reads a set of polygonal obstacles from a file using the function template `read_objects()` (defined in the file `read_objects.h`; see Section 2.3), reads the polygonal robot from a second input file, and constructs the vertical decomposition of the arrangement that represents the configuration space. It then reads a set of queries from a third file, and answers each one independently.

```cpp
int main(int argc, char* argv[])
{
  if (argc < 4) {
    std::cout << "Usage: " << argv[0]
              << " <obstacles_file> <polygon_file> <queries_file>" << std::endl;
    return -1;
  }

  // Read the polygonal obstacles from the first input file.
  std::list<Polygon>    obstacles;
  read_objects<Polygon>(argv[1], std::back_inserter(obstacles));

  // Read the polygonal robot from the second input file.
  std::ifstream         frobot(argv[2]);
  if (! frobot.is_open()) {
    std::cerr << "Failed to open " << argv[2] << std::endl;
    return -1;
  }
  Polygon robot;
  frobot >> robot;
  frobot.close();

  // Reflect the robot about the origin by negating each vertex.
  Polygon::Vertex_const_iterator  vit;
  Polygon                          rot_robot;
  for (vit = robot.vertices_begin(); vit != robot.vertices_end(); ++vit)
```

```
        rot_robot.push_back(Point(−vit−>x(), −vit−>y()));
    if (rot_robot.orientation() == CGAL::CLOCKWISE)
        rot_robot.reverse_orientation();

    // Compute the Minkowski sum of each obstacle with the rotated robot.
    std::list<Polygon>::const_iterator   obs_it;
    std::list<Polygon_with_holes>           c_obstacles;
    for (obs_it = obstacles.begin(); obs_it != obstacles.end(); ++obs_it)
        c_obstacles.push_back(CGAL::minkowski_sum_2(*obs_it, rot_robot));

//Compute forbidden configuraion−space and extract the underlying arrangement.
    // The observer keeps the information updated in the new faces.
    Polygon_set::Traits_2 traits;
    Polygon_set   cforb(traits);
    cforb.join(c_obstacles.begin(), c_obstacles.end());
    Arrangement arr = cforb.arrangement();
    Vertical_decomposition_polygon_set_observer observer;
    observer.attach(arr);
    Kernel* ker = &traits;
    vertical_decomposition(arr, *ker);
    observer.detach();

    // Read the query points, and plan the paths.
    Landmarks_pl   pl(arr);
    std::list<Point_pair> queries;
    read_objects<Point_pair>(argv[3], std::back_inserter(queries));
    std::list<Point_pair>::const_iterator it;
    for (it = queries.begin(); it != queries.end(); ++it) {
        std::list<Point> path;
        plan_path(arr, pl, it−>first, it−>second, std::back_inserter(path), *ker);

        // Print the results.
        std::cout << "Query:_(" << it−>first << ")_and_(" << it−>second
                    << ")_are" << (path.empty() ? "_NOT_" : "_") << "reachable."
                    << std::endl;
        if (path.empty()) continue;
        std::cout << "Path_is:";
        std::list<Point>::const_iterator pit;
        for (pit = path.begin(); pit != path.end(); ++pit)
            std::cout << "_(" << *pit << ")";
        std::cout << std::endl;
    }

    return 0;
}
```

Figure 9.6a shows the obstacles specified in the **house.dat** data file, and the robot specified in the file **diamond.dat**. Figure 9.6b shows the configuration space, the queries specified in the **house_queries.dat** file, and the corresponding paths if they exist. The first two queries result in collision-free paths. For the third query no free path exists.

We exploit the **vertical_decomposition()** function template listed on Page 61. It does not decompose the entire unbounded face. In other words, if there is no arrangement vertex or edge lying below (respectively above) some vertex, it does not add a downward-directed (respectively upward-) vertical segment from this vertex. This suits us, as we assume that the robot is moving in some area enclosed by polygonal obstacles. However, the process can be further optimized to produce only a partial vertical decomposition, where forbidden faces are not decomposed at all.

We maintain the containment flag for a new face created when an existing face is split as a result of a vertical-segment insertion using the observer `Vertical_decomposition_polygon_set_observer` listed below and defined in `mp_tr_polygon.cpp`. Notice that only the `after_split_face()` member function of the observer is populated; see Section 6.1 for further details.

```
class Vertical_decomposition_polygon_set_observer :
  public CGAL::Arr_observer<Arrangement>
{
  void after_split_face(Face_handle f1, Face_handle f2, bool)
  { f2->set_contained(f1->contained()); }
};
```

The function template **plan_path()**, listed below, and defined in `mp_tr_polygon.cpp`, checks whether there exists a collision-free path between two given configurations identified as `ps` and `pg`. If there is, it plans such a path, which is given as a sequence of points defining a piecewise-linear curve. It utilizes the BGL generic implementation of the BFS algorithm to find a collision-free path that intersects the smallest number of free faces.

```
template <typename Output_iterator>
Output_iterator plan_path(Arrangement& arr, Landmarks_pl& pl,
                          const Point& ps, const Point& pg,
                          Output_iterator path, Kernel& ker)
{
  typedef CGAL::Dual<Arrangement>                        Dual_arrangement;
  typedef CGAL::Arr_face_index_map<Arrangement>          Face_index_map;
  typedef std::map<Face_handle, Face_handle, Less_than_handle>     Preds_map;
  typedef std::map<Face_handle, Halfedge_handle, Less_than_handle> Edges_map;

  // Get the free-space faces that contain the source and target points
  // (checking that they are contained in the free space).
  Face_const_handle       fh;
  fh = get_free_face(pl, ps);
  if (fh == Face_const_handle()) return path;
  Face_handle fs = arr.non_const_handle(fh);

  fh = get_free_face(pl, pg);
  if (fh == Face_const_handle()) return path;
  Face_handle fg = arr.non_const_handle(fh);

  // The Boost BFS records the predecessor faces and the predecessor edges,
  // and stops when the goal face discovered.
  Face_index_map          index_map(arr);
  Preds_map               preds_map;
  Edges_map               edges_map;

  // Define a visitor to stop the search when we reach the goal.
  Find_vertex_visitor<Face_handle> find_vertex_visitor(fg);

  // We use a filter graph to traverse only the free faces.
  Dual_arrangement        dual(arr);
  boost::filtered_graph<Dual_arrangement, boost::keep_all, Is_face_free>
    graph(dual, boost::keep_all());

  try {
```

```
        preds_map[fs] = Face_handle(); edges_map[fs] = Halfedge_handle();
        boost::breadth_first_search(graph, fs,
                                    boost::vertex_index_map(index_map).visitor
                                    (make_bfs_visitor(
                                        make_pair(
                                            find_vertex_visitor,
                                            make_pair(
                                                record_predecessors(
                                                    make_assoc_property_map(preds_map),
                                                    boost::on_tree_edge()),
                                                record_edge_predecessors(
                                                    make_assoc_property_map(edges_map),
                                                    boost::on_tree_edge())))))));

        // If there is a path then an exception should have been thrown.
        return path;
    }
    catch(Found_vertex_exception e) {}

    // We arrive here only if an exception was thrown and a path exists.

    // Reconstruct the path from the BFS results going backward.
    Kernel::Construct_midpoint_2  midp = ker.construct_midpoint_2_object();
    *path++ = pg;

    Face_handle f = fg;
    do {
        *path++ = point_in_vertical_trapezoid(f, arr, ker);
        Halfedge_handle he = edges_map[f];
        if (he != Halfedge_handle())
            // Add the midpoint of the associated vertical segment.
            *path++ = midp(he->source()->point(), he->target()->point());
        f = preds_map[f];
    } while (f != Face_handle());
    *path++ = ps;

    return path;
}
```

 Try: As mentioned above, the resulting collision-free paths may contain redundant face-interior points. Alter the code of the **plan_path()** function template to selectively add only the necessary face-interior points.

The **plan_path()** function template utilizes a BGL graph filter to restrict the graph traversal to free faces only. The graph filter creates a filtered view of a graph, where forbidden faces are filtered out using the functor **Is_face_free** listed below, and defined in **mp_tr_polygon.cpp**.

```
struct Is_face_free {
    bool operator()(Face_handle f) const
    { return !f->is_unbounded() && !f->contained(); }
    bool operator()(Face_const_handle f) const
    { return !f->is_unbounded() && !f->contained(); }
};
```

When a collision-free path is found, the exception **Found_vertex_exception**, listed below, and defined in **mp_tr_polygon.cpp**, is thrown, terminating the graph traversal. Observe that a vertex of the traversed graph is in fact a face handle of the arrangement.

class Found_vertex_exception : **public** std :: exception {};

An appropriate instance of the `Find_vertex_visitor` class template is used as a BGL visitor. It is applied at each iteration of the traversal to check whether the goal vertex has been reached.

```
template <typename Vertex> class Find_vertex_visitor {
private:
  Vertex m_goal;
public:
  typedef boost :: on_finish_vertex                event_filter;

  Find_vertex_visitor(Vertex v) : m_goal(v) {}

  template <class Graph> void operator()(Vertex v, const Graph& g)
  { if (v == m_goal) throw Found_vertex_exception(); }
};
```

The function template `get_free_face()`, listed below, and defined in `mp_tr_polygon.cpp`, accepts a point and returns the handle to the free face that contains the point. If the point is not contained in the free space, the function returns an empty handle.

```
Face_const_handle get_free_face(const Landmarks_pl& pl, const Point& p)
{
  // Perform point–location queries and locate the point (configuration).
  CGAL :: Object      obj = pl.locate(p);

  // Check whether the point lies on an edge separating two forbidden faces.
  Halfedge_const_handle  hh;
  if (CGAL :: assign(hh, obj)) {
    if (! hh->face()->contained() && ! hh->twin()->face()->contained())
      return hh->face();
    return Face_const_handle();
  }

  // Check whether the point is contained inside a free bounded face.
  Face_const_handle      fh;
  if (! CGAL :: assign(fh, obj) || fh->is_unbounded() || fh->contained())
    return Face_const_handle();
  return fh;
}
```

Finally, the function template `point_in_vertical_trapezoid()`, listed below, and defined in the header file `point_in_vertical_trapezoid.h`, computes a point located in the interior of a given vertical trapezoid lying as far as possible from the face boundary.

```
template <typename Arrangement, typename Kernel>
typename Kernel :: Point_2
point_in_vertical_trapezoid(typename Arrangement::Face_const_handle f,
                            const Arrangement& arr, const Kernel& ker)
{
  const typename Arrangement :: Traits_2 :: Is_vertical_2 is_vertical =
    arr.traits()->is_vertical_2_object();
  const typename Kernel :: Construct_midpoint_2 midpoint =
    ker.construct_midpoint_2_object();

  // Locate the two edges along the face boundary that are not associated
```

```
// with vertical segments, such that one lies on the upper boundary of the
// face and the other on its lower boundary. Note that these two edges must
// have opposite directions.
typename Arrangement::Halfedge_const_handle he1, he2;
CGAL::Arr_halfedge_direction direction;
bool found = false;
typename Arrangement::Ccb_halfedge_const_circulator first = f->outer_ccb();
typename Arrangement::Ccb_halfedge_const_circulator circ = first;
do {
  if (!is_vertical(circ->curve())) {
    // The current edge is not vertical: assign it as either he1 or he2.
    if (!found) {
      he1 = circ;
      direction = he1->direction();
      found = true;
      continue;
    }
    if (circ->direction() != direction) {
      he2 = circ;
      break;
    }
  }
} while (++circ != first);

// Take the midpoint of the midpoints of he1 and he2.
return midpoint(midpoint(he1->source()->point(), he1->target()->point()),
               midpoint(he2->source()->point(), he2->target()->point()));
}
```

9.3.2   Application: Coordinating Two Disc Robots

The two-disc coordination problem: *Consider two disc robots $B_{r^{(1)}}$ and $B_{r^{(2)}}$ with radii $r^{(1)}$ and $r^{(2)}$, respectively, moving in a room cluttered with polygonal obstacles. Devise a data structure that can efficiently answer queries of the following form: Given two free[7] start positions $s^{(1)}$ and $s^{(2)}$ of the robots and two free goal positions $g^{(1)}$ and $g^{(2)}$, respectively, plan a collision-free motion of the first robot from $s^{(1)}$ to $g^{(1)}$ and of the second robot from $s^{(2)}$ to $g^{(2)}$ such that the robots do not collide during their motions.*

The figure above depicts a simple

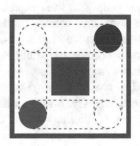

(a) The workspace.

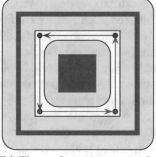

(b) The configuration space of a single robot.

Fig. 9.7: An instance of the two-disc coordination problem.

instance of the two-disc coordination problem. The workspace consists of two obstacles and two congruent-disc robots. A square-shaped obstacle is located inside the hole of another obstacle, which encloses the interesting part of the workspace. The goal position of one robot is the starting position of the other; namely, the robots have to exchange places. As the robot discs have the same radii, their forbidden two-dimensional configuration spaces are identical. A possible solution to

---

[7]In the coordination problem a free position means a position of the robot where it avoids the obstacles as well as the other robot.

the problem, that is, a pair of (simultaneous) collision-free paths for the two robots, respectively, is illustrated in both the workspace and the configuration space (of a single robot).

Consider a single disc robot B_r of radius r and a single obstacle O for a moment. If the disc center, which we choose as its reference point, is located at some point whose distance from O is less than r, then the robot will intersect the obstacle. The corresponding configuration-space obstacle is thus $C(O) = O \oplus B_r$, namely the Minkowski sum of the obstacle and a disc of radius r centered at the origin. Let us denote the forbidden-configuration spaces for the independent robots by $\mathcal{C}_{\text{forb}}^{(1)}$ and $\mathcal{C}_{\text{forb}}^{(2)}$, respectively. We then have

$$\mathcal{C}_{\text{forb}}^{(k)} = \bigcup_{i=1}^{m} P_i \oplus B_{r^{(k)}} \qquad (k = 1, 2) \ .$$

Planning the independent motion of a single disc robot can be done using a process similar to the process described in Section 9.3.1, as we have an explicit representation of the forbidden two-dimensional configuration space, and thus of its complement, which is the corresponding free space. However, the complete configuration space has four degrees of freedom (dofs), as each configuration must represent the locations q_1 and q_2 of the two robots, respectively. Recall that q_i denotes the location of the reference point of the robot $B_{r^{(i)}}$, which is the center of the disc. We denote such a configuration by $\langle q_1, q_2 \rangle$. A configuration is forbidden not only if one of the robots intersects some obstacle, but also if the two robots intersect each other. The free-configuration space in our case can therefore be formulated as follows:

$$\mathcal{C}_{\text{free}} = \left\{ \langle q_1, q_2 \rangle \mid q_1 \notin \mathcal{C}_{\text{forb}}^{(1)} , q_2 \notin \mathcal{C}_{\text{forb}}^{(2)} , B_{r^{(1)}}(q_1) \cap B_{r^{(2)}}(q_2) = \emptyset \right\} \ .$$

Obtaining an explicit representation of the free-configuration space is difficult, mainly due to the problem dimensionality. We therefore solve this motion-planning problem by using a *probabilistic roadmap*—a PRM for short—but still employing the exact two-dimensional arrangements capabilities that we have demonstrated so far. The PRM approach has proved to be successful in numerous motion-planning applications having a high-dimensional configuration space. The roadmap is a graph that attempts to capture the connectivity of the free-configuration space. It is constructed by first picking random configurations, and filtering out forbidden configurations, until some prescribed number N of free configurations is obtained. These configurations compose the roadmap vertices. We then add edges to the graph by considering pairs of vertices in the graph associated with nearby configurations, and connecting them if the (straight) line segment between the configurations constitutes a collision-free path. As we pick only pairs of nearby vertices, figuring out whether the line segment connecting them is contained in the free space is a relatively simple task, carried out by a simple *local planner*.

Once the PRM is constructed, it can be queried to efficiently solve rather complicated motion-planning problems. Given a query, namely, a start s and a goal g configurations, we simply add s and g as graph vertices, and try to connect them to nearby vertices. We only have to test whether the two vertices are reachable from one another in the graph. If so, we can also output a collision-free motion-path, which consists of the intermediate configurations along the path from s to g in the PRM. Note that when a path is not found in the PRM, it does not necessarily mean that a path does not exist in reality. Such false reports are often referred to as *false negatives*. If the configuration space is sufficiently densely sampled though, the probability of false negatives is low.

How do we construct a PRM for the two-disc problem? Note that we have to evaluate two predicates: (i) determine whether a given configuration $\langle q_1, q_2 \rangle$ is free, and (ii) determine whether two given configurations can be connected by a (straight) line segment contained in the free space. When using the PRM method, these two predicates are assumed to be provided by *oracles*. We give concrete implementations of these oracles using an approach that combines the PRM sampling method with techniques that use exact representation (and seek complete solutions).

The first predicate is easy to evaluate, as we can construct an explicit representation of $\mathcal{C}_{\text{forb}}^{(1)}$ and $\mathcal{C}_{\text{forb}}^{(2)}$. Once it is constructed, we check whether $q_1 \notin \mathcal{C}_{\text{forb}}^{(1)}$ and $q_2 \notin \mathcal{C}_{\text{forb}}^{(2)}$. If this is the case, we also check whether $r^{(1)} + r^{(2)} < \|q_2 - q_1\|$, namely whether the distance between the two robot centers is larger than the sum of their radii. This is a sufficient and necessary condition for the two robots to be disjoint. In this particular variant we do not allow the robots to touch the obstacles nor one another.

The second predicate is a bit more involved, as we have to consider the continuous motion of the two discs, each along a straight-line segment. A common approach for handling such predicates is sampling the path at a finite number of intermediate configurations and verifying each one separately. In our case, if we wish to check whether the motion of the disc robots from $\langle u_1, u_2 \rangle$ to $\langle v_1, v_2 \rangle$ is collision-free, we consider the intermediate configurations $\langle u_1 + \frac{j}{K}(v_1 - u1), u_2 + \frac{j}{K}(v_2 - u2) \rangle$ for $j = 0, 1, \ldots, K$, where K is chosen such that $\frac{1}{K}\|v_1 - u_1\|$ and $\frac{1}{K}\|v_2 - u_2\|$ are both smaller than some prescribed step-size $\delta > 0$. As we check for collisions with the workspace obstacles only at a discrete number of points, we have to treat the robot as if it is slightly larger than its actual size. It is possible to show that if the motion path for a robot of radius $r + \frac{\delta}{2}$ is collision-free at a discrete number of sampled points, then the continuous path is collision-free for the original robot.

To this end we define the "inflated" forbidden-configuration spaces as follows:

$$\mathcal{C}_{\text{forb}}^{(k)+} = \bigcup_{i=1}^{m} P_i \oplus B_{r^{(k)} + \frac{1}{2}\delta} \qquad (k = 1, 2) \ .$$

Checking whether a straight path is collision-free reduces to a sequence of point-location queries on $\mathcal{C}_{\text{forb}}^{(1)+}$ and $\mathcal{C}_{\text{forb}}^{(2)+}$. In addition, for every intermediate configuration we check whether the distance of the robot centers is larger than the sum of their inflated radii, to account for their motions.

Instances of the (pure) class template `Two_disc_prm<Kernel>`, defined in the header file `Two_disc_prm.h`, serve as base classes for different possible classes that implement the PRM method. We provide two concrete class-templates, namely, `Two_disc_simple_prm<Kernel>` and `Two_disc_offset_prm<Kernel>`, defined in the header files `Two_disc_simple_prm.h` and `Two_disc_offset_prm.h`, respectively, that derive from the `Two_disc_prm` base class-template. Each of the concrete class templates implements a different variant. The class template `Two_disc_simple_prm<Kernel>` implements the solution described above, while the class template `Two_disc_offset_prm<Kernel>` implements a slightly improved version, which further exploits our exact computation capabilities. It is described later on in this section.

The header file `Two_disc_prm.h` also contains the definition of the `Configuration_4<Kernel>` struct template, an instance of which is used as the type of objects that represent four-dimensional configurations. Its listing follows.

```
template <typename Kernel> struct Configuration_4 {
    typename Kernel::Point_2 pos1;        // the location of the first robot
    typename Kernel::Point_2 pos2;        // the location of the second robot
};
```

The base class template `Two_disc_prm<Kernel>` contains member functions that are common to all different concrete implementations. It is parameterized with a kernel, used to define the representation of the polygonal obstacles. It uses the approximated offset procedure to compute the dilated obstacles and uses a generalized-polygon set to represent the forbidden-configuration spaces it computes. Note that we compute an overestimate of the forbidden-configuration spaces, but this has only a marginal effect on the applicability of the PRM approach, which is probabilistic and introduces false negatives in any case. Note further that the solution is *conservative* in the sense that if the program finds coordinated-motion paths for the robots then these are indeed valid collision-free paths.

The PRM graph is defined as a BOOST adjacency list structure—a structure well-suited for sparse graphs. Our PRM graph has undirected weighted edges. An adjacency-list representation of a graph maintains for each vertex in the graph a sequence of outgoing edges. Recall that a vertex of the PRM graph represents a configuration $\langle q, p \rangle$, which is associated with the locations q and p of the two robots, respectively, in the workspace. The weight of an edge connecting two configurations $\langle q_1, p_1 \rangle$ and $\langle q_2, p_2 \rangle$ is $\|q_2 - q_1\| + \|p_2 - p_1\|$, calculated using floating-point arithmetic, that is, the sum of the (approximate) Euclidean respective distances. The public types nested in the `Two_disc_prm` class template, including the PRM-graph type, are listed next, and the protected types immediately follow.

public:
```
typedef typename Kernel::FT                         Number_type;
typedef typename Kernel::Point_2                     Position_2;
typedef CGAL::Polygon_2<Kernel>                      Obstacle_2;
typedef Configuration_4<Kernel>                      Conf_4;


// Representation of the PRM graph.
typedef boost::adjacency_list<boost::vecS, boost::vecS, boost::undirectedS,
                              boost::no_property,
                              boost::property<boost::edge_weight_t, double>>
                                                     Graph;
typedef boost::graph_traits<Graph>::edge_descriptor  Edge;
typedef boost::property_map<Graph, boost::edge_weight_t>::type Weight_map;
```

Observe that the nested type `CSpace_forb_2` is defined to be a generalized-polygon set containing polygons bounded by circular arcs and line segments.

protected:
```
typedef CGAL::Gps_circle_segment_traits_2<Kernel>   Traits_2;
typedef typename Traits_2::Point_2                   Point_2;
typedef typename Traits_2::Polygon_with_holes_2      CSpace_obstacle_2;
typedef CGAL::General_polygon_set_2<Traits_2>        CSpace_forb_2;
```

Next we list the protected data members.

```
CGAL::Bbox_2        m_bbox;       // the bounding box of the scene
Number_type         m_rad1;       // the radius of the first robot
Number_type         m_rad2;       // the radius of the second robot
CSpace_forb_2       m_cforb1;     // forbidden C-space for the first robot
CSpace_forb_2       m_cforb2;     // forbidden C-space for the second robot
double              m_eps;        // error bound for offset approximation
unsigned int                      m_num_verts; // number of PRM vertices
unsigned int                      m_num_edges; // number of PRM edges
std::vector<Conf_4>               m_confs;     // configurations of PRM vertices
Graph                             m_prm;       // the PRM graph
Weight_map                        m_wgt_map;   // a weight map for the PRM edges
```

The bounding rectangle of the scene is used to restrict the generation of the random points to this rectangle.

The member function-template `query()` listed below accepts a start configuration s, a goal configuration g, and an output iterator. It plans a collision-free motion path from s to g. If a path exists, it is inserted as a sequence of configurations into the container associated with the given output iterator. The function returns a past-the-end iterator for the sequence. If a path is not found, the container remains unchanged.

```
template <typename Output_iterator>
Output_iterator query(const Conf_4& s, const Conf_4& g, Output_iterator oi)
{
  // Clear any edges incident to the PRM vertices representing the source
  // configuration and the goal configuration.
  const unsigned int  s_index = 0;
  const unsigned int  g_index = m_num_verts + 1;

  boost::clear_vertex(s_index, m_prm);
```

```
boost :: clear_vertex(g_index, m_prm);

// Set the source and goal configurations.
m_confs[s_index] = s;
m_confs[g_index] = g;

// If the source or the goal configurations are not free, indicate that
// there is no possible collision-free path.
if (! is_free(m_confs[s_index]) || ! is_free(m_confs[g_index])) return oi;

// Try to connect the source and goal configurations directly.
double len;
if (can_connect(s_index, g_index, len, false)) {
  *oi++ = m_confs[s_index];
  *oi++ = m_confs[g_index];
  return oi;
}

//Try to connect the source and goal configurations with the PRM vertices.
unsigned int v, sc = 0, gc = 0;
for (v = 1; v <= m_num_verts; ++v) {
  // Try to connect the source configuration with v.
  if (can_connect(s_index, v, len)) {
    std :: pair<Edge, bool>  e = boost :: add_edge(s_index, v, m_prm);
    m_wgt_map[e.first] = len;
    ++sc;
  }

  // Try to connect the goal configuration with v.
  if (can_connect(g_index, v, len)) {
    std :: pair<Edge, bool>  e = boost :: add_edge(g_index, v, m_prm);
    m_wgt_map[e.first] = len;
    ++gc;
  }
}

// Perform a Dijkstra search from the vertex representing the source
// configuration and look for the shortest path to the goal configuration.
std :: vector<unsigned int>  preds(boost :: num_vertices(m_prm));

boost :: dijkstra_shortest_paths(m_prm, s_index,
                                 boost :: predecessor_map(&preds[0]));

// The goal configuration is unreachable from the source:
if (preds[g_index] == g_index) return oi;

// According to the predecessor information, prepare a list of vertices
// that occur on the shortest path from the source to the goal.
std :: list<unsigned int>  path_verts;
for (v = g_index; v != s_index; v = preds[v])
  path_verts.push_front(v);
path_verts.push_front(s_index);

// Output the sequence of intermediate configurations that constitute the
```

```
  // shortest collision-free motion-path we have computed.
  std::list<unsigned int>::iterator  it;
  for (it = path_verts.begin(); it != path_verts.end(); ++it)
    *oi++ = m_confs[*it];
  return oi;
}
```

Checking whether a given point is in some forbidden-configuration space is carried out by the member function **is_in_cforb()**, listed next.

```
bool is_in_cforb(const CSpace_forb_2& cforb, const Position_2& q) const
{
  return (cforb.oriented_side(Point_2(q.x(), q.y())) !=
          CGAL::ON_NEGATIVE_SIDE);
}
```

Checking whether two disc robots of given radii intersect when placed at given positions is carried out by the member function **do_intersect()**, listed next.

```
bool do_intersect(const Position_2& pos1, const Number_type& radius1,
                  const Position_2& pos2, const Number_type& radius2) const
{
  // If the distance between the two robot centers is not larger than the
  // sum of their radii, then the robots intersect. We consider the squared
  // values to avoid a square-root operation.
  const Number_type dx = pos2.x() - pos1.x();
  const Number_type dy = pos2.y() - pos1.y();
  return (CGAL::compare(dx*dx + dy*dy, CGAL::square(radius1 + radius2)) !=
          CGAL::LARGER);
}
```

Finally, checking whether a four-dimensional configuration is free is carried out by the member function **is_free()**, listed next.

```
bool is_free(const Conf_4& conf) const
{
  return (! is_in_cforb(m_cforb1, conf.pos1) &&
          ! is_in_cforb(m_cforb2, conf.pos2) &&
          ! do_intersect(conf.pos1, m_rad1, conf.pos2, m_rad2));
}
```

The excerpts of code of the class template **Two_disc_prm** listed above are the highlights of the class. We omit the code listing of the class template **Two_disc_simple_prm** altogether, and continue with a suggested improvement.

─────── *advanced* ───────

The main drawback in the suggested implementation of the oracles lies in the sampling of intermediate configurations. Using a large value for the step size δ decreases the reliability of the oracle, as it increases the probability of having false negatives. On the other hand, using a small δ value increases the number of intermediate configurations, along with the overall time consumption of the process.

Let a robot $B^{(i)}$ move from $p^{(i)}$ to $q^{(i)}$ in a straight line. This path is collision-free iff the straight-line segment $p^{(i)}q^{(i)}$ does not intersect the forbidden-configuration space $\mathcal{C}_{\text{forb}}^{(i)}$. Luckily, the configuration spaces $\mathcal{C}_{\text{forb}}^{(1)}$ and $\mathcal{C}_{\text{forb}}^{(2)}$ are explicitly represented as generalized-polygon sets. Testing whether a generalized-polygon set and a line segment intersect is a relatively simple task.

Try: One efficient method to test whether a line segment intersects a generalized-polygon set is to directly operate on the underlying arrangement of the generalized-polygon set. Alter the code to obtain the two underlying arrangements $\mathcal{A}^{(1)}$ and $\mathcal{A}^{(2)}$ of the two forbidden-configuration spaces $\mathcal{C}_{\text{forb}}^{(1)}$ and $\mathcal{C}_{\text{forb}}^{(2)}$, respectively. Then, for each arrangement $\mathcal{A}^{(i)}$, $i = 1, 2$, locate the arrangement feature containing one end of the line segment $p^{(i)}q^{(i)}$, say $p^{(i)}$. If $p^{(i)}$ coincides with a vertex or lies on an edge, conclude that the path is not collision-free. Otherwise, let f be the face containing $p^{(i)}$. Iterate over the bounding curves of f, and test whether each intersects $p^{(i)}q^{(i)}$. If one does, also conclude that the path is not collision-free.

Sometimes it is necessary to keep a small amount of clearance c between a robot and an obstacle. To this end, we offset each line segment $p^{(i)}q^{(i)}$ by $c^{(i)}$, and check whether $(p^{(i)}q^{(i)} \oplus c^{(i)}) \cap \mathcal{C}_{\text{forb}}^{(i)} = \emptyset$. We set $c^{(i)}$ to be a small fraction of the radius $r^{(i)}$ of the robot $B_{r^{(i)}}$. The `Two_disc_offset_prm<Kernel>` class template, defined in the header file `Two_disc_offset_prm.h`, is an improved version of the class template `Two_disc_simple_prm<Kernel>`. It is based on the techniques proposed in the previous and next paragraphs. The constructor of the class template sets $c^{(i)}$ to be 1% of $r^{(i)}$; see Line 10 and Line 11 in the code excerpt below.

```
1   template <typename ObstacleIterator>
2   Two_disc_offset_prm(const CGAL::Bbox_2& bbox,
3                       ObstacleIterator begin, ObstacleIterator end,
4                       const Number_type& radius1, const Number_type& radius2,
5                       unsigned int n_vertices, unsigned int n_edges) :
6     Base(bbox, begin, end, radius1, radius2, n_vertices, n_edges)
7   {
8     // Compute the desired clearance for each robot, given as a fraction of
9     // its radius.
10    m_clear1 = radius1 / PRM_INV_RELATIVE_CLEARANCE;
11    m_clear2 = radius2 / PRM_INV_RELATIVE_CLEARANCE;
12
13    this->generate_prm(n_vertices, n_edges);
14  }
```

So, it is possible to verify the individual path for each robot using a single operation. But how can we verify that the two robots do not intersect each other? Let us denote the end configurations by $\langle (x_1^{(1)}, y_1^{(1)}), (x_1^{(2)}, y_1^{(2)}) \rangle$ and $\langle (x_2^{(1)}, y_2^{(1)}), (x_2^{(2)}, y_2^{(2)}) \rangle$, and let us assume that each of the robots moves at a fixed velocity. In time $0 \le t \le 1$ the first robot is located at $(x_1^{(1)} + (x_2^{(1)} - x_1^{(1)}) \cdot t, y_1^{(1)} + (y_2^{(1)} - y_1^{(1)}) \cdot t)$, while the second robot is located at $(x_1^{(2)} + (x_2^{(2)} - x_1^{(2)}) \cdot t, y_1^{(2)} + (y_2^{(2)} - y_1^{(2)}) \cdot t)$. It is not difficult to see that $D(t)$, the squared distance between the two robots at time t, is a quadratic function in t, and thus has a single minimum that can be found by solving the linear equation $D'(t_{\min}) = 0$. Now, if $0 < t_{\min} < 1$ we substitute it into the expression $D(t)$ and check whether the squared distance is greater than $(r^{(1)} + r^{(2)})^2$.

The `can_connect()` method of the class template `Two_disc_offset_prm` listed below tests whether the straight path between two given configurations $\langle p^{(1)}, p^{(2)} \rangle$ and $\langle q^{(1)}, q^{(2)} \rangle$ is collision-free. If it is, the total length of the paths of the two robots is computed and returned. It applies the `approximated_offset_2()` function template to the line segments $p^{(1)}q^{(1)}$ and $p^{(2)}q^{(2)}$, and tests whether the resulting approximated offset polygons intersect with the corresponding forbidden-configuration spaces. Then it tests whether the two robots collide.

```
bool can_connect(unsigned int u, unsigned int v, double& len,
                 bool /* bound_steps = true */) const
{
  // Is the straight path plus clearance for the 1st robot collision-free?
  Obstacle path1;
  path1.push_back(this->m_confs[u].pos1);
  path1.push_back(this->m_confs[v].pos1);
```

```
if (this->m_cforb1.do_intersect(approximated_offset_2(path1, m_clear1,
                                                      this->m_eps)))
    return false;

// Is the straight path plus clearance for the 2nd robot collision-free?
Obstacle path2;
path2.push_back(this->m_confs[u].pos2);
path2.push_back(this->m_confs[v].pos2);
if (this->m_cforb2.do_intersect(approximated_offset_2(path2, m_clear2,
                                                      this->m_eps)))
    return false;

// Assuming the first robot moves at constant speed from u_pos1 to v_pos1
// while the second one moves at constant speed from u_pos2 to v_pos2
// from time t = 0 to time t = 1, find the time when the distance between
// them is minimal.
// The squared distance D between the robots at time t is given by:
//    D(t) = alpha*t^2 + 2*beta*t + gamma.
const Number_type ux1 = this->m_confs[u].pos1.x();
const Number_type uy1 = this->m_confs[u].pos1.y();
const Number_type vx1 = this->m_confs[v].pos1.x();
const Number_type vy1 = this->m_confs[v].pos1.y();
const Number_type ux2 = this->m_confs[u].pos2.x();
const Number_type uy2 = this->m_confs[u].pos2.y();
const Number_type vx2 = this->m_confs[v].pos2.x();
const Number_type vy2 = this->m_confs[v].pos2.y();
const Number_type dx = ux2 - ux1, dy = uy2 - uy1;
const Number_type rx1 = vx1 - ux1, rx2 = vx2 - ux2;
const Number_type ry1 = vy1 - uy1, ry2 = vy2 - uy2;
const Number_type dlx = rx2 - rx1, dly = ry2 - ry1;
const Number_type alpha = dlx*dlx + dly*dly;
const Number_type beta = dx*dlx + dy*dly;
const Number_type gamma = dx*dx + dy*dy;
const Number_type t_min = - beta / alpha;

if ((CGAL::sign(t_min) == CGAL::POSITIVE) &&
    (CGAL::compare(t_min, 1) == CGAL::SMALLER))
{
  // We found a valid value 0 < t_min < 1, so the minimal squared
  // distance is given by D(t_min).
  const Number_type    min_sqr_dist = (alpha*t_min + 2*beta)*t_min + gamma;
  bool rc = (CGAL::compare(CGAL::square(this->m_rad1 + this->m_rad2),
                           min_sqr_dist) == CGAL::SMALLER);
  if (rc) {
    const double dist1 = CGAL::sqrt(CGAL::square(CGAL::to_double(rx1)) +
                                    CGAL::square(CGAL::to_double(ry1)));
    const double dist2 = CGAL::sqrt(CGAL::square(CGAL::to_double(rx2)) +
                                    CGAL::square(CGAL::to_double(ry2)));
    len = dist1 + dist2;
  }
  return rc;
}

// The straight-line connection is collision-free and its length equals
```

```
      // the square root of the sum of pairwise distances.
      const double dist1 = CGAL::sqrt(CGAL::square(CGAL::to_double(rx1)) +
                                      CGAL::square(CGAL::to_double(ry1)));
      const double dist2 = CGAL::sqrt(CGAL::square(CGAL::to_double(rx2)) +
                                      CGAL::square(CGAL::to_double(ry2)));
      len = dist1 + dist2;
      return true;
    }
```

───────── *advanced* ─────────

The program below uses an instance of the function template **read_objects()** explained in Section 4.2.1 to read the obstacles from an input file. It then uses the function template **bbox()** to compute the bounding rectangle of the obstacles. The program also reads two numbers that indicate the radii of the two robots, respectively, the number of free configurations to obtain, and the maximum number of free connections between free configurations to establish. The program then constructs a PRM passing all the relevant data to the PRM constructor. Once the PRM is constructed, the program uses it to answer motion-planning queries, which it reads from a second input file using a second instance of the function template **read_objects()**.

```cpp
// File: mp_two_discs.cpp

#include <iostream>
#include <sstream>
#include <string>
#include <list>
#include <boost/timer.hpp>
#include <boost/lexical_cast.hpp>

#include <CGAL/Cartesian.h>
#include <CGAL/Gmpq.h>

#include "read_objects.h"
#include "bbox.h"
#include "Two_disc_offset_prm.h"

typedef CGAL::Cartesian<CGAL::Gmpq>         Kernel;
typedef Kernel::FT                          Number_type;
typedef Kernel::Point_2                     Position;
typedef CGAL::Polygon_2<Kernel>             Obstacle;
typedef std::list<Obstacle>                 Obstacle_set;
typedef Two_disc_offset_prm<Kernel>         PRM;

// The query type.
struct Query { Configuration_4<Kernel> m_s, m_g; };

// Read a single query.
std::istream& operator>>(std::istream& is, Query& q)
{ return is >> q.m_s.pos1 >> q.m_s.pos2 >> q.m_g.pos1 >> q.m_g.pos2; }

// The main application:
int main(int argc, char* argv[])
{
  if (argc < 7) {
    std::cout << "Usage:_" << argv[0]
```

```cpp
                    << "_<obstacle_file>_<radius_1>_<radius_2>"
                    << "_<#_PRM_vertices>_<#_PRM_edges>_<query_file>" << std::endl;
      return -1;
  }

  // Read the polygonal obstacles from the input file.
  Obstacle_set   obstacles;
  read_objects<Obstacle>(argv[1], std::back_inserter(obstacles));

  // Construct the bounding box obstacles.
  CGAL::Bbox_2  bb = bbox(obstacles.begin(), obstacles.end());

  // Construct the PRM according to the command-line parameters.
  Number_type   radius1, radius2;
  std::istringstream iss1(argv[2], std::istringstream::in); iss1 >> radius1;
  std::istringstream iss2(argv[3], std::istringstream::in); iss2 >> radius2;
  const unsigned int n_vertices = boost::lexical_cast<unsigned int>(argv[4]);
  const unsigned int n_edges = boost::lexical_cast<unsigned int>(argv[5]);
  std::cout << "Constructing_the_PRM_..._" << std::flush;
  boost::timer timer;
  PRM prm(bb, obstacles.begin(), obstacles.end(), radius1, radius2,
          n_vertices, n_edges);
  double secs = timer.elapsed();
  std::cout << "Done!" << std::endl
            << "_PRM_has_" << prm.number_of_vertices() << "_vertices_and_"
            << prm.number_of_edges() << "_edges." << std::endl
            << "__Construction_took_" << secs << "_sec." << std::endl;

  // Read the queries.
  std::list<Query> queries;
  read_objects<Query>(argv[6], std::back_inserter(queries));

  std::list<Query>::const_iterator it;
  for (it = queries.begin(); it != queries.end(); ++it) {
    // Use the PRM to plan a collision-free motion path.
    std::list<Configuration_4<Kernel> > path;
    prm.query(it->m_s, it->m_g, std::back_inserter(path));
    if (path.empty()) {
      std::cout << "__No_path_found!" << std::endl;
      continue;
    }

    // Print the collision-free path.
    std::cout << "__Found_a_path_with_" << path.size() - 2
              << "_intermediate_configuration(s)." << std::endl;
    path.clear();
  }
  return 0;
}
```

The class templates `Two_disc_simple_prm` and `Two_disc_offset_prm` have the exact same interface. Thus, you can easily exchange between them in the code above.

The command line that executes the program consists of five arguments, which follow the executable name:

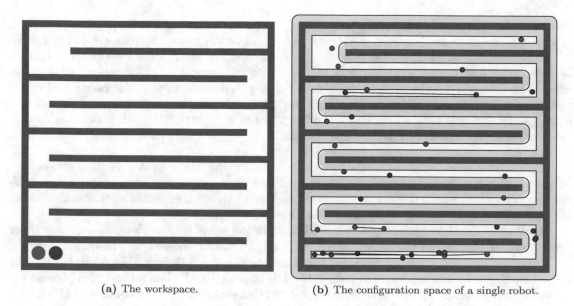

(a) The workspace.   (b) The configuration space of a single robot.

Fig. 9.8: A maze instance of the two-disc coordination problem.

mp_two_discs <obstacle-file> <rad 1> <rad 2> <# vertices> <# edges> <query-file>

The command line below executes the application in an attempt to obtain a collision-free motion for the instance depicted in Figure 9.7a. Typically, limiting the number of vertices and edges in the PRM to 32 and 1,024, respectively, for this particular instance is sufficient to obtain a collision-free motion. Constructing the corresponding PRM takes just a few seconds on a computer clocked at 2 GHz.

mp_two_discs obstacles_simple.dat 3 3 32 1024 queries_simple.dat

The command line below executes the application in an attempt to obtain a collision-free motion for the instance depicted in Figure 9.8a. As the number of samples is very small (16), only few edges are typically established. As a consequence the probability of obtaining a collision-free motion is low.

mp_two_discs obstacles_maze.dat 1 1 16 256 queries_maze.dat

The two-disc coordination application is far from trivial. For complicated scenes, such as the one depicted in Figure 9.8, even the improved concrete implementation of the PRM approach fails to find a solution in reasonable time.

9.4 Bibliographic Notes and Remarks

Computing Minkowski sums is a recurring operation in many geometric applications. Problems involving moving an object amidst obstacles, laying out objects in a cell or on a shop floor, defining a tolerance zone around an object, morphing shapes, sweeping an object along a curve, cutting and packing problems, and more, call for computing the Minkowski sums of polygons or of a polygon and a disc. See [16, 28, 59, 125, 138] for a small sample of applications.

The combinatorial complexity of planar polygonal Minkowski sums, namely, the number of vertices and edges on the boundary of the sum, is well understood. Consider the Minkowski sum of two polygons, one with m vertices and another with n vertices. If both polygons are convex then the maximum complexity of their Minkowski sum is $O(m+n)$ and the sum can be computed in time $O(m+n)$. If both are non-convex then the maximum complexity of the sum is $O(m^2n^2)$

and it can be computed in time $O(m^2n^2\log(mn))$. If one is convex and the other is non-convex, then the complexity of their sum is $O(mn)$ and it can be computed in time $O(mn\log^2(mn))$. All the combinatorial bounds mentioned in this paragraph are tight in the worst case. All these bounds (lower and upper) are fairly easy to show, with the exception of the upper bound in the case of the sum of a convex polygon and a non-convex polygon. A proof of this bound and of the rest can be found in [45, Chapter 13], which is a gentle introduction to the topic. A proof of a more general theorem [127] shows that the offset polygon of a polygon P with n vertices, namely the Minkowski sum of P with a disc, has complexity $O(n)$. The offset can be computed in time $O(n\log^2 n)$.

The running time cited for the algorithm computing the Minkowski sum of a non-convex polygon and a convex object (either a convex polygon or a disc) refers to a simple divide-and-conquer algorithm, where the merge step is sweep-based. Faster algorithms are known, reducing the log-squared factor to log, using generalized Voronoi diagrams [143].

Besides the CGAL software for Minkowski sums of polygons we are aware of only one other exact implementation—in the LEDA library; see Section 1.2.4 for some notes about this library.

The decomposition approach to computing Minkowski sums is described in detail in [1]. The convolution approach is described in [208–210]. The paper [1] and Flato's thesis [68] give an extensive coverage of the related literature on polygon decomposition in the context of Minkowski sums. For a general review of polygon decomposition, see [128]. Specifically, we mention in this chapter four approaches: Two approaches by Greene [86], an optimal dynamic-programming-based approach and an approximation scheme; an approximation scheme by Hertel and Mehlhorn [116]; and a variant by Flato [68] of an approximation scheme by Chazelle and Dobkin [35]. The *2D Minkowski Sums* package of CGAL [212] includes four models of the PolygonConvexDecomposition concept, which are the implementations of the four approaches above. The first three are classes that wrap the decomposition functions included in the *2D Polygon Partitioning* package of CGAL [114].

For more details on the convolution approach to computing Minkowski sums, see, e.g., [93]. Other aspects of planar Minkowski sums include Minkowski sums of algebraic curves [13] and their approximate construction [142].

In three dimensions, the Minkowski sum of two convex polytopes is also well understood. The combinatorial bound in this case is $O(mn)$, where the two polytopes have m and n vertices, respectively. Its complexity is understood even exactly, namely without resorting to asymptotic notation [74]. It can be computed robustly and efficiently using software based on CGAL arrangements. See the papers [71, 74] and the thesis by Fogel [70] for more details.

The case of two non-convex polyhedra is easy on the combinatorics side, where a tight bound of $O(m^3n^3)$ is known, but hard on the robust implementation side. A significant achievement in this direction is the work of Hachenberger [95], who was the first to devise an exact construction for the non-convex polyhedral case, based on CGAL's Nef polyhedra and following the decomposition approach. Producing code that is at once robust and fast remains a challenge. A number of authors describe approximate solutions to this case; see, for example, [144, 145, 205].

The case of one convex polyhedron and one non-convex polyhedron is intriguing from a combinatorial perspective [196]. A sharper combinatorial bound is known for the special case of a non-convex polyhedron and a box [106].

Motion planning is a very broad topic that has attracted intensive research for several decades. We just point out major books [37, 138, 139] and several survey papers [97, 99, 147, 195] on the topic and its relatives.

The two-disc motion-planning problem discussed in Section 9.3.2 was investigated by Hirsch and Halperin [117], who give a more elaborate and comprehensive solution than the one developed in this chapter, based on exact construction of arrangements. Their solution can find paths even for certain settings that incur *tight* passages. The sampling-based method used in the two-disc example, PRM, is described in [126] and [37, Chapter 7].

9.5   Exercises

9.1 Write a function that computes the Minkowski sum of two linear polygons that may contain holes utilizing the decomposition approach. Develop the decomposition methods described below, and compare the overall performance of the corresponding Minkowski sum.

- Develop a decomposition method that is based on the vertical decomposition function-template, which decomposes the faces of a given arrangement into pseudo trapezoids; see Section 3.6.1.

- Develop a decomposition method that is based on constrained triangulation [220].

9.2 Optimize the "Translating Polygonal Robot" application so that only a selective vertical decomposition is applied. More precisely, develop a function template, say `free_vertical_decomposition()`, that accepts an arrangement object and decomposes only the arrangement faces that comprise the free-configuration space. Page 61 contains the listing of the function template `vertical_decomposition()`. Replace the call to the instance of this function template with a call to an instance of the newly developed function template `free_vertical_decomposition()`.

9.3 Alter the motion-planning application presented in Section 9.3.1 to produce shorter collision-free paths.

Naively replacing the application of the breadth-first search (BFS) procedure with an application of a procedure that finds the shortest path in a weighted graph (where the weights are Euclidean lengths), such as the BGL implementation of Dijkstra's shortest path algorithm, may yield a shorter path but not necessarily the shortest path. A more effective algorithm is to compute the visibility graph first and to apply Dijkstra's shortest path algorithm on it; see [45, Chapter 15].

9.4 With respect to the **two-disc coordination** problem consider two free configurations, $\langle u_1, u_2 \rangle$ and $\langle v_1, v_2 \rangle$. Prove that if the motion path for a robot of radius $r + \frac{\delta}{2}$ is collision-free at a discrete number K of sampled points along the straight-line segment connecting $\langle u_1, u_2 \rangle$ and $\langle v_1, v_2 \rangle$, such that $\frac{1}{K}\|v_1 - u_1\| < \delta$ and $\frac{1}{K}\|v_2 - u_2\| < \delta$, then the continuous path is collision-free for the original robot.

9.5 **[Project]** Develop an application that solves a variant of the motion-planning problem of a polygonal robot translating amidst polygonal obstacles, where the robot is allowed to touch the obstacles. That is, where the forbidden-configuration space is open and the free-configuration space is closed.

Hint: In this variant, where touching is allowed, the free-configuration space may contain vertices and edges incident only to faces contained in the forbidden-configuration space. A collision-free path may span more than a single face of the arrangement. Regularized Boolean set operations are useless here, because they may erase critical vertices and edges. Thus, you may need to directly operate on the arrangement that represents the configuration space instead, as suggested by Flato [68].

9.6 **[Project]**

(a) Develop an application that generalizes the two-disc coordination application presented in Section 9.3.2 such that it handles a fixed number $k \geq 2$ of disc robots, where k is set at compile time.

(b) Develop an application that generalizes the two-disc coordination application such that it handles two *translating* polygonal robots.

(c) Combine the two generalizations above in a single application.

9.7 Develop a script in some scripting language (e.g., Perl or Python) that accepts two positive integers m and n and generates two polygons P and Q with $O(m)$ and $O(n)$ vertices, respectively, such that the Minkowski sum $P \oplus Q$ has $O(m^2 n^2)$ vertices.

Chapter 10

Envelopes

It is sometimes sufficient to consider only a *sub-structure* of the arrangement (that is, some portion of the arrangement) to solve a computational problem at hand, rather than the entire arrangement. In this chapter we examine one useful sub-structure called the *envelope* of a set of curves (or a set of surfaces). We discuss envelopes both in $\mathbb{R}^2$ and in $\mathbb{R}^3$. Informally, the envelope of a collection of surfaces in, say, $\mathbb{R}^3$ is what one sees of the arrangement of the surfaces when looking at it from sufficiently far away. (Formal definitions are given below.) In the planar case, the envelope is a one-dimensional structure and can be efficiently (and easily) computed without having to construct the entire arrangement. Still, we use the various models of the traits-class concepts, as defined in Section 5.1, to compute envelopes of arbitrary planar curves.

The envelope computation of a collection of surfaces in $\mathbb{R}^3$ exploits the ability to represent the envelope as a planar arrangement, as the envelope is only two-dimensional. We use geometric operations in the plane as much as possible, but we also have to use a model of a refined traits-class concept that is capable of operating on three-dimensional entities as well. This is therefore a small step into three-dimensional space.

10.1 Envelopes of Curves in the Plane

Given a set $\mathcal{C} = \{C_1, C_2, \ldots, C_n\}$ of continuous x-monotone curves in the plane, their *lower envelope* is defined as the point-wise minimum of all curves in $\mathcal{C}$. Formally, as the curves in $\mathcal{C}$ are continuous x-monotone, we regard each curve $C_k \in \mathcal{C}$ as a univariate function $y = C_k(x)$, defined over some continuous range $R_k \subseteq \mathbb{R}$. The lower envelope is given by the following function:

$$\mathcal{L}_{\mathcal{C}}(x) = \min_{1 \leq k \leq n} \overline{C}_k(x) \ ,$$

where $\overline{C}_k(x) = C_k(x)$ for $x \in R_k$, and $\overline{C}_k(x) = \infty$ otherwise.

Similarly, the *upper envelope* of $\mathcal{C}$ is the point-wise maximum of the x-monotone curves in $\mathcal{C}$, and is given the following function:

$$\mathcal{U}_{\mathcal{C}}(x) = \max_{1 \leq k \leq n} \underline{C}_k(x) \ ,$$

where $\underline{C}_k(x) = -\infty$ for $x \notin R_k$.

10.1.1 Representing the Envelope

The *minimization diagram* of the set $\mathcal{C}$ is the subdivision of the x-axis into 0-dimensional cells called *vertices* and 1-dimensional maximal relatively-open cells called *edges*, such that a particular function that represents a curve of $\mathcal{C}$ (or a particular set of functions that represent a subset of $\mathcal{C}$) attains the lower envelope over all points of a specific cell of the subdivision. Each vertex of the

E. Fogel et al., *CGAL Arrangements and Their Applications*, Geometry and Computing 7, DOI 10.1007/978-3-642-17283-0_10, © Springer-Verlag Berlin Heidelberg 2012

Fig. 10.1: The lower envelope of eight line segments,
labeled $A, \ldots, H$, as constructed in `ex_envelope_`
`segments.cpp`. The mini-
mization diagram is shown
at the bottom, where each
diagram vertex points to
the planar point of the
lower envelope associated
with it. The labels of the
segments that induce a diagram edge are displayed below the edge. Note that
there exists one edge that
represents an overlap, and
there are four edges that
represent empty intervals.

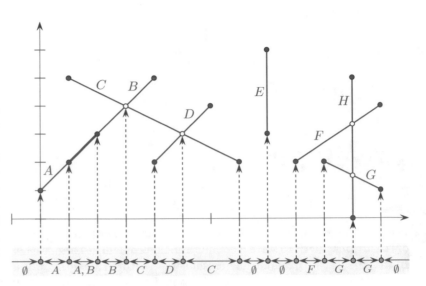

minimization diagram is incident to two edges, one lying to its left and the other to its right. An
edge in the envelope diagram represents a continuous interval on the x-axis.

In non-degenerate situations, an edge is induced by a single curve (or by no curves at all, if
there are no x-monotone curves defined over the interval), and a vertex is induced either by a
single curve, and corresponds to one of its endpoints, or by two curves, and corresponds to their
intersection point. In a degenerate situation a vertex v could be induced by more than two curves
that intersect at a point p_v, which attains the lower envelope over x_{p_v}, and an edge e could be
induced by several overlapping curves defined over the interval represented by e. Thus, every
diagram vertex v is associated with a set of x-monotone curves that induce the envelope over the
x-coordinate represented by v, and every diagram edge e is associated with a set of x-monotone
curves that induce the lower envelope over the interval represented by e. This set is empty if no
x-monotone curves are defined over this interval. We emphasize that the identity of the curves
that induce the envelope over each cell is fixed. Figure 10.1 shows the lower envelope of a set of
eight line segments, and the structure of their minimization diagram. The *maximization diagram*
is similarly defined for upper envelopes. We refer to either diagram as the *envelope diagram*
hereafter.

10.1.2   Constructing the Envelope Diagram

The output of the lower-envelope (respectively, upper envelope) computation is a minimization diagram (respectively, maximization diagram) represented as a one-dimensional arrangement, where
each feature f (i.e., vertex or edge) of the arrangement is labeled with the set of x-monotone curves
that attain the minimum (respectively, maximum) over f. The label can contain a single curve or
several curves, or it can be the empty set.

Lower and upper envelopes can be efficiently computed using a divide-and-conquer approach.
Computing the envelope diagram of a single x-monotone curve C_k is trivial; we take the boundary
of its range of definition R_k and label the features it induces accordingly. Given a set $\hat{C}$ of (not
necessarily x-monotone) curves in $\mathbb{R}^2$, we subdivide each curve into a finite number of weakly
x-monotone curves, and obtain the set C. Then, we split the set into two disjoint subsets C_1 and
C_2 of roughly the same size, and we compute their respective envelope diagrams M_1 and M_2
recursively. Finally, we merge M_1 and M_2 to obtain the final envelope diagram M. We merge
M_1 and M_2 in linear time (linear in the complexity of M_1 and M_2), traversing the diagrams in
parallel.

The merge step is carried out in three sub-steps described below and illustrated in Figure 10.2.

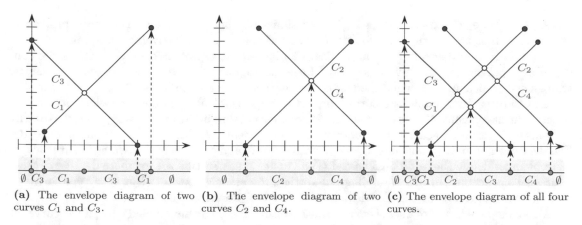

(a) The envelope diagram of two curves C_1 and C_3.

(b) The envelope diagram of two curves C_2 and C_4.

(c) The envelope diagram of all four curves.

Fig. 10.2: The merge step of the divide-and-conquer algorithm that computes envelope diagrams.

1. We overlay the two (one-dimensional) arrangements that represent the envelope diagrams $\mathcal{M}_1$ and $\mathcal{M}_2$ to obtain the arrangement $\mathcal{A}''$. For each feature in $\mathcal{A}''$ we maintain references to the inducing features in $\mathcal{M}_1$ and $\mathcal{M}_2$.

2. We determine the structure of the minimization diagram over each feature in $\mathcal{A}''$ to obtain the arrangement $\mathcal{A}'$, which is a refinement of the arrangement representing $\mathcal{M}$. We then label each feature f of $\mathcal{A}'$ with the correct envelope curve. For each feature f we must consider the two relevant features in $\mathcal{M}_1$ and $\mathcal{M}_2$ and their corresponding labels l_1 and l_2. If both sets are empty, there is no curve defined over f, and we label it with the empty set. If exactly one of these labels is the empty set, we label f with the nonempty set among the two. Handling the remaining case, where both labels are nonempty sets, is a bit more complicated, at least for edges. The x-monotone curves in the sets l_1 and l_2 are defined over f, and their envelope over f is the envelope of $C_1 \cup C_2$ there. Since all the x-monotone curves of one label l_i overlap over the current feature f, it is sufficient to consider a representative curve C_i from each label, and compute their minimization diagram over f, denoted by $\mathcal{M}'$. If f is an edge, we must consider the intersections between C_1 and C_2, or more precisely, the projection onto the x-axis of the intersections, which may split the edge f. We then label all the features of the arrangement that represents $\mathcal{M}'$ restricted to f (where f is considered relatively open) with the correct labels, which might be either one of the labels l_1 and l_2, or $l_1 \cup l_2$ in the case of an overlap.

3. Finally, we apply a cleanup step, which removes redundant vertices of $\mathcal{A}'$, and obtain the final arrangement that represents the envelope diagram $\mathcal{M}$ of $\mathcal{C}$. Vertices incident to pairs of edges with the same label are redundant. We remove such vertices and merge the incident two edges into a single edge.

This relatively simple process is efficient and leads to an algorithm with near-optimal running time. The complexity of the lower envelope of a set of well-behaved curves (see Section 1.3.3) is linear or very slightly super-linear in the number of input curves. (Envelopes of curves are combinatorially equivalent to a special family of words called *Davenport-Schinzel* sequences, and the complexity bounds are derived from bounds on the maximal length of such words.) The algorithm above, in turn, runs in near-linear time in the number of input curves. See Section 10.5 for running times of this and of a theoretically better algorithm.

10.1.3 The Envelope-Software Components

The *Envelopes of Curves in 2D* package of CGAL contains two sets of free functions, as follows: `lower_envelope_x_monotone_2(begin, end, diag)` (and similarly, `upper_envelope_x_monotone_2()`) accepts a range of x-monotone curves and computes their envelope diagram;

`lower_envelope_2(begin, end, diag)` (and similarly, `upper_envelope_2()`) accepts a range of *arbitrary* (not necessarily x-monotone) curves and computes their envelope diagram.

A minimization diagram (or a maximization diagram) is represented by a model of the concept EnvelopeDiagram_1. Any model of the EnvelopeDiagram_1 concept must define a geometric traits-class, which in turn must model the ArrangementXMonotoneTraits_2 concept if the diagram is induced by x-monotone curves, or the ArrangementTraits_2 concept (which refines the concept ArrangementXMonotoneTraits_2) if the diagram is induced by arbitrary (not necessarily x-monotone) curves. Recall that a model of the ArrangementXMonotoneTraits_2 concept must define the `Point_2` and `X_monotone_curve_2` nested types, and a model of the ArrangementTraits_2 concept must define the additional nested type `Curve_2`; see Section 5.1 for more details on these traits-class concepts. In any case, diagram vertices are associated with objects of type `Point_2` and diagram edges are associated with objects of type `X_monotone_curve_2`.

An envelope diagram of a set of curves consists either of a single unbounded edge, if the curve set is empty or if the envelope contains a single unbounded curve that is in front of (or below or above) all other curves, or of at least one vertex and two unbounded edges, while each additional vertex comes with an additional edge. The *leftmost* and *rightmost* edges of an envelope diagram represent the unbounded intervals that start at $-\infty$ and end at ∞, respectively. Let `diag` be an envelope-diagram object, the type of which, say `Diagram_1`, models the concept EnvelopeDiagram_1. The nested types `Diagram_1::Vertex_const_handle` and `Diagram_1::Edge_const_handle` serve as pointers to a vertex and an edge of the diagram, respectively. The *leftmost* and *rightmost* edges of `diag` can be obtained with the calls `diag.leftmost()` and `diag.rightmost()`, respectively. In the example depicted in Figure 10.1 we have only bounded curves, so the leftmost and rightmost edges represent empty intervals. This is not the case when we deal, for example, with envelopes of sets of lines.

Let v and e be handles to a vertex and an edge of an envelope diagram, respectively. The calls `v->right()` and `v->left()` return the edge to the right and the edge to left of the vertex v, respectively. Similarly, the calls `e->right()` and `e->left()` return the vertex to the right and the vertex to left of the edge e. The calls `v->curves_begin()` and `v->curves_end()` return iterators of the nested `Diagram_1::Curve_const_iterator` type that define the valid range of curves that attain the envelope over v. The call `v->curve()` returns a representative curve from the range. Similarly, the calls `e->curves_begin()` and `e->curves_end()` return iterators that define the valid range of curves that attain the envelope over e. The call `e->curve()` returns a representative curve from the range.

10.1.4   Using the Traits Classes

In this section we demonstrate the usage of various traits classes for constructing the envelopes of sets of planar curves. In particular, we show how to attach auxiliary data to the curves.

Example: The program below demonstrates how to compute and traverse the minimization diagram of line segments illustrated in Figure 10.1. We attach a label (a `char` in this case) to each input segment, and use these labels when we print the minimization diagram. We instantiate the `Arr_curve_data_traits_2<BaseTraits, XMonotoneCurveData>` class template, substituting the `BaseTraits` template parameter with a traits class that handles segments, as defined in the header file `arr_exact_construction_segments.h`; see Section 3.4.1.

```
// File: ex_envelope_segments.cpp

#include <list>
#include <iostream>

#include <CGAL/basic.h>
#include <CGAL/Arr_curve_data_traits_2.h>
#include <CGAL/Envelope_diagram_1.h>
#include <CGAL/envelope_2.h>
```

```cpp
#include "arr_exact_construction_segments.h"

typedef CGAL::Arr_curve_data_traits_2<Traits, char>      Data_traits;
typedef Data_traits::X_monotone_curve_2                  Label_seg;
typedef CGAL::Envelope_diagram_1<Data_traits>            Diagram;

int main()
{
  // Consrtuct the input segments and label them 'A' ... 'H'.
  std::list<Label_seg> segs;
  segs.push_back(Label_seg(Segment(Point(0, 1), Point(2, 3)), 'A'));
  segs.push_back(Label_seg(Segment(Point(1, 2), Point(4, 5)), 'B'));
  segs.push_back(Label_seg(Segment(Point(1, 5), Point(7, 2)), 'C'));
  segs.push_back(Label_seg(Segment(Point(4, 2), Point(6, 4)), 'D'));
  segs.push_back(Label_seg(Segment(Point(8, 3), Point(8, 6)), 'E'));
  segs.push_back(Label_seg(Segment(Point(9, 2), Point(12, 4)), 'F'));
  segs.push_back(Label_seg(Segment(Point(10, 2), Point(12, 1)), 'G'));
  segs.push_back(Label_seg(Segment(Point(11, 0), Point(11, 5)), 'H'));

  // Compute the minimization diagram that represents their lower envelope.
  Diagram min_diag;
  CGAL::lower_envelope_x_monotone_2(segs.begin(), segs.end(), min_diag);

  // Print the minimization diagram.
  Diagram::Edge_const_handle       e = min_diag.leftmost();
  while (e != min_diag.rightmost()) {
    Diagram::Curve_const_iterator   cit;
    std::cout << "Edge:";
    if (! e->is_empty())
      for (cit = e->curves_begin(); cit != e->curves_end(); ++cit)
        std::cout << '␣' << cit->data();
    else std::cout << "␣[empty]";
    std::cout << std::endl;

    Diagram::Vertex_const_handle v = e->right();
    std::cout << "Vertex(" << v->point() << "):␣";
    for (cit = v->curves_begin(); cit != v->curves_end(); ++cit)
      std::cout << '␣' << cit->data();
    std::cout << std::endl;

    e = v->right();
  }
  CGAL_assertion(e->is_empty());
  std::cout << "Edge:␣[empty]" << std::endl;
  return 0;
}
```

When executed, the program above inserts the lines below into the standard output-stream. (The line numbers are not part of the output). Compare with Figure 10.1.

```
1 Edge: [empty]         10 Vertex(4 2):  D       18 Vertex(9 2):  F
2 Vertex(0 1):  A       11 Edge: D                19 Edge: F
3 Edge: A               12 Vertex(5 3):  C D      20 Vertex(10 2):  G
4 Vertex(1 2):  B A     13 Edge: C                21 Edge: G
```

```
 5 Edge: A B              14 Vertex(7 2):  C        22 Vertex(11 0):  H
 6 Vertex(2 3):  A B      15 Edge: [empty]          23 Edge: G
 7 Edge: B                16 Vertex(8 3):  E        24 Vertex(12 1):  G
 8 Vertex(3 4):  B C      17 Edge: [empty]          25 Edge: [empty]
 9 Edge: C
```

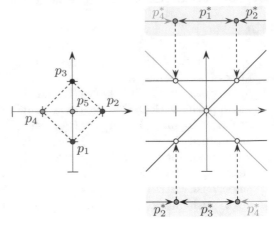

Example: The next program computes the convex hull of a set of input points by constructing envelopes of (unbounded) lines, which are dual to the input points. We use the **Arr_linear_traits_2** class template as instantiated in the header file **arr_linear.h** (see Section 4.1.1) to compute the lower and upper envelopes of the set of dual lines. We read a set of points $\mathcal{P} = p_1, \ldots, p_n$ from an input file, and compute the corresponding dual lines $\mathcal{P}^* = p_1^*, \ldots, p_n^*$, where the line p^* dual to a point $p = (p_x, p_y)$ is given by $y = p_x x - p_y$; see Section 4.2. Each line in the dual plane is associated with the index of the point inducing it. We then compute the lower and upper envelopes of $\mathcal{P}^*$. Observe that the lines that form the lower envelope of $\mathcal{P}^*$ are dual to the points along the *upper* part of the convex hull of $\mathcal{P}$, and the lines that form the upper envelope of $\mathcal{P}^*$ are dual to the points along the *lower* part of the convex hull. The left part of the figure above depicts a simple case of five input points. The dual lines and their envelopes are depicted in the right part; see, e.g., [45, Section 11.4] for more details. Finally, we traverse the lower envelope and continue with the upper envelope from left to right to obtain the convex hull of the point set $\mathcal{P}$.

Note that in non-degenerate cases the leftmost edge of the minimization diagram is associated with the same line as is the rightmost edge of the maximization diagram, and the rightmost edge of the minimization diagram is associated with the same line as is the leftmost edge of the maximization diagram. Thus, we skip the rightmost edges of both diagrams in the general case.

```
// File: ex_convex_hull.cpp

#include <vector>

#include <CGAL/basic.h>
#include <CGAL/Arr_curve_data_traits_2.h>
#include <CGAL/Envelope_diagram_1.h>
#include <CGAL/envelope_2.h>

#include "arr_linear.h"
#include "read_objects.h"
#include "dual_plane.h"

typedef CGAL::Arr_curve_data_traits_2<Traits, unsigned int>   Data_traits;
typedef Data_traits::X_monotone_curve_2                       Dual_line;
typedef CGAL::Envelope_diagram_1<Data_traits>                Diagram;

using CGAL::lower_envelope_x_monotone_2;
using CGAL::upper_envelope_x_monotone_2;

int main(int argc, char* argv[])
{
```

```cpp
  // Read the points from the input file.
  const char * filename = (argc > 1) ? argv[1] : "ch_points.dat";
  std::vector<Point>    points;
  read_objects<Point>(filename, std::back_inserter(points));

  // Consturct the lines dual to the points.
  std::list<Dual_line>  dual_lines;
  std::vector<Point>::const_iterator it;
  unsigned int          k = 0;
  for (it = points.begin(); it != points.end(); ++it) {
    dual_lines.push_back(Dual_line(dual_line<Traits>(*it), k++));
  }

  // Compute the lower envelope of dual lines, which corresponds to the upper
  // part of the convex hull, and their upper envelope, which corresponds to
  // the lower part of the convex hull.
  Diagram               min_diag, max_diag;
  lower_envelope_x_monotone_2(dual_lines.begin(), dual_lines.end(), min_diag);
  upper_envelope_x_monotone_2(dual_lines.begin(), dual_lines.end(), max_diag);

  // Output the points along the boundary convex hull in a counterclockwise
  // order. We start by traversing the minimization diagram from right to
  // left, and then traverse the maximization diagram from left to right.
  std::cout << "The convex hull of " << points.size() << " input points:";
  Diagram::Edge_const_handle  e = min_diag.leftmost();
  while (e != min_diag.rightmost()) {
    std::cout << " (" << points[e->curve().data()] << ')';
    e = e->right()->right();
  }
  // Handle the degenerate case of a vertical convex hull edge.
  if (e->curve().data() != max_diag.leftmost()->curve().data())
    std::cout << " (" << points[e->curve().data()] << ')';

  e = max_diag.leftmost();
  while (e != max_diag.rightmost()) {
    std::cout << " (" << points[e->curve().data()] << ')';
    e = e->right()->right();
  }
  // Handle the degenerate case of a vertical convex hull edge.
  if (e->curve().data() != min_diag.leftmost()->curve().data())
    std::cout << " (" << points[e->curve().data()] << ')';

  std::cout << std::endl;
  return 0;
}
```

The example above is given for illustrative purposes, in order to demonstrate how to construct and traverse the lower and upper envelopes of unbounded curves. Computing the convex hull of a set of points can be performed much more efficiently by the *2D Convex Hulls and Extreme Points* package of CGAL.[1]

---

[1]Even though the procedure we present for convex hull computation has running time of $O(n \log n)$, which is asymptotically as good as the running time of other convex hull procedures, it has one practical drawback: Our procedure uses geometric constructions, as it computes intersections between the dual lines. Other convex hull algorithms only require exact evaluation of geometric predicates. They can use geometric kernels that do not

10.2   Application: Nearest Jeep over Time

The nearest jeep over time problem: *We are given a set of jeeps* $\mathcal{J} = \{J_1, J_2, \ldots, J_n\}$ *that move in the desert on a planar terrain with constant speed. Each jeep is therefore given by* $J_k = \langle p_k, \vec{v}_k \rangle$, *where* $p_k = (x_k, y_k) \in \mathbb{R}^2$ *is the initial location of the jeep at time* $t = 0$, *and* $\vec{v}_k$ *is its velocity vector. Determine, at each time* $t \geq 0$, *the nearest jeep to some base station located at* $b = (x_0, y_0)$.

At first glance, you may wonder how the problem stated above is related to lower envelopes. As we are interested in the nearest jeep at each moment in time, namely the jeep whose distance from the base station b is minimal, the following idea comes to mind: If we express the distance of each jeep J_k from b as a function of time $D_k(t)$, the minimization diagram of $D_1(t), \ldots, D_n(t)$ will give us exactly what we need.

Let us assume that the velocity vectors are given as $\vec{v}_k = (v_{k,x}, v_{k,y})$, namely by the velocity in the x-direction and the velocity in the y-direction; see the illustration on the right. The location of J_k at time $t \geq 0$ is therefore given by

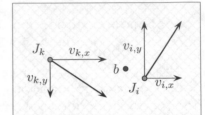

$$p_k + \vec{v}_k t = (x_k + v_{k,x}t, \; y_k + v_{k,y}t) \,.$$

Instead of considering the distance of the jeep to the base station, we prefer to express its *squared distance*. This way we avoid square root operations, while the nature of the minimization problem remains unchanged. Let us therefore define

$$
\begin{aligned}
D_k(t) &= \; \| p_k + \vec{v}_k t - b \|^2 = \\
&= \; (x_k + v_{k,x}t - x_0)^2 + (y_k + v_{k,y}t - y_0)^2 \\
&= \; (v_{k,x}^2 + v_{k,y}^2) \cdot t^2 + 2\left[(x_k - x_0)v_{k,x} + (y_k - y_0)v_{k,y}\right] \cdot t + \left[(x_k - x_0)^2 + (y_k - y_0)^2\right]
\end{aligned}
$$

The function above is a parabola $D_k(t) = \alpha_k t^2 + \beta_k t + \gamma_k$, where

$$
\begin{aligned}
\alpha_k &= \; v_{k,x}^2 + v_{k,y}^2 \,, \\
\beta_k &= \; 2\left[(x_k - x_0)v_{k,x} + (y_k - y_0)v_{k,y}\right] \,, \\
\gamma_k &= \; (x_k - x_0)^2 + (y_k - y_0)^2 \,.
\end{aligned}
$$

We can therefore use the traits class for arcs of rational functions (see Section 5.4.3) and construct the parabolic arcs of $D_1(t), \ldots, D_n(t)$ defined over $[0, \infty)$. (We prefer using the traits class for arcs of rational functions rather then the traits class for conic arcs as the former supports unbounded curves.) By computing the lower envelope of these parabolic arcs and obtaining their minimization diagram, we can identify the time intervals where the identity of the nearest jeep remains the same through the edges of the diagram, and the events where two (or more) jeeps become equally distant from the base station through the vertices of the diagram.

The program below solves the **nearest jeep over time problem**. It reads the location of the base station $b = (x_0, y_0)$ from the command line. The number of jeeps, their initial locations, and their velocities are read from an input file, which is also specified on the command line. The function uses the instantiation of the traits class for arcs of rational functions, as given in the header file `arr_rat_functions.h` and, attaches an integer to each parabolic arc it constructs, which represents the jeep index. It is then straightforward to construct the minimization diagram and determine which jeep is closest to the base station at each time interval.

```
// File: nearest_jeep.cpp

#include <list>
#include <vector>
```

---

construct geometric objects in an exact manner, thus saving time in practice; see Section 1.4.4.

```
#include <iostream>
#include <boost/lexical_cast.hpp>
#include <boost/tuple/tuple.hpp>
#include <boost/tuple/tuple_io.hpp>

#include <CGAL/basic.h>
#include <CGAL/Arr_curve_data_traits_2.h>
#include <CGAL/Envelope_diagram_1.h>
#include <CGAL/envelope_2.h>

#include "arr_rat_functions.h"
#include "read_objects.h"

typedef CGAL::Arr_curve_data_traits_2<Traits, unsigned int> Data_traits;
typedef Data_traits::Curve_2                                 Indexed_curve;
typedef CGAL::Envelope_diagram_1<Data_traits>                Diagram;
typedef boost::tuple<int, int, int, int>                     Jeep;

int main(int argc, char* argv[])
{
  if (argc < 4) {
    std::cerr << "Usage:_" << argv[0] << "_<input_file>_<x_0>_<y_0>"
              << std::endl;
    return -1;
  }


  // Define a traits class object and a constructor for rational functions.
  Traits traits;
  Traits::Construct_curve_2 construct = traits.construct_curve_2_object();

  // Analyze the command line and obtain the name of the input file
  // and the location of the base station b = (x_0, y_0).
  const char*    filename = argv[1];
  const Traits::Rational  x_0(boost::lexical_cast<int>(argv[2]));
  const Traits::Rational  y_0(boost::lexical_cast<int>(argv[3]));

  // Read the jeeps from the file, and construct the parabolic arcs
  // representing the functions of their squared distance from the base b.
  std::vector<Jeep> jeeps;
  read_objects<Jeep>(filename, std::back_inserter(jeeps));
  std::list<Indexed_curve>        arcs;
  std::vector<Jeep>::const_iterator it;
  unsigned int k = 0;
  for (it = jeeps.begin(); it != jeeps.end(); ++it) {
    Traits::Rational p_x = Traits::Rational(boost::get<0>(*it));
    Traits::Rational p_y = Traits::Rational(boost::get<1>(*it));
    Traits::Rational v_x = Traits::Rational(boost::get<2>(*it));
    Traits::Rational v_y = Traits::Rational(boost::get<3>(*it));

    // Construct the parabolic arc whose supporting conic is
    // alpha*x^2 + beta*x - y + gamma = 0.
    // Note that we also associate the parabolic arc with its jeep index.
    std::vector<Traits::Rational> coefficients(3);
    coefficients[2] = CGAL::square(v_x) + CGAL::square(v_y);
```

```
  coefficients [1] = 2*((p_x − x_0)*v_x + (p_y − y_0)*v_y);
  coefficients [0] = CGAL::square (p_x − x_0) + CGAL::square (p_y − y_0);

  // Construct an arc over the interval [0, infinity).
  Traits::Curve_2 arc = construct(coefficients.begin(), coefficients.end(),
                                  Alg_real(0), true);
  arcs.push_back(Indexed_curve(arc, k++));
}


// Compute the minimization diagram that represents the lower envelope
// of the parabolic arcs.
Diagram   min_diag;
CGAL::lower_envelope_2(arcs.begin(), arcs.end(), min_diag);

// Every left bounded edge in the diagram represents a time interval over
// which the identity of the nearest jeep does not change. Print these
// time intervals and the index of the nearest jeep for each interval.
Diagram::Edge_const_handle     e = min_diag.leftmost()−>right()−>right ();
while (e != min_diag.rightmost()) {
  CGAL_assertion(! e−>is_empty ());
  std::cout << "From time " << CGAL::to_double(e−>left()−>point ().x())
            << " to time " << CGAL::to_double(e−>right()−>point ().x())
            << " the nearest jeep no. is " << e−>curve ().data () << '.'
            << std::endl;
  e = e−>right()−>right ();
}
std::cout << "From time " << CGAL::to_double(e−>left()−>point ().x())
          << " to infinity the nearest jeep no. is " << e−>curve ().data()
          << '.' << std::endl;
return 0;
}
```

10.3   Envelopes of Surfaces in 3-Space

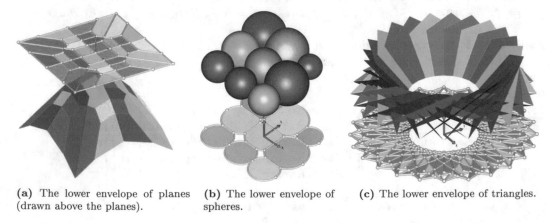

(a) The lower envelope of planes (drawn above the planes).   (b) The lower envelope of spheres.   (c) The lower envelope of triangles.

Fig. 10.3: Lower envelopes of surfaces in 3-space.

A surface S in $\mathbb{R}^3$ is called *xy-monotone* if every line parallel to the z-axis intersects it at most once. A horizontal triangle in space is an example of an xy-monotone surface.[2] On the other hand, the sphere $x^2 + y^2 + z^2 = 1$ is *not* xy-monotone as the z-axis intersects it at $(0, 0, -1)$ and at $(0, 0, 1)$. The xy-plane subdivides it into an upper hemisphere and a lower hemisphere, which are xy-monotone. A continuous xy-monotone surface can therefore be viewed as the graph of a bivariate function $z = S(x, y)$, defined over some continuous range $R \subseteq \mathbb{R}^2$. Given a set $\mathcal{S} = \{S_1, S_2, \ldots, S_n\}$ of xy-monotone surfaces, their *lower envelope* is defined as the graph of the point-wise minimum of all the corresponding functions. Namely, the lower envelope of the set $\mathcal{S}$ is defined as the graph of the following function:

$$\mathcal{L}_\mathcal{S}(x, y) = \min_{1 \leq k \leq n} \overline{S}_k(x, y) \ ,$$

where we define $\overline{S}_k(x, y) = S_k(x, y)$ for (x, y) in the range R_k of definition of S_k, and $\overline{S}_k(x, y) = \infty$ otherwise.

Similarly, the *upper envelope* of $\mathcal{S}$ is the graph of the point-wise maximum of the functions:

$$\mathcal{U}_\mathcal{S}(x, y) = \max_{1 \leq k \leq n} \underline{S}_k(x, y) \ ,$$

where in this case $\underline{S}_k(x, y) = -\infty$ for $(x, y) \notin R_k$.

We consider *vertical* surfaces, namely patches of planes that are perpendicular to the xy-plane, as *weakly xy-monotone*, to handle degenerate inputs properly. Weakly xy-monotone surfaces are given special treatment.

Given a set of xy-monotone surfaces $\mathcal{S}$, the *minimization diagram* of $\mathcal{S}$ is a subdivision of the xy-plane into cells such that the identities of the surfaces that induce the lower envelope over a specific cell of the subdivision (be it a face, an edge, or a vertex) are the same. In non-degenerate situations a face is induced by a single surface (or by no surfaces at all if there are no xy-monotone surfaces defined over it); an edge is induced by a single surface and corresponds to the projection of a maximal continuous portion of its boundary that does not meet any other surface, or by two surfaces and corresponds to the projection of a maximal continuous portion of their intersection curve that does not meet any other surface; and a vertex is induced by a single surface and corresponds to the projection of a boundary point, or by two surfaces and corresponds to the projection of an intersection point of one surface boundary with the other surface, or by three surfaces and corresponds to the projection of an intersection point. The *maximization diagram* is symmetrically defined for upper envelopes. Recall that we refer to both these diagrams as *envelope diagrams*.

As an envelope diagram is no more than a planar subdivision, it is possible to represent it as an arrangement object. An instance of the class template `Envelope_diagram_2<Traits>` represents an envelope-diagram object. It derives from an instance of the class template `Arrangement_2<Traits,Dcel>`, where the `Dcel` template parameter is substituted with an extended DCEL data structure (see Section 6.2), such that each face, halfedge, and vertex is extended to store its *originators*, that is, the xy-monotone surfaces that induce this feature. Specifically, each record is extended with a container of surfaces, the value type of which is `Traits::Xy_monotone_surface_3`. Let v, e, and f be handles to a vertex, an edge, and a face of an envelope diagram, respectively. The calls `v->surfaces_begin()` and `v->surfaces_end()` return iterators of the nested `Envelope_diagram_2::Surface_const_iterator` type that define the valid range of surfaces that attain the envelope over v. Similarly, `e->surfaces_begin()` and `e->surfaces_end()` and `f->surfaces_begin()` and `f->surfaces_end()` return iterators that define the valid range of surfaces that attain the envelope over e and f, respectively. Each of the calls `v->surface()`, `e->surface()`, and `f->surface()` returns a representative surface from the corresponding range.

Lower and upper envelopes of surfaces in three dimensions can be efficiently computed using a generalization of the divide-and-conquer approach taken in the two-dimensional case. First, note that computing the envelope diagram of a single xy-monotone surface S_k is trivial; we project its boundary onto the xy-plane, and label the faces it induces accordingly. For example, if S_k is a

---

[2] A triangle in 3-space is a *surface patch* but we use the word *surface* for short.

lower hemisphere, its projected boundary is a circle that subdivides the plane into two faces; the face lying inside the circle is labeled $\{S_k\}$, and the outer unbounded face, is labeled the empty set. Note that S_k may have no boundary at all, for example, if it is a non-vertical plane. In this case we construct an arrangement that contains just a single unbounded face and label it $\{S_k\}$.

Given a set $\hat{\mathcal{S}}$ of (not necessarily xy-monotone) surfaces in $\mathbb{R}^3$, we subdivide each surface into a finite number of weakly xy-monotone surfaces, thus obtaining the set $\mathcal{S}$. Then, we split the set $\mathcal{S}$ into two disjoint subsets $\mathcal{S}_1$ and $\mathcal{S}_2$ of roughly the same size, and compute their envelope diagrams $\mathcal{M}_1$ and $\mathcal{M}_2$ recursively. Finally, we merge the diagrams in three steps similar to the steps used to merge one-dimensional envelope diagrams; see Section 10.1.2. We overlay the arrangements that represent the envelope diagrams $\mathcal{M}_1$ and $\mathcal{M}_2$, and then refine the resulting arrangement and clean up the refined arrangement, removing redundant edges and vertices to obtain the final arrangement that represents the envelope diagram of $\mathcal{S}$. While in principle the process is fairly straightforward, its implementation is rather intricate, as special attention has been devoted to its performance. For example, the number of algebraic operations was reduced as much as possible. These operation are especially costly when non-linear algebraic surfaces are involved, and the software is perfectly capable of handling a variety of such surfaces. Wherever possible, combinatorial knowledge for a diagram feature is propagated to its neighbors instead of performing algebraic operations; see Section 10.5 for more details and references.

10.3.1   The Envelope-Traits Concept

The implementation of the envelope-computation algorithm is generic and can handle arbitrary surfaces. It is parameterized with a traits class, which defines the geometry of the surfaces it handles, and supports all the necessary functionality on these surfaces, and on their projections onto the xy-plane. The traits class must model the concept EnvelopeTraits_3 described below.

The EnvelopeTraits_3 concept refines the concept ArrangementXMonotoneTraits_2. Namely, a model of this concept must define the planar types `Point_2` and `X_monotone_curve_2` and support basic operations on them, as listed in Section 5.1. These types and operations enable the representation of envelope diagrams as arrangements, and the

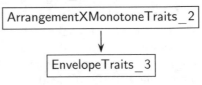

employment of arrangement overlay-computations by the envelope-construction algorithm. In addition, the EnvelopeTraits_3 concept must define the spatial types `Surface_3` and `Xy_monotone_surface_3`. Any model of the envelope-traits concept must also support the following operations on these spatial types:

Make_xy_monotone_3: Subdivide a given surface into continuous xy-monotone surfaces. It is possible at this stage to disregard xy-monotone surfaces that do not contribute to the surface envelope. (For example, if we are given a sphere, it is possible to return just its lower hemisphere if we are interested in the lower envelope; the upper hemisphere is obviously redundant in such a case.)

Construct_projected_boundary_2: Given an xy-monotone surface S, construct all the planar x-monotone curves that form the boundary of the vertical projection of the boundary of S onto the xy-plane. This operation is used at the bottom of the recursion to build the minimization diagram of a single xy-monotone surface.

Construct_projected_intersections_2: Construct all geometric entities that comprise the projection (onto the xy-plane) of the intersection between two xy-monotone surfaces S_1 and S_2. These entities may be:

- A planar curve that is the projection of a 3D intersection curve of S_1 and S_2 (for example, the intersection curve between two spheres is a 3D circle, which becomes an ellipse when projected onto the xy-plane).

  In many cases it is also possible to indicate whether the two surfaces transversely intersect and change their relative z-positions on either side of the intersection curve, or whether they maintain their relative z-positions. Providing this information is optional. When provided, it is used by the algorithm to determine the relative order of S_1 and S_2 on one side of their intersection curve when their order on the other side of that curve is known, thus improving the performance of the algorithm.

- A point, induced by the projection of a tangency point of S_1 and S_2, *or* by the projection of a vertical intersection curve onto the xy-plane.

The set of entities is empty if S_1 and S_2 do not intersect.

Compare_z_at_xy_3: Given two xy-monotone surfaces S_1 and S_2, and a planar point $p = (x_0, y_0)$ that lies in their common xy-definition range, determine the z-order of S_1 and S_2 over p; namely, compare $S_1(x_0, y_0)$ and $S_2(x_0, y_0)$.

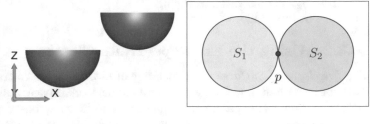

This operation is used only in degenerate situations, in order to determine the surface inducing the envelope over a vertex. The figure above illustrates a situation where this operation is applied. The hemispheres S_1 and S_2 have only a single two-dimensional point (p) in their common xy-definition range, while S_2 is above S_1 at p.

Compare_z_at_xy_above_3: Given two xy-monotone surfaces S_1 and S_2, and a planar x-monotone curve c that is the projected intersection of S_1 and S_2 or part of it, determine the z-order of S_1 and S_2 immediately above (respectively below) the curve c. Note that c is a planar x-monotone curve, and we refer to the region above (or below) it *in the xy-plane*. The region above a curve is the region to the left of the curve when going from the lexicographically smaller end of the curve towards its lexicographically larger end. In the figure to the right the z-coordinate of S_1 is larger than the z-coordinate of S_2 immediately above pq and smaller than it immediately below pq.

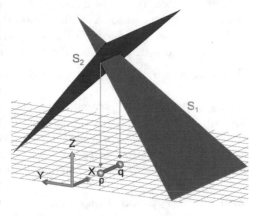

This operation is used by the algorithm to determine which surface induces the envelope over a face incident to c.

Compare_z_at_xy_below_3: Given two xy-monotone surfaces S_1 and S_2, and a planar x-monotone curve c that fully lies in their common xy-definition range, and such that S_1 and S_2 do not intersect over the interior of c, determine the relative z-order of s_1 and s_2 over the interior of c. Namely, compare $S_1(x', y')$ and $S_2(x', y')$ for some point (x', y') in c.

This operation is used to determine which surfaces induce the envelope over an edge associated with the x-monotone curve c, or a face incident to c, in situations where the previous predicate cannot be used, since c is *not* an intersection curve of S_1 and S_2, as illustrated in the figure to the right.

 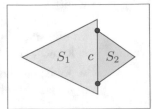

10.3.2 Using the Envelope-Traits Classes

The *Envelopes of Surfaces in 3D* package of CGAL contains a traits class `Env_triangle_traits_3<Kernel>` for handling triangles in three-dimensional space, and another traits class `Env_sphere_traits_3<ConicTraits>` for 3D spheres. The latter is based on geometric operations on conic curves (ellipses). In addition, the package includes a traits-class decorator that enables users to attach external (non-geometric) data to surfaces. In this section we demonstrate how to use the two traits classes that handle *bounded* surfaces, and show how to attach auxiliary data to these surfaces using the decorator.

The *Envelopes of Surfaces in 3D* package also includes a traits class named `Env_plane_traits_3<Kernel>` that handles planes in $\mathbb{R}^3$. Section 10.4 introduces an application that utilizes the upper envelope of a set of planes.

The function template `print_diagram()` listed next, and defined in the header file `print_diagram.h`, prints either a minimization diagram or a maximization diagram. It is used by some of the examples introduced in this chapter. It shows how to traverse the envelope diagram using the arrangement traversal-methods; see Section 2.2.1.

```cpp
#include <boost/iterator/transform_iterator.hpp>

using boost::make_transform_iterator;

// Obtain the data attached to a surface by a traits decorator.
template <typename Surface>
const char& data(const Surface& surface) { return surface.data(); }

template <typename Diagram>
void print_diagram(const Diagram& diag)
{
  typedef typename Diagram::Xy_monotone_surface_3        Surf;

  // Go over all arrangement faces.
  typename Diagram::Face_const_iterator fit;
  for (fit = diag.faces_begin(); fit != diag.faces_end(); ++fit) {
    // Print the face boundary.
    if (fit->is_unbounded()) std::cout << "[Unbounded_face]";
    else {
      // Print the vertices along the outer boundary of the face.
      std::cout << "[Face]__";
      typename Diagram::Ccb_halfedge_const_circulator ccb = fit->outer_ccb();
      do std::cout << '(' << ccb->target()->point() << ")__";
      while (++ccb != fit->outer_ccb());
    }

    // Print the labels of the surfaces that induce the envelope on this face.
```

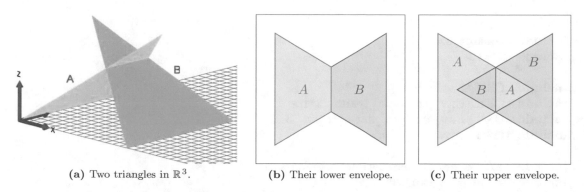

(a) Two triangles in $\mathbb{R}^3$.   (b) Their lower envelope.   (c) Their upper envelope.

Fig. 10.4: The lower and upper envelopes of two triangles.

```
    std::cout << "—>␣␣" << fit–>number_of_surfaces() << "␣surface(s):␣";
    std::copy(make_transform_iterator(fit–>surfaces_begin(), &data<Surf>),
              make_transform_iterator(fit–>surfaces_end(), &data<Surf>),
              std::ostream_iterator<char>(std::cout, ",␣"));
    std::cout << std::endl;
}

// Go over all arrangement edges.
typename Diagram::Edge_const_iterator eit;
for (eit = diag.edges_begin(); eit != diag.edges_end(); ++eit) {
    // Print the labels of the surfaces that induce the envelope on this edge.
    std::cout << "[Edge]␣␣(" << eit–>source()–>point() << ")␣␣("
              << eit–>target()–>point() << ")␣␣—>␣␣"
              << eit–>number_of_surfaces() << "␣surface(s):␣";
    std::copy(make_transform_iterator(eit–>surfaces_begin(), &data<Surf>),
              make_transform_iterator(eit–>surfaces_end(), &data<Surf>),
              std::ostream_iterator<char>(std::cout, ",␣"));
    std::cout << std::endl;
}

// Go over all arrangement vertices.
typename Diagram::Vertex_const_iterator vit;
for (vit = diag.vertices_begin(); vit != diag.vertices_end(); ++vit) {
    //Print the labels of the surfaces that induce the envelope on the vertex.
    std::cout << "[Vertex]␣␣(" << vit–>point() << ")␣␣—>␣␣"
              << vit–>number_of_surfaces() << "␣surface(s):␣";
    std::copy(make_transform_iterator(vit–>surfaces_begin(), &data<Surf>),
              make_transform_iterator(vit–>surfaces_end(), &data<Surf>),
              std::ostream_iterator<char>(std::cout, ",␣"));
    std::cout << std::endl;
}
}
```

Example: The example program listed below shows how to use the envelope-traits class for 3D triangles. It constructs the lower and upper envelopes of the two triangles, depicted in Figure 10.4a, and inserts the triangle labels that induce each cell in the output diagrams into the standard output-stream. For convenience, we use the traits-class decorator `Env_surface_data_traits_3` to label the triangles. Printing the triangles of the diagrams translates to printing their labels.

```
// File: ex_envelope_triangles.cpp
```

```cpp
#include <iostream>
#include <list>

#include <CGAL/basic.h>
#include <CGAL/Env_triangle_traits_3.h>
#include <CGAL/Env_surface_data_traits_3.h>
#include <CGAL/envelope_3.h>

#include "arr_exact_construction_segments.h"
#include "print_diagram.h"

typedef CGAL::Env_triangle_traits_3<Kernel>              Traits_3;
typedef Kernel::Point_3                                  Point_3;
typedef Traits_3::Surface_3                              Triangle_3;
typedef CGAL::Env_surface_data_traits_3<Traits_3, char> Data_traits;
typedef Data_traits::Surface_3                           Data_triangle;
typedef CGAL::Envelope_diagram_2<Data_traits>            Envelope_diagram;

int main()
{
  // Construct the input triangles, marked A and B.
  std::list<Data_triangle>    triangles;
  Triangle_3 ta(Point_3(0, 0, 0), Point_3(0, 6, 0), Point_3(5, 3, 4));
  Triangle_3 tb(Point_3(6, 0, 0), Point_3(6, 6, 0), Point_3(1, 3, 4));
  triangles.push_back(Data_triangle(ta, 'A'));
  triangles.push_back(Data_triangle(tb, 'B'));

  // Compute and print the minimization diagram.
  Envelope_diagram  min_diag;
  CGAL::lower_envelope_3(triangles.begin(), triangles.end(), min_diag);
  std::cout << std::endl << "The minimization diagram:" << std::endl;
  print_diagram(min_diag);

  // Compute and print the maximization diagram.
  Envelope_diagram  max_diag;
  CGAL::upper_envelope_3(triangles.begin(), triangles.end(), max_diag);
  std::cout << std::endl << "The maximization diagram:" << std::endl;
  print_diagram(max_diag);

  return 0;
}
```

Example: The next program demonstrates how to instantiate and use the envelope-traits class for spheres, based on the **Arr_conic_traits_2** class template (as defined in the header file **arr_conics.h**), which handles the projected intersection curves. The program reads a set of spheres from an input file and constructs their lower envelope.

```cpp
// File: ex_envelope_spheres.cpp

#include <iostream>
#include <list>
#include <boost/timer.hpp>
```

```
#include <CGAL/basic.h>
#include <CGAL/Env_sphere_traits_3.h>
#include <CGAL/envelope_3.h>

#include "arr_print.h"
#include "arr_conics.h"
#include "read_objects.h"

typedef CGAL::Env_sphere_traits_3<Traits>          Traits_3;
typedef Traits_3::Surface_3                        Sphere;
typedef CGAL::Envelope_diagram_2<Traits_3>         Envelope_diagram;

int main(int argc, char* argv[])
{
  const char* filename = (argc > 1) ? argv[1] : "spheres.dat";
  std::list<Sphere>  spheres;
  read_objects<Sphere>(filename, std::back_inserter(spheres));
  Envelope_diagram    min_diag;
  std::cout << "Constructing_the_lower_envelope_of_" << spheres.size()
            << "_spheres." << std::endl;

  // Compute the envelope.
  boost::timer timer;
  CGAL::lower_envelope_3(spheres.begin(), spheres.end(), min_diag);
  double secs = timer.elapsed();

  // Print the minimization diagram.
  print_arrangement_size(min_diag);
  std::cout << "Construction_took_" << secs << "_seconds." << std::endl;
  return 0;
}
```

10.4 Application: Locating the Farthest Point

The farthest point problem: *Given a set of points $P = \{p_1, \ldots, p_n\}$ in the plane, devise a data structure that can efficiently answer queries of the following form: Locate the farthest point (or points in case of a tie) in P from a given query point q.*

For each point $p_i = (x_i, y_i) \in P$, we can define the distance function of any point $(x, y) \in \mathbb{R}^2$ from p_i. As we wish to avoid unnecessary square root operations, we consider the squared distance. Hence, our function is

$$z = (x - x_i)^2 + (y - y_i)^2 .$$

This function defines a paraboloid S_i, namely, a surface of degree 2 in $\mathbb{R}^3$. It turns out that the maximization diagram of $S_1, \ldots, S_n$ is exactly the data structure we need, as the paraboloid that induces the upper envelope at a query point q identifies the farthest point from q. (In the case of a tie several paraboloids induce the upper envelope.) Recall, that the maximization diagram is represented as an arrangement. Once the maximization diagram is constructed, all we have to do is locate the point q in the diagram, an operation conveniently supported for arrangements (Section 3.1.1).

The following observation simplifies the construction of the maximization diagram. It enables the use of linear objects rather than degree 2 surfaces, considerably saving time while constructing the diagram. Note that the equation of each of the paraboloids contains the term $x^2 + y^2$. As we are interested in the relative z-positions of the paraboloids, and *not* in their absolute positions, it is possible to omit these quadratic terms and obtain the equation of a plane:

$$T_i : \ 2x_i \cdot x + 2y_i \cdot y + z - (x_i^2 + y_i^2) = 0 \ .$$

Computing the upper envelope of $T_1, \ldots, T_n$ results in a maximization diagram that can be used to efficiently answer farthest-point queries.

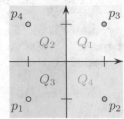

The program below applies the technique described above to answer farthest-point queries. It first reads the set of points P from an input file, constructs the corresponding planes $T_1, \ldots, T_n$, and computes their upper envelope. In the figure to the right the farthest Voronoi-cell of a point p_i, which is contained in the quadrant $Q_{(i+2)\%4}$, is the opposite quadrant Q_i. Notice the usage of the surface-data traits to associate each plane T_i with the corresponding point p_i. The program then reads a sequence of query points from a second input file, and answers each query by issuing a point-location query on the constructed maximization diagram.

```cpp
// File: farthest_point.cpp

#include <fstream>
#include <list>

#include <CGAL/basic.h>
#include <CGAL/Exact_predicates_exact_constructions_kernel.h>
#include <CGAL/Env_plane_traits_3.h>
#include <CGAL/Env_surface_data_traits_3.h>
#include <CGAL/envelope_3.h>
#include <CGAL/Arr_naive_point_location.h>

#include "read_objects.h"
#include "arr_print.h"

typedef CGAL::Exact_predicates_exact_constructions_kernel  Kernel;
typedef Kernel::FT                                         Number_type;
typedef Kernel::Point_2                                    Point;
typedef Kernel::Plane_3                                    Plane;
typedef CGAL::Env_plane_traits_3<Kernel>                   Traits;
typedef CGAL::Env_surface_data_traits_3<Traits, Point>     Data_traits;
typedef Data_traits::Surface_3                             Data_plane;
typedef CGAL::Envelope_diagram_2<Data_traits>              Envelope_diagram;
typedef CGAL::Arr_naive_point_location<Envelope_diagram>   Naive_pl;

int main(int argc, char* argv[])
{
  const char* filename1 = (argc > 1) ? argv[1] : "fp_points.dat";
  const char* filename2 = (argc > 2) ? argv[2] : "fp_queries.dat";
  std::list<Point> points;
  read_objects<Point>(filename1, std::back_inserter(points));

  std::list<Data_plane> planes;
  std::list<Point>::const_iterator it;
  for (it = points.begin(); it != points.end(); ++it) {
    // The surface induced by the point (p_x, p_y) is the plane:
    //   2*p_x*x + 2*p_y*y + z - (p_x^2 + p_y^2) = 0.
    Number_type px = it->x(), py = it->y();
```

```
    Plane plane(2*px, 2*py, 1, -(px*px + py*py));
    planes.push_back(Data_plane(plane, *it));
  }

  // Compute the maximization diagram of the planes.
  Envelope_diagram      max_diag;
  CGAL::upper_envelope_3(planes.begin(), planes.end(), max_diag);

  // Perform the queries.
  Naive_pl pl(max_diag);
  std::list<Point> queries;
  read_objects<Point>(filename2, std::back_inserter(queries));
  for (it = queries.begin(); it != queries.end(); ++it) {
    CGAL::Object obj = pl.locate(*it);

    // Print the query result according to the location of the query point.
    Envelope_diagram::Surface_const_iterator sit, sit_end;
    Envelope_diagram::Halfedge_const_handle  he;
    Envelope_diagram::Face_const_handle       f;
    Envelope_diagram::Vertex_const_handle     v;
    if (CGAL::assign(f, obj)) {
      sit = f->surfaces_begin();
      sit_end = f->surfaces_end();
    }
    else if (CGAL::assign(he, obj)) {
      sit = he->surfaces_begin();
      sit_end = he->surfaces_end();
    }
    else if (CGAL::assign(v, obj)) {
      sit = v->surfaces_begin();
      sit_end = v->surfaces_end();
    }
    else CGAL_error();

    std::cout << "The farthest point(s) from (" << *it << "):";
    while (sit != sit_end) std::cout << " (" << sit++->data() << ")";
    std::cout << std::endl;
  }
  return 0;
}
```

Note that it is possible to compute the lower envelope of the planes $T_1, \ldots, T_n$ and to use the resulting minimization diagram to locate the point nearest to each query point. This minimization diagram is the well-known *Voronoi diagram* of the input points under the Euclidean metric. This specific task can be accomplished more effectively by constructing the Voronoi diagram of the input points with alternative algorithms. CGAL provides a package that can efficiently construct and query nearest-neighbor Voronoi diagrams. However, Voronoi diagrams are defined more generally, and the *Envelopes of Surfaces in 3D* package can be used to compute fairly general types of Voronoi diagrams in the plane.

10.5   Bibliographic Notes and Remarks

Envelopes of curves and surfaces have been intensively studied in computational geometry. The underlying theory is covered in depth in the book by Sharir and Agarwal [196]. You can compute

envelopes of curves using operations provided by the *Envelopes of Curves in 2D* package of
CGAL [211]. Operations that compute envelopes of surfaces are provided by the *Envelopes of
Surfaces in 3D* package of CGAL [156].

The complexity of envelopes of well-behaved curves (see Section 1.3.3) in the plane is near-
linear in the number of curves. Envelopes of curves are combinatorially equivalent to a special
family of words called *Davenport-Schinzel* sequences. This connection has led to a variety of
intriguing results on the combinatorics of envelopes, like the proof that the maximum combinatorial
complexity of the envelope of a set of n line segments in the plane is $\Theta(n\alpha(n))$.[3] For more details,
see [196].

No similar equivalence is known in three-dimensional space. However, tight or near-tight
bounds are known for envelopes of well-behaved surfaces. In general, the complexity of the envelopes of n well-behaved surfaces in three-dimensional space is near-quadratic in the number of
surfaces [104, 193]. In some special cases the complexity is near-linear [188], or even sharply linear,
as in the case of planes.

A straightforward divide-and-conquer approach, as described in Section 10.1, yields an algorithm for constructing the envelope of curves in the plane with running time that has a multi-
plicative log factor over the worst-case complexity of the envelope. Hershberger [112] improved
this algorithm slightly, such that for segments, for example, it runs in worst-case optimal time
$O(n \log n)$. (Without Hershberger's improvement, the standard divide-and-conquer algorithm runs
in time $O(n\alpha(n) \log n)$.)

For envelopes in three-dimensional space the divide-and-conquer approach remains fairly simple, and it runs in time that is close to the worst-case combinatorial complexity of the envelope.
Namely, in general it runs in time that is near-quadratic in the number of surfaces. Its efficiency,
however, is not at all obvious and stems from a combinatorial result due to Agarwal et al. [3] on
the complexity of the overlay of minimization diagrams.

Alternatively, one can construct the envelopes incrementally; see, e.g., [30]. In three dimensions
the algorithm is conceptually much more complex than the divide-and-conquer algorithm.

Both two-dimensional and three-dimensional envelopes are computed in CGAL using divide-
and-conquer. In the two-dimensional case the implementation is fairly straightforward and the
conquer step amounts to a careful merge of the lists of vertices of two minimization diagrams,
with the detecting and adding of extra vertices during the merge step. Implementation-wise,
the three-dimensional case is tremendously more complex. The code, implemented mostly by
Meyerovitch [154, 155], is heavily based on the two-dimensional arrangement software and adds
to it close to 15, 000 lines of code. The conquer step is very intricate when preparing for arbitrary surfaces and handling all degeneracies. Also, major effort has been invested in propagating
combinatorial information throughout the process in order to save in costly algebraic operations;
see [154, 155] for more details.

Envelopes of curves in two-dimensional space and envelopes of surfaces in three-dimensional
space can be used to solve a variety of problems. For example, they can be used as building blocks of
collision-free motion paths planners for multi-axis numerically-controlled machines (NC-machine);
see [210, 215] for the exploitation of envelopes of curves in a hybrid approach to plan collision-free
motion paths in 5-axis NC-machining. Envelopes of surfaces can be used to compute fairly general
Voronoi diagrams in the plane (in terms of sites and distance metrics) [29, Chapter 2], [54], [191].
Also, they can be used to compute the so-called sandwich region (see Exercises 10.1 and 10.3),
which in turn has its own line of applications; see, e.g., [123].

10.6   Exercises

10.1 We are given a set H of halfplanes in the plane, and we wish to compute their intersection
$\bigcap H = \bigcap_{h_i \in H} h_i$. We can compute $\bigcap H$ with the help of the two-dimensional envelope code.
Partition H into three subsets as follows: H_1 is the set of halfplanes bounded from above,
H_2 is the set of halfplanes bounded from below, and H_3 is the set of halfplanes bounded by

---

[3]Recall that $\alpha()$ is the inverse of Ackermann's function.

a vertical line. Let L_i denote the lines bounding the halfplanes in H_i, for example, if the halfplane $a_i x + b_i y \leq c_i$ is in H_j, then the line $a_i x + b_i y = c_i$ is in L_j. The intersection $\bigcap H$ is obtained by computing the *sandwich region* (see the Bibliographic Notes and Remarks section above) between the lower envelope of the lines in L_1 and the upper envelope of the lines in L_2. This sandwich region needs to be further intersected with the vertical slab defined by the halfplanes in H_3.

10.2 An alternative way to compute the intersection of a given set of halfplanes in the plane is to first obtain the lower and upper envelopes computed in the previous exercise using duality and a convex hull procedure, in the spirit of the algorithm in Section 10.1.4 for computing the convex hull of a set of points. (Notice that although the convex hull always exists and may have many vertices on its boundary, the corresponding intersection of halfplanes may be empty.) Choose a convex hull procedure in CGAL and compare the running time of this option versus the procedure you developed in Exercise 10.1 above on large inputs. Explain the comparison results.

10.3 Given two sets of conic curves C_1 and C_2 compute the sandwich region between the lower envelope of C_1 and the upper envelope of C_2. The major difference between this sandwich region and the sandwich region for lines is that the former may comprise several connected components.

10.4 Write an efficient program that accepts a polyhedron and a set of points in the plane. For each input point p it computes the point in the lower envelope of the polyhedron vertically above p. **Hint:** Triangulate the facets of the polyhedron.

10.5 The power distance is measured from a point $x \in \mathbb{R}^2$ to a disc D with center c and positive radius r, and is defined to be $\rho(x, D) = (x - c)^2 - r^2$. The power diagram of a set of discs is the nearest-site Voronoi diagram induced by the power distance.

Write a program that computes the power diagram of a set of discs using lower envelopes of planes. Represent the power diagrams as minimization diagrams.

10.6 **[Project]** The *Envelopes of Curves in 2D* package supports only lower and upper envelope computations. Develop a function template with the prototype below that computes the approximate arbitrary envelope of segments. This exercise is analogous to Exercise 3.2, where you are asked to develop code that supports approximate arbitrary ray-shooting.

```
template <typename Input_iterator, typename Direction, typename Rational,
          typename Envelope_diagram_1>
void approximate_envelope_of_segment_2(Input_iterator begin,
                                       Input_iterator end,
                                       Direction_2& direction,
                                       Rational epsilon,
                                       Envelope_diagram_1& diagram)
```

Like the `lower_envelope_x_monotone_2()` and `upper_envelope_x_monotone_2()` function templates, the new function template accepts a range of input segments and a reference to an envelope diagram to store the result. It accepts, in addition, a direction d in $\mathbb{R}^2$ and a rational number ϵ. Denote the angle defined by the positive x-direction and d as α.

The function computes the envelope of the segments as seen when looking at the segments from a point far away lying on a vector in the d direction. First, it computes an angle β' that approximates $\beta = -\alpha + \pi/2$ such that $|\sin \beta' - \sin \beta| \leq \epsilon$ and $|\cos \beta' - \cos \beta| \leq \epsilon$. Then, it rotates the input segments by β' and computes the upper envelope of the transformed segments.

You may use the free function `rational_rotation_approximation()` provided by CGAL to obtain the desired $\sin \beta'$ and $\cos \beta'$. The implementation of this function is based on a method presented by Canny, Donald, and Ressler [33].

Chapter 11

Prospects

In this closing chapter, we briefly discuss ongoing development as well as future directions for possible extension of the *2D Arrangements* package and its relatives. As in most of the chapters, we refer to work that is described elsewhere; the bibliographic remarks are scattered throughout the entire chapter and are not compiled in a special section at the end.

11.1 Arrangements on Curved Surfaces

We are given a surface S in $\mathbb{R}^3$ and a set $\mathcal{C}$ of curves embedded on this surface. The curves subdivide S into cells of dimension 0 (*vertices*), 1 (*edges*), and 2 (*faces*). This subdivision is the *arrangement* $\mathcal{A}(\mathcal{C})$ induced by $\mathcal{C}$ on S. Arrangements embedded on curved surfaces in $\mathbb{R}^3$ are generalizations of arrangements embedded in the plane. While this is easily stated, the mechanism to support such generalizations is not trivially implemented. However, the reward for having generic software that supports generalized arrangements is tremendous. In this section we provide a glimpse into a practical framework for the construction, maintenance, and manipulation of arrangements embedded on two-dimensional orientable parametric surfaces such as planes, cylinders, spheres, and tori, and surfaces homeomorphic to them; see Figure 11.1. Such arrangements have many theoretical and practical applications; see, e.g., [4, 20, 22, 73, 98].

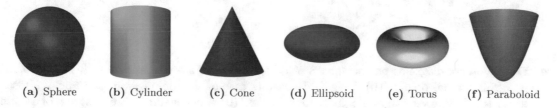

(a) Sphere   **(b)** Cylinder   **(c)** Cone   **(d)** Ellipsoid   **(e)** Torus   **(f)** Paraboloid

Fig. 11.1: Various two-dimensional parametric surfaces, arrangements on which are supported by the new framework.

The *2D Arrangements* package of the latest version of CGAL contains an implementation of the framework, albeit only partially documented.[1] Concretizations for several types of surfaces and several types of curves embedded on the surfaces have already been implemented. Some of them are also included in the *2D Arrangements* package; others are included in experimental packages, which are also expected to be included in future releases of CGAL. Examples are arrangements of arcs of great circles on the sphere [75, 76], of intersection curves between quadric surfaces and a fixed quadric [21], and of intersection curves between arbitrary algebraic surfaces and a fixed Dupin cyclide [25]. The torus is a Dupin cyclide. This is the first implementation of a generic algorithm

---

[1]The manual of the *2D Arrangements* package in CGAL's latest version does yet not include material for non-planar surfaces.

E. Fogel et al., *CGAL Arrangements and Their Applications*, Geometry and Computing 7,
DOI 10.1007/978-3-642-17283-0_11, © Springer-Verlag Berlin Heidelberg 2012

that can handle arrangements on a large class of parametric surfaces. Cazals and Loriot [34, 46] have developed a software package that computes exact arrangements of circles on a sphere. Their software is specialized for the spherical case though.

The starting point for the new framework (arrangements on surfaces) was the CGAL package for constructing and maintaining arrangements of bounded curves in the plane [213]. It could not handle unbounded curves directly. Rather, unbounded curves had to be clipped by the user in a preprocessing phase, so that no essential information about the arrangements (e.g., a finite intersection point) was lost. This solution was clearly inconvenient. The initial goal was to extend the package so that unbounded curves in the plane could be handled within the package, and hence the full functionality of the package would also be available for arrangements of such curves. Evidently, this goal was achieved; see Chapter 4. The solution carries further than that; it can also deal with arrangements on surfaces in $\mathbb{R}^3$.

The framework is implemented as a class template `Arrangement_on_surface_2` parameterized by template parameters *geometry traits* and *topology traits*, that is, `Arrangement_on_surface_2<GeoTraits, TopTraits>`. A concretization is obtained by instantiating the class template, where models of the geometry-traits and the topology-traits concepts substitute the two template parameters, respectively. The topology-traits class deals with the topology of the parametric surface. A parametric surface S is given by a continuous function $\phi_S : \Phi \to \mathbb{R}^3$, where the domain $\Phi = U \times V$ is a rectangular two-dimensional parameter space; $S = \phi_S(\Phi)$. U and V are open, half-open, or closed intervals with endpoints in $\overline{\mathbb{R}}$. The (unbounded) plane is modeled with ϕ_S being the identity mapping. The geometry-traits class introduces the C++ type names of the basic geometric objects (i.e., point, curve, and u-monotone curve) and a small set of operations on objects of these types, such as comparing two points in uv-lexicographic order and computing intersections of curves. The traits-concept hierarchy described in Chapter 5 was kept the same (except for the renaming of the x and y coordinates to u and v, respectively), and a few new refined concepts were introduced; see [22] for a complete description of the new full traits-concept hierarchy.

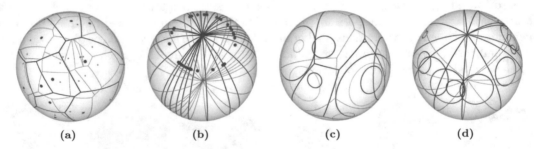

(a)   (b)   (c)   (d)

Fig. 11.2: Voronoi diagrams on the sphere. All diagram edges are geodesic arcs. (a) The Voronoi diagram of 32 random points. (b) A highly degenerate case of Voronoi diagram of 30 point sites on the sphere. (c) The power diagram of 10 random circles. (d) A degenerate power diagram of 14 sites on the sphere.

The effort needed to generalize existing algorithms, such as the plane sweep and zone traversal algorithms, originally designed for arrangements of bounded curves in the plane, was reduced by extensive reuse of code. The implementation of the generalization benefits from all software tools described in this book. There is a well-known connection between Voronoi diagrams in $\mathbb{R}^d$ and lower envelopes of the corresponding distance functions to the sites in $\mathbb{R}^{d+1}$ [54]. This observation also holds for the spherical case. For example, given a set $P = \{p_1, \ldots, p_n\}$ of points on the sphere, if we take $f_i(x) = \rho(x, p_i)$ to be the geodesic distance between x and p_i on the sphere, for $1 \leq i \leq n$, then the minimization diagram of $\{f_1, \ldots, f_n\}$ over $\mathbb{S}^2$ corresponds to the Voronoi diagram of P over $\mathbb{S}^2$. Setter implemented a framework for constructing various two-dimensional Voronoi diagrams [191, 192], exploiting the generic envelope-code of CGAL; see Chapter 10. Figure 11.2 shows various Voronoi diagrams the bisectors of which are geodesic arcs. The diagrams are computed using the aforementioned new framework. They are represented as arrangements induced by geodesic arcs embedded on the sphere.

11.2 Higher-Dimensional Arrangements

In earlier chapters you have seen a variety of usages of two-dimensional arrangements for pattern matching, GIS, robot motion-planning, and more. Could there be similar interest in and use for three- or higher-dimensional arrangements? The answer is a definite yes. There are numerous uses for higher-dimensional arrangements in general and three-dimensional arrangements in particular. Consider for example the computation of the minimum-area triangle presented in Section 4.2.1. A similar problem can be asked in higher dimensions, *the minimum-volume simplex problem*: Given a collection P of n points in $\mathbb{R}^d$, find a subset of $d + 1$ of the points such that the simplex, which is their convex hull, has the smallest volume over all possible selections of such subsets of P. The desired simplex can be found by constructing an arrangement of n hyperplanes in $\mathbb{R}^d$ (each hyperplane dual to a point in P), and searching over it, in total time $O(n^d)$ [53].

Most applications in the book have interesting and relevant three-dimensional counterparts, including the computation of Minkowksi sums or offset polyhedra and finding the largest cardinality matching among points sets.

In this section we discuss representational and (combinatorial/topological) algorithmic issues of higher-dimensional arrangements. Additional aspects are discussed in the Section 11.3 and Section 11.4.

We are given a set Σ of n well-behaved algebraic surfaces[2] in $\mathbb{R}^3$. Assume we have at our disposal the software for computing arrangements of algebraic curves embedded on a parametric surface, as described in Section 11.1 above. What would then be the shortest path to producing information about the three-dimensional arrangement $\mathcal{A}(\Sigma)$? Here is a simple solution that produces partial yet meaningful information about $\mathcal{A}(\Sigma)$. Take one surface $\sigma \in \Sigma$ and intersect it with all the other surfaces in Σ. You obtain a set of curves embedded on σ; let's call this set Γ_σ. Using the two-dimensional machinery, compute the arrangement $\mathcal{A}_\sigma(\Gamma_\sigma)$ embedded on σ. Repeat the same process for every surface of Σ, to get n two-dimensional arrangements. This already results in significant data: You have just produced all the low (0-, 1-, or 2-) dimensional cells of the arrangement. You are only missing the three-dimensional cells. Moreover, you have, through the respective DCEL structures, the connectivity information on each surface separately. With additional moderate effort you can obtain the information on the connectivity among all the lower-dimensional features embedded on different surfaces and thus obtain a partial representation of the three-dimensional cells as well. (See the handling of the union of spheres in [102, 105] to get an idea how this can be done.)

The advantage of the arrangement-per-surface approach is that it uses only two-dimensional machinery, and it barely adds other features to the representation beyond the two-dimensional arrangements on the surfaces. (Recall that for now we only concern ourselves with combinatorial algorithmic aspects and postpone the discussion of algebra.) However, it still falls short of providing us with information about the three-dimensional cells of the arrangement and their connectivity, in particular the connection between a 3D cell and 3D "holes" it may contain. Such information may be of central importance in applications like robot motion planning. We therefore describe an alternative approach. It is (conceptually) as simple, relies primarily on computing two-dimensional arrangements, but has a very different flavor: It adds many extra features to the subdivision of space, from which the three-dimensional cells and their connectivity is easily deduced. The approach is based on Collins's Cylindrical Algebraic Decomposition (or C.A.D. for short) [39], and we describe it here for a collection $\mathcal{T}$ of n (possibly intersecting) triangles in $\mathbb{R}^3$. Let $\mathcal{T}$ be a collection of n triangles in $\mathbb{R}^3$. Let B be an axis-parallel box bounding all the triangles in $\mathcal{T}$. The goal is to produce a refinement of the arrangement $\mathcal{A}(\mathcal{T})$ within the box B. Compute the pairwise intersections of the triangles in $\mathcal{T}$, assuming for simplicity of exposition that each such intersection is a line segment. More generally, the description here assumes general position

---

[2]For an elaborate discussion on what constitute well-behaved surfaces see [4].

of the triangles.[3] Then take these intersection segments together with the boundary edges of the triangles and project all of them onto the xy-plane. Let us denote the set of projected line segments, together with the boundary of the projection $\hat{B}$ of the box B, by $\mathcal{S}$.

Next, compute the planar arrangement $\mathcal{A}(\mathcal{S})$, and add a vertical line-segment through every vertex in the arrangement $\mathcal{A}(\mathcal{S})$, spanning the height of $\hat{B}$. Denote the set $\mathcal{S}$ augmented with these vertical line segments by $\mathcal{S}'$. The faces of $\mathcal{A}(\mathcal{S}')$ are either triangles or trapezoids. We extend each edge of the arrangement $\mathcal{A}(\mathcal{S}')$ in the z-direction into a vertical rectangle spanning the height of the box B, obtaining a collection of vertical prisms. Take each such prism and slice it into smaller prisms by cutting it with the triangles of $\mathcal{T}$ that cross it.

We have thus refined the arrangement $\mathcal{A}(\mathcal{T})$. Each cell in this subdivision is simply shaped (is convex, in particular), and has at most six facets. Furthermore, as is easy to verify, each cell is contained in a unique three-dimensional cell of $\mathcal{A}(\mathcal{T})$. This refinement of $\mathcal{A}(\mathcal{T})$ can be conveniently represented by a *connectivity graph* as follows: Each three-dimensional cell of the subdivision is represented by a node in the graph, and two nodes in the graph are connected by an arc if the corresponding cells share a facet that is *not* contained in an input triangle. The connected components of this graph are in one-to-one correspondence with the three-dimensional cells of $\mathcal{A}(\mathcal{T})$. We add information to each node to describe the geometric information of the corresponding cell (say the coordinates of the vertices of the prism). We can further add specially marked arcs to the graph for pairs of cells that share a facet contained in a triangle to capture the entire connectivity of the three-dimensional arrangement.

The algorithm and resulting refinement can be made more efficient if instead of adding vertical lines through every vertex of the planar arrangement $\mathcal{A}(\mathcal{S})$, we apply vertical decomposition (see Section 3.6). The major difficulty in implementing such an algorithm is the handling of all possible degeneracies once we relax the general position assumption.

Moving from triangles to algebraic surfaces is very complex due to numerous factors, and due primarily to the algebraic machinery that is needed to support the construction. Still, a significant advancement in this direction for three-dimensional arrangements of surfaces is reported in [26], using the CGAL *2D Arrangements* package for constructing the projected arrangement. So one may think that the problem of computing three-dimensional arrangements in practice is solved. But here is the catch: We started with n triangles and ended up with a representation that has $\Omega(n^7)$ cells in the worst case, while the underlying arrangement itself can have complexity at most $O(n^3)$. If we are more careful and decompose the planar arrangement of projected segments with planar vertical decomposition, the number of cells drops from $\Omega(n^7)$ to $\Omega(n^5)$. This means that the C.A.D.-style decomposition will only be effective for very small values of n.

There are alternative representations for three- (or higher-) dimensional arrangements. A relative of the DCEL has been extended to a so-called cell-tuple structure in high dimensions [31]. Already, in the usage of the DCEL for two-dimensional arrangements, the treatment of holes is often tedious and time-consuming—a problem that is alleviated by refined representations like the C.A.D.-style decomposition we have described earlier.

This discussion leads us to what we believe would be a suitable representation of three-dimensional arrangements: The vertical decomposition of the arrangement or lighter variants of it. The vertical decomposition of a three-dimensional arrangement is an extension of the well-known trapezoidal decomposition in the plane; see Section 3.6. Just as trapezoidal decomposition applies to general arrangements of curves, its higher-dimensional counterpart, the vertical decomposition, applies very broadly to arrangements of hypersurfaces in higher dimensions; see, e.g., [98, 196]. Vertical decomposition is much more economical than C.A.D. For example, for an arrangement of n triangles it comprises at most $O(n^3)$ simply shaped cells (see [42]; notice that such a tight bound is not known in general for arrangements of well-behaved surfaces in $\mathbb{R}^3$, but almost-tight bounds are known). The shape of the cells is similar to the shape of the cells produced by C.A.D. However, their connectivity across the triangles (or more generally, the inducing surfaces) is more

---

[3]By the general position for arrangements of triangles in 3-space we mean that no four triangles share a common point, no three triangles intersect at more than a point, no two triangles intersect at more than a segment, no two edges of two different triangles intersect, no vertex of one triangle lies in another triangle, and no triangle edge intersects another triangle at more than a point [11].

complicated to describe; see the discussion in [42, Section 4.3]. We have gained some practical experience with such decompositions for arrangements of triangles and arrangements of polyhedral surfaces in $\mathbb{R}^3$ [197].

While asymptotically the vertical decomposition is an efficient method, it can hide rather significant constants in the big-Oh notation; these constants are not very big but they are often larger than is practically desirable. An attempt at reducing the number of cells is via the so-called partial vertical decomposition; see [197].

In view of the anticipated hardships of representing and manipulating arrangements in higher dimensions, a possibly effective approach of a completely different flavor is to hybridize exact lower-dimensional representations with sampling-based methods rooted in robot motion planning (see Section 9.4). Sampling-based methods represent connectivity of cells in an arrangement by randomly sampling points in space, and connecting nearby points that belong to the same cell by straight line edges. We have already applied such a hybrid approach to certain four-dimensional arrangements (see Section 9.3.2 and [117]).

11.3   Fixed-Precision Geometric Computing

The approach to geometric computing taken throughout the book, and more generally in the *2D Arrangements* package and in CGAL as a whole, is the so-called Exact Geometric Computation (EGC) paradigm; see Section 1.3.2. It requires that predicates never err, and it is achieved by using various software-implemented arithmetic tools to emulate arbitrary-precision calculations of the predicates.

It is rare in scientific or engineering applications that we need to produce exact results. Good approximations will in most cases do the job, since the input to the geometric computation comes from modeling or measurements that are approximate to begin with. Moreover, using special tools for exact geometric computing is in general more costly than using machine arithmetic, both in running time and storage space. We nevertheless adapt EGC, since to date this is the only approach that is at once correct and sufficiently general for implementing geometric algorithms.

If you implement a selective geometric algorithm (namely an algorithm that does not produce new geometric objects) in the plane, such as an algorithm to compute the convex hull of a set of planar points, and you use state-of-the art software for that, you will barely feel the difference in running time between using exact computation and using machine floating-point arithmetic. (Assuming that your program will terminate and produce correct results while using floating-point arithmetic.) This is due to big headways in developing speedup tools for exact computation. The same holds even for constructive geometric algorithms (namely algorithms that compute new geometric entities, such as intersection points) for planar linear objects, such as those for arrangements of lines in the plane. If, however, you implement the construction of an arrangement of algebraic surfaces (even of moderate degree) in three-dimensional space, you will notice a significant slowdown when switching from machine floating-point arithmetic to exact arithmetic.

Several attempts have been made over the years to do consistent geometric computation with limited-precision arithmetic (see Section 4 of [187] for a survey). These attempts are mostly ad-hoc and rather intricate.

The use of limited precision in the computation of arrangements has been investigated in the studies on arrangements of lines [80], on arrangements of algebraic curves [158, 160], and (our own work) on arrangements of circles [101] and spheres [105]. We call our approach *controlled perturbation*, or CP for short, as it relies on carefully perturbing the input objects that induce the arrangement. (We remark that CP has been studied more broadly; see [81, 152, 175, 177, 184].) The goal in developing CP was twofold: (i) allow for consistent geometric computing while using fixed-precision arithmetic, and interdependently (ii) remove all degeneracies in the constructed arrangement.

This brings about another issue with exact geometric computation. It implicitly assumes that we handle all possible degeneracies, which makes programming extremely complex. In fact, this is what we do for two-dimensional arrangements: We handle all conceivable degeneracies, and report all the degeneracies in the topological output. We anticipate that doing the same in higher dimensions, that is, coping with all possible degeneracies in higher-dimensional arrangements of algebraic surfaces, will be a tremendously difficult task.

Therefore, a major challenge we foresee on the way to developing robust and efficient software for arrangements of surfaces in higher dimensions is devising a systematic approximation approach to computing such arrangements with fixed-precision arithmetic, while removing degeneracies.

Another issue that comes up when working with arrangements of curves computed exactly, or more generally with any other nontrivial geometric structure that was computed exactly, is the representation of the coordinates. The exact coordinates of an arrangement of curves may be either too big (easily hundreds of bits per coordinate for natural real-life inputs) in the case of rational computation, or impossible to store numerically with finite precision when computing with algebraic numbers. How do we pass on our results to an engineer for further computation? What we would like to do is to cut short the coordinates such that they fit in some fixed-precision computer words, while preserving the geometry and topology of the structure that we computed.

A satisfactory rounding solution exists for polygonal objects in the plane. It is called *snap rounding*, has been intensively studied [43, 85, 92, 103, 113, 118, 176], and is frequently used. The package *2D Snap Rounding* of CGAL [178] offers, as its name implies, snap rounding for arrangements of segments in the plane. It is based on so-called *iterated* snap rounding [103].

Snap rounding seems, however, difficult to extend to other settings, although results exist for Bézier curves [58], for geodesic arcs on the sphere [136, 137], and for polyhedral subdivisions [79]. Alternative rounding schemes appear, for example, in [49, 87, 159]. Another major challenge in fixed-precision approximation of arrangements is to devise a consistent and efficient rounding scheme for arrangements of curves in the plane and for arrangements in higher dimensions, both linear and curved.

11.4 More Applications

The diversity and range of uses of arrangements is still greater than surveyed in the book so far, and in this section we briefly touch on a few more applications. In particular we mention a few additional problems that were solved using the CGAL *2D Arrangements* package.

Movable separability of sets [203] is an area abounding in problems that can be efficiently solved using arrangements. It is concerned with questions of how to move objects away from one another so as to unlock a compound object or disassemble the parts of a product. For a large family of disassembly and assembly problems, there is a general-solution approach that relies on constructing arrangements in the space of all possible motions of the parts [100, 218]. Previously, the method has only been implemented for very restricted types of motions, primarily infinitesimal motions, which give rise to arrangements of polytopes [91]. Recently, with the development of arrangements on the sphere (Section 11.1), we have been able to apply the method to finite translational motions [72]. The area of mechanical assembly planning offers many more interesting geometric problems on arrangements and otherwise.

Another set of problems, of a somewhat similar flavor, arises in *casting and molding*. In casting, hot liquid is poured into a mold, and after it cools down the cast object is removed from the mold. Typical (though not necessarily) molds consist of two parts and are designed for repeated use. One needs to design the two parts so that they will allow for taking out the object without breaking the mold. Arrangements come in handy in the design of feasible or optimal molds and, in particular, good parting lines between the parts of the mold [5, 134]. In their implementation of the parting-line finding algorithm, Khardekar and McMains [134] use the CGAL *2D Arrangements* package.

There is a whole gamut of *visibility* problems where arrangements play a natural role. An interesting family of visibility problems concerns the construction of *aspect graphs* [180]. Aspect graphs are data structures that encode the different views of a scene. They vary according to the type of objects in the scene (for example, convex polyhedra), the possible view points (which determine what a view, or a projection, of the scene is), and the combinatorial characterization of a view. They are in fact arrangements that subdivide the space of all possible viewpoints into cells such that all the views inside a single cell are combinatorially the same (see, e.g., [44]). A compact encoding of the information over all the cells of the arrangement, proposed by Gigus et al. [84] in the context of aspect graphs, has much broader applicability to efficiently storing extra information over the cells of arrangements. A closely related concept is that of *the visibility complex*; see, e.g., [52, 181]. Another rich area of visibility problems is that of the so-called *art gallery* problem and its many variants [170, 172]. The classical art-gallery problem asks for the minimum number of *guards* that achieve visibility coverage of a simple polygon, and is known to be NP-hard [141]. A recent solution based on linear programming [15] uses the CGAL *2D Arrangements* package in its implementation.

In Section 7.3 you have studied one optimization problem, that of finding the largest common subset between two point sets under translation. Many optimization problems are solved efficiently with arrangements [4, Section 14.4]. The solution is often found by traversing appropriate arrangements, where each cell corresponds to one candidate solution, efficiently searching for the best cell/solution. One such problem is maximizing the area of an axially symmetric polygon inscribed in a simple polygon. Its solution involves arrangements of rational arcs, and it uses the CGAL *2D Arrangements* package in its implementation [14, 185]. Another optimization problem that has recently been solved using the CGAL *2D Arrangements* is finding the minimum width annulus of set of discs in the plane [192].

In Section 9.3 we discussed the connection between motion-planning problems and the arrangements of constraints that they induce. It is well known that for a single-query motion-planning problem (to compute a path between the start and goal configurations of a robot), it suffices to construct a single cell in the arrangement of constraint curves or surfaces in order to find a path if one exists: The cell that contains the start configuration of the robot [94, 96]. Recently, Warmat [206] implemented the intricate single-face algorithm for arrangements of line segments, as described in [196], using the CGAL *2D Arrangements* package.

To find out more about applications of arrangements, consult the arrangements books, book chapters, and surveys cited in Section 1.6.

11.5 Exercises

11.1 [**Project**] Devise and implement a data structure to represent arrangements of planes in $\mathbb{R}^3$, based on the arrangement-per-surface idea outlined above in Section 11.2.

Use this structure to solve the following minimum-volume simplex problem: Given a set $P = \{p_1, p_2, \ldots, p_n\}$ of n points in $\mathbb{R}^3$, find four distinct points $p_i, p_j, p_k, p_l \in P$ such that the volume of the simplex $\triangle p_i p_j p_k p_l$ is minimal among all the simplices defined by four distinct points in the set. The machinery described in Section 4.2.1 for the two-dimensional case extends smoothly to the three-dimensional case. The solution is described by Edelsbrunner [53] in any dimension.

11.2 [**Project**] Implement the C.A.D.-style decomposition described in Section 11.2 to represent arrangements of triangles in $\mathbb{R}^3$.

Carry out extensive experiments and record the growth of the complexity of the structure as a function of the number of triangles for various families of triangles: sparse collections, highly intersecting, randomly selected inside a cube (choose each coordinate of each corner of a triangle uniformly at random in the range $[0, 1]$).

11.3 [**Project**] Adapt the C.A.D.-style decomposition you have developed in Exercise 11.2 to handle arrangements of planes.

Experimentally compare the two approaches to constructing arrangements of planes, the arrangement-per-surface and the C.A.D.-style decomposition, in terms of time and storage space requirements.

Bibliography

[1] P. K. Agarwal, E. Flato, and D. Halperin. Polygon decomposition for efficient construction of Minkowski sums. *Computational Geometry: Theory and Applications*, 21:39–61, 2002. 238

[2] P. K. Agarwal, J. Pach, and M. Sharir. State of the union (of geometric objects): A review. In J. E. Goodman, J. Pach, and R. Pollack, editors, *Computational Geometry: Twenty Years Later*, pages 9–48. American Mathematical Society, Providence, 2008. 206

[3] P. K. Agarwal, O. Schwarzkopf, and M. Sharir. The overlay of lower envelopes and its applications. *Discrete & Computational Geometry*, 15:1–13, 1996. 260

[4] P. K. Agarwal and M. Sharir. Arrangements and their applications. In J.-R. Sack and J. Urrutia, editors, *Handbook of Computational Geometry*, chapter 2, pages 49–119. Elsevier Science Publishers, B.V. North-Holland, Amsterdam, 2000. 17, 263, 265, 269

[5] H.-K. Ahn, M. de Berg, P. Bose, S.-W. Cheng, D. Halperin, J. Matoušek, and O. Schwarzkopf. Separating an object from its cast. *Computer-Aided Design*, 34(8):547–559, 2002. 268

[6] O. Aichholzer and H. Krasser. Abstract order type extension and new results on the rectilinear crossing number. *Computational Geometry: Theory and Applications*, 36(1):2–15, 2006. Special Issue on the 21st European Workshop on Computational Geometry. 65

[7] A. Alexandrescu. *Modern C++ Design: Generic Programming And Design Patterns Applied*. Addison-Wesley, Boston, MA, USA, 2001. 17, 125, 158

[8] N. Alon, D. Halperin, O. Nechushtan, and M. Sharir. The complexity of the outer face in arrangements of random segments. In *Proceedings of the 24th Annual ACM Symposium on Computational Geometry (SoCG)*, pages 69–78. Association for Computing Machinery (ACM) Press, 2008. 65

[9] H. Alt, U. Fuchs, G. Rote, and G. Weber. Matching convex shapes with respect to the symmetric difference. *Algorithmica*, 21(1):89–103, 1998. 206

[10] C. Ambühl, S. Chakraborty, and B. Gärtner. Computing largest common point sets under approximate congruence. In *Proceedings of the 8th Annual European Symposium on Algorithms (ESA)*, volume 1879 of *LNCS*, pages 52–64, 2000. 171

[11] B. Aronov and M. Sharir. Triangles in space or building (and analyzing) castles in the air. *Combinatorica*, 10(2):137–173, 1990. 266

[12] M. H. Austern. *Generic Programming and the STL*. Addison-Wesley, Boston, MA, USA, 1999. 2, 5, 17, 125

[13] C. L. Bajaj and M.-S. Kim. Generation of configuration space obstacles: The case of moving algebraic curves. *Algorithmica*, 4(2):157–172, 1989. 238

[14] G. Barequet and V. Rogol. Maximizing the area of an axially symmetric polygon inscribed in a simple polygon. *Computers & Graphics*, 31(1):127–136, 2007. 269

[15] T. Baumgartner, S. P. Fekete, A. Kröller, and C. Schmidt. Exact solutions and bounds for general art gallery problems. In *Workshop on Algorithm Engineering and Experiments*, pages 11–22, 2010. 269

[16] J. A. Bennell and X. Song. A comprehensive and robust procedure for obtaining the nofit polygon using Minkowski sums. *Computers & Operations Research*, 35(1):267–281, 2008. 237

[17] J. L. Bentley. Multidimensional binary search trees used for associative searching. *Communications of the ACM*, 18(9):509–517, September 1975. 63

[18] J. L. Bentley and T. Ottmann. Algorithms for reporting and counting geometric intersections. *IEEE Transactions on Computers*, 28(9):643–647, 1979. 64

[19] E. Berberich, A. Eigenwillig, M. Hemmer, S. Hert, K. Mehlhorn, and E. Schömer. A computational basis for conic arcs and Boolean operations on conic polygons. In *Proceedings of the 10th Annual European Symposium on Algorithms (ESA)*, volume 2461 of *LNCS*, pages 174–186. Springer, 2002. 124, 125, 157

[20] E. Berberich, E. Fogel, D. Halperin, M. Kerber, and O. Setter. Arrangements on parametric surfaces II: Concretizations and applications. *Mathematics in Computer Science*, 4:67–91, 2010. 157, 263

[21] E. Berberich, E. Fogel, D. Halperin, K. Mehlhorn, and R. Wein. Sweeping and maintaining two-dimensional arrangements on surfaces: A first step. In *Proceedings of the 15th Annual European Symposium on Algorithms (ESA)*, pages 645–656, 2007. 81, 124, 263

[22] E. Berberich, E. Fogel, D. Halperin, K. Mehlhorn, and R. Wein. Arrangements on parametric surfaces I: General framework and infrastructure. *Mathematics in Computer Science*, 4:45–66, 2010. 81, 124, 263, 264

[23] E. Berberich, M. Hemmer, and M. Kerber. A generic algebraic kernel for non-linear geometric applications. Technical Report 7274, Inria, 2010. 157

[24] E. Berberich, M. Hemmer, L. Kettner, E. Schömer, and N. Wolpert. An exact, complete and efficient implementation for computing planar maps of quadric intersection curves. In *Proceedings of the 21st Annual ACM Symposium on Computational Geometry (SoCG)*, pages 99–106. Association for Computing Machinery (ACM) Press, 2005. 124

[25] E. Berberich and M. Kerber. Exact arrangements on tori and Dupin cyclides. In *Proceedings of 2008 ACM Symposium on Solid and Physical Modeling (SPM)*, pages 59–66. Association for Computing Machinery (ACM) Press, 2008. 263

[26] E. Berberich, M. Kerber, and M. Sagraloff. Exact geometric-topological analysis of algebraic surfaces. In *Proceedings of the 24th Annual ACM Symposium on Computational Geometry (SoCG)*, pages 164–173. Association for Computing Machinery (ACM) Press, 2008. 266

[27] K.-F. Böhringer, B. R. Donald, and D. Halperin. On the area bisectors of a polygon. *Discrete & Computational Geometry*, 22(2):269–285, 1999. 206

[28] J.-D. Boissonnat, E. de Lange, and M. Teillaud. Slicing Minkowski sums for satellite antenna layout. *Computer-Aided Design*, 30(4):255–265, 1998. 237

[29] J.-D. Boissonnat and M. Teillaud, editors. *Effective Computational Geometry for Curves and Surfaces*. Springer, 2006. xi, 17, 124, 125, 260

[30] J.-D. Boissonnat and M. Yvinec. *Algorithmic Geometry.* Cambridge University Press, Cambridge, UK, 1998. Translated by H. Brönnimann. 17, 260

[31] E. Brisson. Representing geometric structures in d dimensions: Topology and order. *Discrete & Computational Geometry*, 9:387–426, 1993. 266

[32] H. Brönnimann, C. Burnikel, and S. Pion. Interval arithmetic yields efficient dynamic filters for computational geometry. *Discrete Applied Mathematics*, 109:25–47, 2001. 13

[33] J. Canny, B. Donald, and E. K. Ressler. A rational rotation method for robust geometric algorithms. In *Proceedings of the 8th Annual ACM Symposium on Computational Geometry (SoCG)*, pages 251–260. Association for Computing Machinery (ACM) Press, 1992. 65, 128, 261

[34] F. Cazals and S. Loriot. Computing the arrangement of circles on a sphere, with applications in structural biology. *Computational Geometry: Theory and Applications*, 42(6–7):551–565, 2009. 264

[35] B. Chazelle and D. P. Dobkin. Optimal convex decompositions. In G. T. Toussaint, editor, *Computational Geometry*, pages 63–133. North-Holland, Amsterdam, 1985. 238

[36] B. Chazelle, L. J. Guibas, and D.-T. Lee. The power of geometric duality. *BIT*, 25:76–90, 1985. 81

[37] H. Choset, K. M. Lynch, S. Hutchinson, G. Kantor, W. Burgard, L. E. Kavraki, and S. Thrun. *Principles of Robot Motion: Theory, Algorithms, and Implementations.* MIT Press, Boston, MA, 2005. 238

[38] L. D. Christophe, C. Lemaire, and J.-M. Moreau. Fast Delaunay point-location with search structures. In *Proceedings of the 11th Canadian Conference on Computational Geometry*, pages 136–141, 1999. 63

[39] G. E. Collins. Quantifier elimination for real closed fields by cylindrical algebraic decomposition. In *Proceedings of the 2nd GI Conference on Automata Theory and Formal Languages*, volume 33 of *LNCS*, pages 134–183. Springer, 1975. 265

[40] F. Comellas and J. L. A. Yebra. New lower bounds for Heilbronn numbers. *Electronic Journal of Combinatorics*, 9(1), 2002. 81

[41] T. H. Cormen, C. E. Leiserson, R. L. Rivest, and C. Stein. *Introduction to Algorithms.* MIT Press, 3rd edition, 2009. 171

[42] M. de Berg, L. J. Guibas, and D. Halperin. Vertical decompositions for triangles in 3-space. *Discrete & Computational Geometry*, 15(1):35–61, 1996. 266, 267

[43] M. de Berg, D. Halperin, and M. H. Overmars. An intersection-sensitive algorithm for snap rounding. *Computational Geometry: Theory and Applications*, 36(3):159–165, 2007. 268

[44] M. de Berg, D. Halperin, M. H. Overmars, and M. J. van Kreveld. Sparse arrangements and the number of views of polyhedral scenes. *International Journal of Computational Geometry and Applications*, 7(3):175–195, 1997. 269

[45] M. de Berg, M. van Kreveld, M. H. Overmars, and O. Cheong. *Computational Geometry: Algorithms and Applications.* Springer, Berlin, Germany, 3rd edition, 2008. 17, 40, 59, 64, 81, 157, 206, 238, 239, 246

[46] P. M. de Castro, F. Cazals, S. Loriot, and M. Teillaud. Design of the CGAL 3D spherical kernel and application to arrangements of circles on a sphere. *Computational Geometry: Theory and Applications*, 42(6–7):536–550, 2009. 264

[47] P. M. M. de Castro, S. Pion, and M. Teillaud. Exact and efficient computations on circles in CGAL. In *Abstracts of 23rd European Workshop on Computational Geometry*, pages 219–222, 2007. 124

[48] O. Deussen and B. Lintermann. *Digital Design of Nature: Computer Generated Plants and Organics*. Springer, Berlin, Heidelberg, Germany, 2005. 127

[49] O. Devillers and P. Guigue. Inner and outer rounding of Boolean operations on lattice polygonal regions. *Computational Geometry: Theory and Applications*, 33(1–2):3–17, 2006. 268

[50] O. Devillers, S. Pion, and M. Teillaud. Walking in a triangulation. *International Journal of Foundations of Computer Science*, 13:181–199, 2002. 63

[51] L. Devroye, E. P. Mücke, and B. Zhu. A note on point location in Delaunay triangulations of random points. *Algorithmica*, 22:477–482, 1998. 63

[52] F. Durand, G. Drettakis, and C. Puech. The 3D visibility complex. *ACM Transactions on Graphics*, 21(2):176–206, 2002. 269

[53] H. Edelsbrunner. *Algorithms in Combinatorial Geometry*. Springer, Berlin, Germany, 1987. 17, 81, 172, 265, 269

[54] H. Edelsbrunner and R. Seidel. Voronoi diagrams and arrangements. *Discrete & Computational Geometry*, 1:25–44, 1986. 260, 264

[55] J. Edmonds. Paths, trees, and flowers. *Canadian Journal of Mathematics*, 17:449–467, 1965. 172

[56] A. Eigenwillig and M. Kerber. Exact and efficient 2D-arrangements of arbitrary algebraic curves. In *Proceedings of the 19th Annual Symposium on Discrete Algorithms*, pages 122–131, Philadelphia, PA, USA, 2008. Society for Industrial and Applied Mathematics (SIAM). 125

[57] A. Eigenwillig, L. Kettner, E. Schömer, and N. Wolpert. Complete, exact and efficient computations with cubic curves. In *Proceedings of the 20th Annual ACM Symposium on Computational Geometry (SoCG)*, pages 409–418. Association for Computing Machinery (ACM) Press, 2004. 124

[58] A. Eigenwillig, L. Kettner, and N. Wolpert. Snap rounding of Bézier curves. In *Proceedings of the 23rd Annual ACM Symposium on Computational Geometry (SoCG)*, pages 158–167, Gyeongju, South Korea, 2007. Association for Computing Machinery (ACM) Press. 268

[59] G. Elber and M.-S. Kim. Offsets, sweeps, and Minkowski sums. *Computer-Aided Design*, 31(3):163, 1999. 237

[60] I. Z. Emiris, A. Kakargias, S. Pion, M. Teillaud, and E. P. Tsigaridas. Towards an open curved kernel. In *Proceedings of the 20th Annual ACM Symposium on Computational Geometry (SoCG)*, pages 438–446. Association for Computing Machinery (ACM) Press, 2004. 124

[61] J. Erickson. *Lower Bounds for Fundamental Geometric Problems*. Ph.D. dissertation, University of California at Berkeley, 1996. 81

[62] E. Ezra, D. Halperin, and M. Sharir. Speeding up the incremental construction of the union of geometric objects in practice. *Computational Geometry: Theory and Applications*, 27(1):63–85, 2004. 206

[63] E. Ezra and M. Sharir. Output-sensitive construction of the union of triangles. *SIAM Journal on Computing*, 34(6):1331–1351, 2005. 206

[64] A. Fabri, G.-J. Giezeman, L. Kettner, S. Schirra, and S. Schönherr. On the design of CGAL a computational geometry algorithms library. *Software — Practice and Experience*, 30(11):1167–1202, 2000. Special Issue on Discrete Algorithm Engineering. 7, 10, 11

[65] A. Fabri and S. Pion. A generic lazy evaluation scheme for exact geometric computations. In 2nd *Library-Centric Software Design Workshop*, 2006. 13

[66] A. Fabri and L. Rineau. CGAL and the QT graphics view framework. In *CGAL User and Reference Manual*. CGAL Editorial Board, 3.8 edition, 2011. http://www.cgal.org/Manual/3.8/doc_html/cgal_manual/packages.html#Pkg:GraphicsView. 35

[67] U. Finke and K. H. Hinrichs. Overlaying simply connected planar subdivisions in linear time. In *Proceedings of the 11th Annual ACM Symposium on Computational Geometry (SoCG)*, pages 119–126. Association for Computing Machinery (ACM) Press, 1995. 158

[68] E. Flato. Exact and efficient construction of planar Minkowski sums. M.Sc. thesis, The Blavatnik School of Computer Science, Tel-Aviv University, Israel, 2000. http://acg.cs.tau.ac.il/tau-members-area/generalpublications/m.sc.-theses/flato-thesis.pdf. 238, 239

[69] E. Flato, D. Halperin, I. Hanniel, O. Nechushtan, and E. Ezra. The design and implementation of planar maps in CGAL. *The ACM Journal of Experimental Algorithmics*, 5, 2000. 41

[70] E. Fogel. *Minkowski Sum Construction and other Applications of Arrangements of Geodesic Arcs on the Sphere*. Ph.D. dissertation, The Blavatnik School of Computer Science, Tel-Aviv University, Israel, 2009. http://acg.cs.tau.ac.il/tau-members-area/generalpublications/phd-theses/efif-thesis.pdf. 238

[71] E. Fogel and D. Halperin. Exact and efficient construction of Minkowski sums of convex polyhedra with applications. *Computer-Aided Design*, 39(11):929–940, 2007. 157, 238

[72] E. Fogel and D. Halperin. Polyhedral assembly partitioning with infinite translations or the importance of being exact. In H. Choset, M. Morales, and T. D. Murphey, editors, *Algorithmic Foundations of Robotics VIII*, volume 57 of *Springer Tracts in Advanced Robotics*, pages 417–432. Springer, Heidelberg, Germany, 2009. 157, 268

[73] E. Fogel, D. Halperin, L. Kettner, M. Teillaud, R. Wein, and N. Wolpert. Arrangements. In J.-D. Boissonnat and M. Teillaud, editors, *Effective Computational Geometry for Curves and Surfaces*, chapter 1, pages 1–66. Springer, 2007. 263

[74] E. Fogel, D. Halperin, and C. Weibel. On the exact maximum complexity of Minkowski sums of polytopes. *Discrete & Computational Geometry*, 42(4):654–669, 2009. 238

[75] E. Fogel, O. Setter, and D. Halperin. Exact implementation of arrangements of geodesic arcs on the sphere with applications. In *Abstracts of 24th European Workshop on Computational Geometry*, pages 83–86, 2008. 157, 263

[76] E. Fogel, O. Setter, and D. Halperin. Movie: Arrangements of geodesic arcs on the sphere. In *Proceedings of the 24th Annual ACM Symposium on Computational Geometry (SoCG)*, pages 218–219. Association for Computing Machinery (ACM) Press, 2008. 157, 263

[77] E. Fogel, R. Wein, and D. Halperin. Code flexibility and program efficiency by genericity: Improving CGAL's arrangements. In *Proceedings of the 12th Annual European Symposium on Algorithms (ESA)*, volume 3221 of *LNCS*, pages 664–676, 2004. 41

[78] E. Fogel, R. Wein, B. Zukerman, and D. Halperin. 2D regularized Boolean set-operations. In *CGAL User and Reference Manual*. CGAL Editorial Board, 3.8 edition, 2011. http://www.cgal.org/Manual/3.8/doc_html/cgal_manual/packages.html#Pkg:BooleanSetOperations2. 206

[79] S. Fortune. Vertex-rounding a three-dimensional polyhedral subdivision. *Discrete & Computational Geometry*, 22(4):593–618, 1999. 268

[80] S. Fortune and V. J. Milenkovic. Numerical stability of algorithms for line arrangements. In *Proceedings of the 7th Annual ACM Symposium on Computational Geometry (SoCG)*, pages 334–341. Association for Computing Machinery (ACM) Press, 1991. 267

[81] S. Funke, C. Klein, K. Mehlhorn, and S. Schmitt. Controlled perturbation for Delaunay triangulations. In *Proceedings of the 16th Annual Symposium on Discrete Algorithms*, pages 1047–1056, Philadelphia, PA, USA, 2005. Society for Industrial and Applied Mathematics (SIAM). 267

[82] A. Gajentaan and M. H. Overmars. On a class of $O(n^2)$ problems in computational geometry. *Computational Geometry: Theory and Applications*, 5:165–185, 1995. 81, 206

[83] E. Gamma, R. Helm, R. Johnson, and J. M. Vlissides. *Design Patterns — Elements of Reusable Object-Oriented Software*. Addison-Wesley, Boston, MA, USA, 1995. 17, 44, 49, 64, 125, 129, 158

[84] Z. Gigus, J. F. Canny, and R. Seidel. Efficiently computing and representing aspect graphs of polyhedral objects. *IEEE Transactions on Pattern Analysis and Machine Intelligence*, 13(6):542–551, 1991. 269

[85] M. Goodrich, L. J. Guibas, J. Hershberger, and P. Tanenbaum. Snap rounding line segments efficiently in two and three dimensions. In *Proceedings of the 13th Annual ACM Symposium on Computational Geometry (SoCG)*, pages 284–293. Association for Computing Machinery (ACM) Press, 1997. 268

[86] D. H. Greene. The decomposition of polygons into convex parts. In F. P. Preparata, editor, *Computational Geometry*, volume 1 of *Advances in Computing Research*, pages 235–259. JAI Press, Greenwich, Connecticut, 1983. 238

[87] D. H. Greene and F. F. Yao. Finite-resolution computational geometry. In *Proceedings of the 27th Annual IEEE Symposium on the Foundations of Computer Science*, pages 143–152, 1986. 268

[88] B. Grünbaum. *Convex Polytopes*. John Wiley & Sons, New York, NY, 1967. 17

[89] B. Grünbaum. Arrangements of hyperplanes. *Congressus Numerantium*, 3:41–106, 1971. 17

[90] B. Grünbaum. *Arrangements and Spreads*. Number 10 in Regional Conf. Ser. Math. American Mathematical Society, Providence, RI, 1972. 17

[91] L. J. Guibas, D. Halperin, H. Hirukawa, J.-C. Latombe, and R. H. Wilson. Polyhedral assembly partitioning using maximally covered cells in arrangements of convex polytopes. *International Journal of Computational Geometry and Applications*, 8(2):179–200, 1998. 268

[92] L. J. Guibas and D. Marimont. Rounding arrangements dynamically. *International Journal of Computational Geometry and Applications*, 8:157–176, 1998. 268

[93] L. J. Guibas, L. Ramshaw, and J. Stolfi. A kinetic framework for computational geometry. In *Proceedings of the 24th Annual IEEE Symposium on the Foundations of Computer Science*, pages 100–111, 1983. 238

[94] L. J. Guibas, M. Sharir, and S. Sifrony. On the general motion-planning problem with two degrees of freedom. *Discrete & Computational Geometry*, 4:491–521, 1989. 269

[95] P. Hachenberger. Exact Minkowksi sums of polyhedra and exact and efficient decomposition of polyhedra into convex pieces. *Algorithmica*, 55(2):329–345, 2009. 238

[96] D. Halperin. Robot motion planning and the single cell problem in arrangements. *Journal of Intelligent and Robotic Systems*, 11(1–2):45–65, 1994. 269

[97] D. Halperin. Robust geometric computing in motion. *International Journal of Robotics Research*, 21(3):219–232, 2002. 238

[98] D. Halperin. Arrangements. In J. E. Goodman and J. O'Rourke, editors, *Handbook of Discrete and Computational Geometry*, chapter 24, pages 529–562. Chapman & Hall/CRC, Boca Raton, FL, 2nd edition, 2004. 17, 263, 266

[99] D. Halperin, L. E. Kavraki, and J.-C. Latombe. Robotics. In J. E. Goodman and J. O'Rourke, editors, *Handbook of Discrete and Computational Geometry*, chapter 48, pages 1065–1093. Chapman & Hall/CRC, Boca Raton, FL, 2nd edition, 2004. 238

[100] D. Halperin, J.-C. Latombe, and R. H. Wilson. A general framework for assembly planning: The motion space approach. *Algorithmica*, 26:577–601, 2000. 268

[101] D. Halperin and E. Leiserowitz. Controlled perturbation for arrangements of circles. *International Journal of Computational Geometry and Applications*, 14(4–5):277–310, 2004. 267

[102] D. Halperin and M. H. Overmars. Spheres, molecules, and hidden surface removal. *Computational Geometry: Theory and Applications*, 11(2):83–102, 1998. 265

[103] D. Halperin and E. Packer. Iterated snap rounding. *Computational Geometry: Theory and Applications*, 23(2):209–225, 2002. 268

[104] D. Halperin and M. Sharir. New bounds for lower envelopes in three dimensions, with applications to visibility in terrains. *Discrete & Computational Geometry*, 12:313–326, 1994. 260

[105] D. Halperin and C. R. Shelton. A perturbation scheme for spherical arrangements with application to molecular modeling. *Computational Geometry: Theory and Applications*, 10:273–287, 1998. 265, 267

[106] D. Halperin and C.-K. Yap. Combinatorial complexity of translating a box in polyhedral 3-space. *Computational Geometry: Theory and Applications*, 9(3):181–196, 1998. 238

[107] I. Hanniel. The design and implementation of planar arrangements of curves in CGAL. M.Sc. thesis, The Blavatnik School of Computer Science, Tel-Aviv University, Israel, 2000. http://acg.cs.tau.ac.il/tau-members-area/generalpublications/m.sc.-theses/HannielThesis.pdf. 41

[108] I. Hanniel and D. Halperin. Two-dimensional arrangements in CGAL and adaptive point location for parametric curves. In *Proceedings of the 4th Workshop on Algorithm Engineering (WAE)*, volume 1982 of *LNCS*, pages 171–182, 2000. 41

[109] I. Hanniel and R. Wein. An exact, complete and efficient computation of arrangements of Bézier curves. *IEEE Transactions on Automation Science and Engineering*, 6(3):399–408, 2009. 125

[110] I. Haran. Efficient point location in general planar subdivisions using landmarks. M.Sc. thesis, The Blavatnik School of Computer Science, Tel-Aviv University, Israel, 2006. http://acg.cs.tau.ac.il/tau-members-area/generalpublications/m.sc.-theses/IditThesis.pdf. 64

[111] I. Haran and D. Halperin. An experimental study of point location in planar arrangements in CGAL. *The ACM Journal of Experimental Algorithmics*, 13, 2008. 64

[112] J. Hershberger. Finding the upper envelope of n line segments in $O(n \log n)$ time. *Information Processing Letters*, 33:169–174, 1989. 260

[113] J. Hershberger. Improved output-sensitive snap rounding. *Discrete & Computational Geometry*, 39(1):298–318, 2008. 268

[114] S. Hert. 2D polygon partitioning. In CGAL *User and Reference Manual*. CGAL Editorial Board, 3.8 edition, 2011. `http://www.cgal.org/Manual/3.8/doc_html/cgal_manual/packages.html#Pkg:PolygonPartitioning2`. 238

[115] S. Hert, M. Hoffmann, L. Kettner, S. Pion, and M. Seel. An adaptable and extensible geometry kernel. *Computational Geometry: Theory and Applications*, 38(1–2):16–36, 2007. 12, 124

[116] S. Hertel and K. Mehlhorn. Fast triangulation of the plane with respect to simple polygons. *Information and Control*, 64:52–76, 1985. 238

[117] S. Hirsch and D. Halperin. Hybrid motion planning: Coordinating two discs moving among polygonal obstacles in the plane. In J.-D. Boissonnat, J. Burdick, and K. Goldberg, editors, *Algorithmic Foundations of Robotics V*, pages 239–255. Springer, Heidelberg, Germany, 2004. 238, 267

[118] J. Hobby. Practical segment intersection with finite precision output. *Computational Geometry: Theory and Applications*, 13:199–214, 1999. 268

[119] C. M. Hoffmann. *Geometric and Solid Modeling: an Introduction*. Morgan Kaufmann, San Francisco, CA, USA, 1989. 158

[120] C. M. Hoffmann. Solid modeling. In J. E. Goodman and J. O'Rourke, editors, *Handbook of Discrete and Computational Geometry*, chapter 56, pages 1257–1278. Chapman & Hall/CRC, Boca Raton, FL, 2nd edition, 2004. 175

[121] S.-C. Huang and C.-F. Wang. Genetic algorithms for approximation of digital curves with line segments and circular arcs. *Journal of the Chinese Institute of Engineers*, 32(4):437–444, 2009. 124

[122] R. V. Jean. *Phyllotaxis: A Systemic Study in Plant Morphogenesis*. Cambridge University Press, Cambridge, UK, 1994. Cambridge Books Online. 127

[123] H. Kaplan, N. Rubin, and M. Sharir. Line transversals of convex polyhedra in $\mathbb{R}^3$. In *Proceedings of the 20th Annual Symposium on Discrete Algorithms*, pages 170–179, Philadelphia, PA, USA, 2009. Society for Industrial and Applied Mathematics (SIAM). 260

[124] V. Karamcheti, C. Li, I. Pechtchanski, and C. K. Yap. A core library for robust numeric and geometric computation. In *Proceedings of the 15th Annual ACM Symposium on Computational Geometry (SoCG)*, pages 351–359. Association for Computing Machinery (ACM) Press, 1999. 9

[125] A. Kaul and J. Rossignac. Solid-interpolating deformations: Construction and animation of PIPs. *Computers & Graphics*, 16(1):107–115, 1992. 237

[126] L. E. Kavraki, P. Švestka, J.-C. Latombe, and M. H. Overmars. Probabilistic roadmaps for path planning in high dimensional configuration spaces. *IEEE Transactions on Robotics and Automation*, 12:566–580, 1996. 238

[127] K. Kedem, R. Livne, J. Pach, and M. Sharir. On the union of Jordan regions and collision-free translational motion amidst polygonal obstacles. *Discrete & Computational Geometry*, 1:59–70, 1986. 206, 238

[128] M. Keil. Polygon decomposition. In J.-R. Sack and J. Urrutia, editors, *Handbook of Computational Geometry*, chapter 11, pages 491–518. Elsevier Science Publishers, B.V. North-Holland, Amsterdam, 2000. 238

[129] L. Kettner. Using generic programming for designing a data structure for polyhedral surfaces. *Computational Geometry: Theory and Applications*, 13:65–90, 1999. 41

[130] L. Kettner. 3D polyhedral surfaces. In *CGAL User and Reference Manual*. CGAL Editorial Board, 3.8 edition, 2011. http://www.cgal.org/Manual/3.8/doc_html/cgal_manual/packages.html#Pkg:Polyhedron. 41

[131] L. Kettner. Halfedge data structures. In *CGAL User and Reference Manual*. CGAL Editorial Board, 3.8 edition, 2011. http://www.cgal.org/Manual/3.8/doc_html/cgal_manual/packages.html#Pkg:HDS. 41

[132] L. Kettner, K. Mehlhorn, S. Pion, S. Schirra, and C. Yap. Classroom examples of robustness problems in geometric computations. *Computational Geometry: Theory and Applications*, 40(1):61–78, May 2008. 7

[133] L. Kettner and S. Näher. Two computational geometry libraries: LEDA and CGAL. In J. E. Goodman and J. O'Rourke, editors, *Handbook of Discrete and Computational Geometry*, chapter 65, pages 1435–1463. Chapman & Hall/CRC, Boca Raton, FL, 2nd edition, 2004. 7, 10, 17

[134] R. Khardekar and S. McMains. Efficient computation of a near-optimal primary parting line. In *Proceedings of the 2009 ACM Symposium on Solid and Physical Modeling (SPM)*, pages 319–324. Association for Computing Machinery (ACM) Press, 2009. 268

[135] D. G. Kirkpatrick. Optimal search in planar subdivisions. *SIAM Journal on Computing*, 12(1):28–35, 1983. 63

[136] B. Kozorovitzky. Snap rounding on the sphere. M.Sc. thesis, The Blavatnik School of Computer Science, Tel-Aviv University, Israel, 2010. http://acg.cs.tau.ac.il/projects/internal-projects/snap-rounding-on-the-sphere/thesis.pdf. 268

[137] B. Kozorovitzky and D. Halperin. Snap rounding on the sphere. In *Abstracts of 26th European Workshop on Computational Geometry*, pages 213–216, 2010. 268

[138] J.-C. Latombe. *Robot Motion Planning*. Kluwer Academic Publishers, Norwell, Massachusetts, 1991. 237, 238

[139] S. M. LaValle. *Planning Algorithms*. Cambridge University Press, Cambridge, UK, 2006. Available at http://planning.cs.uiuc.edu/. 238

[140] S. Lazard, L. Peñaranda, and E. Tsigaridas. A CGAL based algebraic kernel based on RS and application to arrangements. In *Abstracts of 24th European Workshop on Computational Geometry*, pages 91–94, 2008. 124

[141] D. T. Lee and A. K. Lin. Computational complexity of art gallery problems. *IEEE Transactions on Information Theory*, 32(2):276–282, 1986. 269

[142] I.-K. Lee, M.-S. Kim, and G. Elber. Polynomial/rational approximation of Minkowski sum boundary curves. *Graphical Models and Image Processing*, 60(2):136–165, 1998. 238

[143] D. Leven and M. Sharir. Planning a purely translational motion for a convex object in two-dimensional space using generalized Voronoi diagrams. *Discrete & Computational Geometry*, 2:9–31, 1987. 238

[144] W. Li and S. McMains. A GPU-based voxelization approach to 3D Minkowski sum computation. In *Proceedings of the 2010 ACM Symposium on Solid and Physical Modeling (SPM)*, pages 31–40. Association for Computing Machinery (ACM) Press, 2010. 238

[145] J.-M. Lien. A simple method for computing Minkowski sum boundary in 3D using collision detection. In H. Choset, M. Morales, and T. D. Murphey, editors, *Algorithmic Foundations of Robotics VIII*, volume 57 of *Springer Tracts in Advanced Robotics*, pages 401–415. Springer, 2009. 238

[146] P. Lienhardt. N-dimensional generalized combinatorial maps and cellular quasi-manifolds. *International Journal of Computational Geometry and Applications*, 4(3):275–324, 1994. 41

[147] M. C. Lin and D. Manocha. Collision and proximity queries. In J. E. Goodman and J. O'Rourke, editors, *Handbook of Discrete and Computational Geometry*, chapter 35, pages 787–807. Chapman & Hall/CRC, Boca Raton, FL, 2nd edition, 2004. 238

[148] C. Linhart, D. Halperin, I. Hanniel, and S. Har-Peled. An experimental study of on-line methods for zone construction in arrangements of lines in the plane. *International Journal of Computational Geometry and Applications*, 13(6):463–485, 2003. 206

[149] J. Matoušek. *Lectures on Discrete Geometry*, volume 212 of *Graduate Texts in Mathematics*. Springer, Heidelberg, Germany, 2002. 17

[150] J. Matoušek, J. Pach, M. Sharir, S. Sifrony, and E. Welzl. Fat triangles determine linearly many holes. *SIAM Journal on Computing*, 23(1):154–169, 1994. 206

[151] K. Mehlhorn and S. Näher. LEDA: *A Platform for Combinatorial and Geometric Computing*. Cambridge University Press, Cambridge, UK, 2000. 7, 9, 17, 63

[152] K. Mehlhorn, R. Osbild, and M. Sagraloff. Reliable and efficient computational geometry via controlled perturbation. In M. Bugliesi, B. Preneel, V. Sassone, and I. Wegener, editors, *Proceedings of the 33rd International Colloquium on Automata, Languages and Programming*, volume 4051 of *Lecture Notes in Computer Science*, pages 299–310, Venice, Italy, 2006. Springer. 267

[153] G. Melquiond and S. Pion. Formal certification of arithmetic filters for geometric predicates. In *Proceedings of the 17th IMACS World Congress on Scientific, Applied Mathematics and Simulation*, 2005. 13

[154] M. Meyerovitch. Robust, generic and efficient construction of envelopes of surfaces in three-dimensional space. M.Sc. thesis, The Blavatnik School of Computer Science, Tel-Aviv University, Israel, 2006. http://acg.cs.tau.ac.il/tau-members-area/generalpublications/m.sc.-theses/michalthesis.pdf. 260

[155] M. Meyerovitch. Robust, generic and efficient construction of envelopes of surfaces in three-dimensional space. In *Proceedings of the 14th Annual European Symposium on Algorithms (ESA)*, volume 4168 of *LNCS*, pages 792–803. Springer, 2006. 260

[156] M. Meyerovitch, R. Wein, and B. Zukerman. 3D envelopes. In CGAL *User and Reference Manual*. CGAL Editorial Board, 3.8 edition, 2011. http://www.cgal.org/Manual/3.8/doc_html/cgal_manual/packages.html#Pkg:Envelope3. 260

[157] S. Micali and V. V. Vazirani. An $o(\sqrt{|V|}|e|)$ algorithm for finding maximum matching in general graphs. In *Proceedings of the 21st Annual IEEE Symposium on the Foundations of Computer Science*, pages 17–27, 1980. 172

[158] V. J. Milenkovic. Calculating approximate curve arrangements using rounded arithmetic. In *Proceedings of the 5th Annual ACM Symposium on Computational Geometry (SoCG)*, pages 197–207. Association for Computing Machinery (ACM) Press, 1989. 267

[159] V. J. Milenkovic. Shortest path geometric rounding. *Algorithmica*, 27(1):57–86, 2000. 268

[160] V. J. Milenkovic and E. Sacks. An approximate arrangement algorithm for semi-algebraic curves. *Computational Geometry: Theory and Applications*, 17(2):175–198, 2007. 267

[161] M. Mucha and P. Sankowski. Maximum matchings via Gaussian elimination. In *Proceedings of the 45th Annual IEEE Symposium on the Foundations of Computer Science*, pages 248–255, Washington, DC, USA, 2004. IEEE Computer Society Press. 172

[162] E. P. Mücke, I. Saias, and B. Zhu. Fast randomized point location without preprocessing in two- and three-dimensional Delaunay triangulations. In *Proceedings of the 12th Annual ACM Symposium on Computational Geometry (SoCG)*, pages 274–283. Association for Computing Machinery (ACM) Press, 1996. 63

[163] D. E. Muller and F. P. Preparata. Finding the intersection of two convex polyhedra. *Theoretical Computer Science*, 7:217–236, 1978. 41

[164] K. Mulmuley. *Computational Geometry: An Introduction Through Randomized Algorithms*. Prentice Hall, Englewood Cliffs, NJ, 1993. 17, 63

[165] T. M. Murali and T. A. Funkhouser. Consistent solid and boundary representations from arbitrary polygonal data. In *Proceedings of the 1997 Symposium on Interactive 3D graphics*, I3D '97, pages 155–162, New York, NY, USA, 1997. Association for Computing Machinery (ACM) Press. 158

[166] D. A. Musser and A. A. Stepanov. Generic programming. In *Proceedings of International Conference on Symbolic and Algebraic Computation*, volume 358 of *LNCS*, pages 13–25. Springer, 1988. 2

[167] D. R. Musser, G. J. Derge, and A. Saini. *STL Tutorial and Reference Guide: C++ Programming with the Standard Template Library*. Professional Computing Series. Addison-Wesley, Boston, MA, USA, 2nd edition, 2001. 17

[168] N. Myers. A new and useful template technique: "Traits". In S. B. Lippman, editor, *C++ Gems*, volume 5 of *SIGS Reference Library*, pages 451–458. Cambridge University Press, Cambridge, UK, 1998. 2, 3, 124

[169] M. Odersky and M. Zenger. Independently extensible solutions to the expression problem. In *Proceedings of the 12th International Workshop on Foundations of Object-Oriented Languages*, January 2005. http://homepages.inf.ed.ac.uk/wadler/fool. 5

[170] J. O'Rourke. *Art Gallery Thereoms and Algorithms*. Oxford University Press, New York, NY, 1987. 269

[171] J. O'Rourke. *Computational Geometry in C*. Cambridge University Press, New York, NY, 2nd edition, 1998. 17, 81

[172] J. O'Rourke. Visibility. In J. E. Goodman and J. O'Rourke, editors, *Handbook of Discrete and Computational Geometry*, chapter 28, pages 643–663. Chapman & Hall/CRC, Boca Raton, FL, 2nd edition, 2004. 269

[173] M. H. Overmars. Designing the computational geometry algorithms library CGAL. In *Proceedings of ACM Workshop on Applied Computational Geometry, Towards Geometric Engineering*, volume 1148, pages 53–58, London, UK, 1996. Springer. 10

[174] J. Pach and G. Tardos. On the boundary complexity of the union of fat triangles. *SIAM Journal on Computing*, 31(6):1745–1760, 2002. 206

[175] E. Packer. Finite-precision approximation techniques for planar arrangements of line segments. M.Sc. thesis, The Blavatnik School of Computer Science, Tel-Aviv University, Israel, 2002. `http://acg.cs.tau.ac.il/tau-members-area/generalpublications/m.sc.-theses/EliThesis.pdf`. 267

[176] E. Packer. Iterated snap rounding with bounded drift. *Computational Geometry: Theory and Applications*, 40(3):231–251, 2008. 268

[177] E. Packer. Controlled perturbation of arrangements of line segments in 2D with smart processing order. Accepted for publication in Computational Geometry: Theory and Applications, 2010. 267

[178] E. Packer. 2D snap rounding. In *CGAL User and Reference Manual*. CGAL Editorial Board, 3.8 edition, 2011. `http://www.cgal.org/Manual/3.8/doc_html/cgal_manual/packages.html#Pkg:SnapRounding2`. 268

[179] M. Pellegrini. Ray shooting and lines in space. In J. E. Goodman and J. O'Rourke, editors, *Handbook of Discrete and Computational Geometry*, chapter 37, pages 839–856. Chapman & Hall/CRC, Boca Raton, FL, 2nd edition, 2004. 81

[180] W. H. Plantinga and C. R. Dyer. Visibility, occlusion, and the aspect graph. *International Journal of Computer Vision*, 5(2):137–160, 1990. 268

[181] M. Pocchiola and G. Vegter. The visibility complex. *International Journal of Computational Geometry and Applications*, 6(3):279–308, 1996. 269

[182] F. P. Preparata and M. I. Shamos. *Computational Geometry: An Introduction*. Springer, New York, NY, 3rd edition, 1990. 7

[183] W. Press, S. Teukolsky, W. Vetterling, and B. Flannery. *Numerical Recipes in C++*. Cambridge University Press, Cambridge, UK, 2nd edition, 2002. 124

[184] S. Raab. Controlled perturbation for arrangements of polyhedral surfaces with application to swept volumes. In *Proceedings of the 15th Annual ACM Symposium on Computational Geometry (SoCG)*, pages 163–172. Association for Computing Machinery (ACM) Press, 1999. 267

[185] V. Rogol. Maximizing the area of an axially-symmetric polygon inscribed by a simple polygon. M.Sc. thesis, Technion, Haifa, Israel, 2003. 269

[186] N. Sarnak and R. E. Tarjan. Planar point location using persistent search trees. *Communications of the ACM*, 29(7):669–679, July 1986. 63

[187] S. Schirra. Robustness and precision issues in geometric computation. In J.-R. Sack and J. Urrutia, editors, *Handbook of Computational Geometry*, chapter 14, pages 597–632. Elsevier Science Publishers, B.V. North-Holland, Amsterdam, 2000. 7, 267

[188] J. T. Schwartz and M. Sharir. On the two-dimensional Davenport-Schinzel problem. *Journal of Symbolic Computation*, 10:371–393, 1990. 260

[189] M. Seel. 2D Boolean operations on Nef polygons. In *CGAL User and Reference Manual*. CGAL Editorial Board, 3.8 edition, 2011. `http://www.cgal.org/Manual/3.8/doc_html/cgal_manual/packages.html#Pkg:Nef2`. 206

[190] R. Seidel. A simple and fast incremental randomized algorithm for computing trapezoidal decompositions and for triangulating polygons. *Computational Geometry: Theory and Applications*, 1(1):51–64, 1991. 63

[191] O. Setter. Constructing two-dimensional Voronoi diagrams via divide-and-conquer of envelopes in space. M.Sc. thesis, The Blavatnik School of Computer Science, Tel-Aviv University, Israel, 2009. `http://acg.cs.tau.ac.il/projects/internal-projects/vd-via-dc-of-envelopes/thesis.pdf`. 157, 260, 264

[192] O. Setter, M. Sharir, and D. Halperin. Constructing two-dimensional Voronoi diagrams via divide-and-conquer of envelopes in space. *Transactions on Computational Science*, 9:1–27, 2010. 157, 264, 269

[193] M. Sharir. Almost tight upper bounds for lower envelopes in higher dimensions. *Discrete & Computational Geometry*, 12:327–345, 1994. 260

[194] M. Sharir. A near-linear algorithm for the planar 2-center problem. *Discrete & Computational Geometry*, 18(2):125–134, 1997. 206

[195] M. Sharir. Algorithmic motion planning. In J. E. Goodman and J. O'Rourke, editors, *Handbook of Discrete and Computational Geometry*, chapter 47, pages 1037–1064. Chapman & Hall/CRC, Boca Raton, FL, 2nd edition, 2004. 238

[196] M. Sharir and P. K. Agarwal. *Davenport-Schinzel Sequences and Their Geometric Applications*. Cambridge University Press, New York, NY, 1995. 17, 64, 206, 238, 259, 260, 266, 269

[197] H. Shaul and D. Halperin. Improved construction of vertical decompositions of three-dimensional arrangements. In *Proceedings of the 18th Annual ACM Symposium on Computational Geometry (SoCG)*, pages 283–292. Association for Computing Machinery (ACM) Press, 2002. 267

[198] J. G. Siek, L.-Q. Lee, and A. Lumsdaine. *The BOOST Graph Library*. Addison-Wesley, Boston, MA, USA, 2002. 7, 17, 49, 161, 171

[199] J. Snoeyink. Point location. In J. E. Goodman and J. O'Rourke, editors, *Handbook of Discrete and Computational Geometry*, chapter 34, pages 767–786. Chapman & Hall/CRC, Boca Raton, FL, 2nd edition, 2004. 63

[200] B. Stroustrup. *The C++ Programming Language*. Addison-Wesley, Boston, MA, USA, 3rd edition, 2004. 17

[201] R. E. Tarjan. *Data Structures and Network Algorithms*. Society for Industrial and Applied Mathematics (SIAM), Philadelphia, PA, USA, 1983. 172

[202] The CGAL Project. *CGAL User and Reference Manual*. CGAL Editorial Board, 3.8 edition, 2011. `http://www.cgal.org/Manual/3.8/doc_html/cgal_manual/contents.html`. 10

[203] G. T. Toussaint. Movable separability of sets. In *Computational Geometry*, pages 335–375. North-Holland, 1985. 268

[204] D. Vandevoorde and N. M. Josuttis. *C++ Templates: The Complete Guide*. Addison-Wesley, Boston, MA, USA, November 2002. 17

[205] G. Varadhan and D. Manocha. Accurate Minkowski sum approximation of polyhedral models. *Graphical Models and Image Processing*, 68(4):343–355, 2006. 238

[206] J. Warmat. Computing a single face in an arrangement of line segments with CGAL. M.Sc. thesis, Rheinische Friedrich-Wilhelms-Universität Bonn, Institut für Informatik I, 2009. 269

[207] R. Wein. High-level filtering for arrangements of conic arcs. In *Proceedings of the 10th Annual European Symposium on Algorithms (ESA)*, volume 2461 of *LNCS*, pages 884–895. Springer, 2002. 124

[208] R. Wein. Exact and efficient construction of planar Minkowski sums using the convolution method. In *Proceedings of the 14th Annual European Symposium on Algorithms (ESA)*, pages 829–840, 2006. 238

[209] R. Wein. Exact and approximate construction of offset polygons. *Computer-Aided Design*, 39(6):518–527, 2007. 124

[210] R. Wein. *The Integration of Exact Arrangements with Effective Motion Planning.* Ph.D. dissertation, The Blavatnik School of Computer Science, Tel-Aviv University, Israel, 2007. `http://acg.cs.tau.ac.il/tau-members-area/generalpublications/m.sc.-theses/WeinMscThesis.pdf`. 238, 260

[211] R. Wein. 2D envelopes. In *CGAL User and Reference Manual*. CGAL Editorial Board, 3.8 edition, 2011. `http://www.cgal.org/Manual/3.8/doc_html/cgal_manual/packages.html#Pkg:Envelope2`. 260

[212] R. Wein. 2D Minkowski sums. In *CGAL User and Reference Manual*. CGAL Editorial Board, 3.8 edition, 2011. `http://www.cgal.org/Manual/3.8/doc_html/cgal_manual/packages.html#Pkg:MinkowskiSum2`. 238

[213] R. Wein, E. Fogel, B. Zukerman, and D. Halperin. Advanced programming techniques applied to CGAL's arrangement package. *Computational Geometry: Theory and Applications*, 38(1–2):37–63, 2007. Special issue on CGAL. 41, 124, 264

[214] R. Wein, E. Fogel, B. Zukerman, and D. Halperin. 2D arrangements. In *CGAL User and Reference Manual*. CGAL Editorial Board, 3.8 edition, 2011. `http://www.cgal.org/Manual/3.8/doc_html/cgal_manual/packages.html#Pkg:Arrangement2`. 15

[215] R. Wein, O. Ilushin, G. Elber, and D. Halperin. Continuous path verification in multi-axis NC-machining. *International Journal of Computational Geometry and Applications*, 15(4):351–377, 2005. 260

[216] R. Wein and B. Zukerman. Exact and efficient construction of planar arrangements of circular arcs and line segments with applications. Technical Report ACS-TR-121200-01, The Blavatnik School of Computer Science, Tel-Aviv University, Israel, 2006. 124

[217] D. B. West. *Introduction to Graph Theory*. Prentice Hall, 2nd edition, 1999. 171

[218] R. H. Wilson and J.-C. Latombe. Geometric reasoning about mechanical assembly. *Artificial Intelligence*, 71(2):371–396, 1994. 268

[219] C. K. Yap. Robust geometric computation. In J. E. Goodman and J. O'Rourke, editors, *Handbook of Discrete and Computational Geometry*, chapter 41, pages 927–952. Chapman & Hall/CRC, Boca Raton, FL, 2nd edition, 2004. 7, 9

[220] M. Yvinec. 2D triangulations. In *CGAL User and Reference Manual*. CGAL Editorial Board, 3.8 edition, 2011. `http://www.cgal.org/Manual/3.8/doc_html/cgal_manual/packages.html#Pkg:Triangulation2`. 239

[221] T. Zaslavsky. *Facing up to Arrangements: Face-Count Formulas for Partitions of Space by Hyperplanes*, volume 154 of *Memoirs Amer. Math. Soc.* American Mathematical Society, Providence, RI, 1975. 17

Index

Ackermann function
    inverse, 206, 260
adaptor, 13
    of arrangement, 16, 161, 172, 173
    of iterator, 161
    of output iterator, 63
    of traits, 91, 125, 194–196, 218
`add_vertical_segment()`, 59, 81
algebraic structure, 11, 98
`Algebraic_kernel_d_1`, 106, 111, 114
AlgebraicKernel_d_1, 106, 111, 114
Ann, 63
API, *see* application programming interface
application programming interface, 10
`approximated_offset_2()`, 216, 219, 233
arc
    circular, 16, 53, 87, 98–102, 104, 124–128,
        158, 161, 175, 190, 192–194, 199,
        214–216, 230
    conic, 15, 21, 35, 83, 102–104, 125, 190,
        215, 218, 248
    geodesic, 264
    rational function, xviii, 21, 105–109, 124,
        125, 127
argument-dependent lookup, 15
`Arr_accessor`, 33
`Arr_algebraic_segment_traits_2`, 111, 113,
    114, 125–127
`Arr_Bezier_curve_traits_2`, 109, 110, 125,
    127, 191, 195
`Arr_circle_segment_traits_2`, 98–100, 125,
    126, 158, 191, 194
`Arr_circular_line_arc_traits_2`, 125, 126
`Arr_conic_traits_2`, 102, 103, 109, 125–127,
    191, 218, 256
`Arr_consolidated_curve_data_traits_2`,
    118
`Arr_curve_data_traits_2`, 117, 118, 244
`Arr_dcel_base`, 136
`Arr_default_dcel`, 21, 132, 137, 139
`Arr_extended_dcel`, 134, 136, 137, 140
`Arr_extended_dcel_text_formatter`, 137
`Arr_face_extended_dcel`, 132, 137, 140

`Arr_face_extended_text_formatter`, 137
`Arr_face_index_map`, 164
`Arr_face_overlay_traits`, 140
`Arr_inserter`, 37
`Arr_landmarks_point_location`, 46
`Arr_landmarks_vertices_generator`, 45
`Arr_linear_traits_2`, 21, 35, 94, 125, 246
`Arr_non_caching_segment_basic_traits_2`,
    93, 125
`Arr_non_caching_segment_traits_2`, 21, 92,
    93, 125, 191
`Arr_non_caching_segment_traits_basic_2`,
    93
`Arr_oblivious_side_tag`, 85, 90
`Arr_observer`, 129, 130
`Arr_open_side_tag`, 90
`Arr_polyline_traits_2`, 95, 117, 120, 125,
    191
`Arr_polyline_traits_2>`, 95
`Arr_rational_arc_traits_2`, 35, 126, 127,
    191
`Arr_rational_function_traits_2`, 90, 105,
    106, 125
`Arr_segment_traits_2`, 21, 28, 53, 92, 93, 97,
    118, 125, 191
`Arr_text_formatter`, 35, 137
`Arr_traits_2`, 195
`Arr_trapezoid_ric_point_location`, 46
`Arr_vertex_index_map`, 162, 163
arrangement, xi–xiii, 1–4, 7–13, 15–17, 19–27,
    29, 30, 32–38, 41–62, 64, 65, 67–70,
    72–74, 76, 81, 83–86, 88, 89, 92, 93,
    95–104, 117, 124, 129, 134, 137–140,
    142–148, 150–154, 157–159, 164, 167,
    168, 172, 173, 178, 185, 189, 190, 195,
    203, 206, 207, 209, 214, 220, 222, 223,
    233, 238, 241, 252, 260, 263–268
    convex, 81, 158
    decomposing, 221, 222, 239
    definition, 1, 19
    extending, 16, 35, 129, 130, 132, 137, 140,
        144, 145, 155, 158, 164, 178, 207, 251
    graph, 161, 162, 164, 165, 172, 173

E. Fogel et al., *CGAL Arrangements and Their Applications*, Geometry and Computing 7,
DOI 10.1007/978-3-642-17283-0, © Springer-Verlag Berlin Heidelberg 2012

Printed in the United States
By Bookmasters